Community Policing
A Policing Strategy
for the 21st Century

Michael J. Palmiotto, PhD
Wichita State University
Wichita, Kansas

AN ASPEN PUBLICATION®
Aspen Publishers, Inc.
Gaithersburg, Maryland
2000

This publication is designed to provide accurate and authoritative information in regard to the Subject Matter covered. It is sold with the understanding that the publisher is not engaged in rendering legal, accounting, or other professional service. If legal advice or other expert assistance is required, the service of a competent professional person should be sought. (From a Declaration of Principles jointly adopted by a Committee of the American Bar Association and a Committee of Publishers and Associations.)

Library of Congress Cataloging-in-Publication Data

Palmiotto, Michael.
Community policing: a policing strategy for the 21st century/
Michael J. Palmiotto.
p.   cm.
Includes bibliographical references and index.
ISBN 0-8342-1087-8
1. Community policing.   2. Crime prevention—Citizen participation.   3. Police-community relations.   4. Police administration.   I. Title.
HV7936.C83P35   1999
363.2'3—dc21
99-14791
CIP

Orders: (800) 638-8437
Customer Service: (800) 234-1660

About Aspen Publishers • For more than 35 years, Aspen has been a leading professional publisher in a variety of disciplines. Aspen's vast information resources are available in both print and electronic formats. We are committed to providing the highest quality information available in the most appropriate format for our customers. Visit Aspen's Internet site for more information resources, directories, articles, and a searchable version of Aspen's full catalog, including the most recent publications: **http://www.aspenpublishers.com**
Aspen Publishers, Inc. • The hallmark of quality in publishing
Member of the worldwide Wolters Kluwer group.

Editorial Services: Kate Hawker
Library of Congress Catalog Card Number: 99-14791
ISBN: 0-8342-1087-8

Printed in the United States of America

1   2   3   4   5

Table of Contents

Preface

This book provides an overview on community policing and focuses on the needs of students who wish to learn about community policing in our society. The first chapter covers police history relevant to community policing. Chapters 2 and 3 cover police culture, police discretion, and police misconduct, which must be reviewed in any book on community policing. Crime prevention is examined in Chapter 4, and police-community relations, team policing, directed patrol, response time, and ministations, foot patrols, and the broken windows theory as elements of community policing strategy in Chapter 5. Chapter 6 clarifies the concepts of communities, neighborhoods, and multiculturalism. Chapter 7 explains problem-oriented policing, a major component of community policing, and Chapter 8 explains community-oriented policing. Chapter 9 discusses the administrative transition from traditional policing to community policing. Last, the book covers implementation of community policing (Chapter 10), selected approaches to community policing (Chapter 11), and distinctive community policing programs (Chapter 12).

Pedagogical features of the book include chapter objectives that orient the student to the main objectives of the chapter, a chapter summary that reinforces the major topics in the chapter, key terms for each chapter to assist students in reviewing material, and discussion questions to help students test their knowledge of the chapter.

Acknowledgments

A large number of individuals deserve appreciation for their assistance in getting this book published. I thank all those who provided information that reveals what is actually occurring in community policing, including Frum Himelfarb of the Royal Canadian Mounted Police, Major Dan Reynolds of the Savannah Police Department, Chief Robert E. Deu Pree of the Astoria, Oregon, Police Department, and numerous other police administrators.

A special acknowledgment goes to Michael Birzer for writing Chapter 9, "Organizational Change and Community Policing," and Anna Chandler for writing the section on multiculturalism that appears in Chapter 6. I also thank the Aspen publishing staff who provided guidance in the development of this book, including Susan Beauchamp, acquisitions editor; Kathleen McGuire Gilbert, developmental editor; and Kate Hawker, production editor.

I would like to thank my wife, Emily, for her support and encouragement throughout this project. Without her support, this book would not have been published.

Chapter 1

Police History Relevant to Community Policing

CHAPTER OBJECTIVES

1. Understand why a knowledge of police history is important for understanding policing today.
2. Be familiar with the Peelian reform principles.
3. Be familiar with the crime-fighting image of policing.
4. Be familiar with police issues of the 1960s.
5. Be familiar with governments' influence on policing in the 1960s.
6. Be familiar with the concept of police professionalism.
7. Be familiar with the various eras of policing.

INTRODUCTION

The philosophy of community policing is being advanced as the new policing system for the twenty-first century. How can police administrators, leaders, or students of policing support or even understand the value of and need for community policing in our contemporary world if they lack knowledge of where the police have been? How can effective decisions and policies be implemented if those politicians and police administrators making decisions have no knowledge or understanding of police history? Formal policing as we know it today is relatively new, being less than two centuries old. Knowledge can be gained from familiarity with the historical role the police have played in society. This holds true even though the history of policing is sketchy and the variety of roles the police have played in the past have not all been documented. There does appear to be sufficient information of value to modern-day police. For example, anyone familiar with policing or police officers often hears the claim that they are not social workers, but in the late nineteenth century and early twentieth century, the police performed tasks that are currently performed by social service agencies, including finding jobs for the unemployed, operating soup kitchens, and using the police station as a night shelter.

Politicians and police administrators, along with police supporters and students of policing, should understand the history of policing, including its various strategies and police methods, if they hope to have an impact on the field. Police decision makers with a knowledge of the various police roles and tactics used throughout history should be in a better position than those with no or limited knowledge of police history to improve policing or at least avoid some of the problems that have occurred in the past.

EARLY POLICING

Throughout history, individuals have been expected to follow the rules of the majority. The demand to control the behavior of individuals can be traced to ancient times, when the tribe, clan, or family controlled the behavior of the individual and had the responsibility to enforce its informal rules or customs. This was the first form of community policing: the tribe, clan, or family, as an early form of community, was responsible for policing itself. Eventually, rules became laws, and rule breaking was formalized as an act committed against the state.

POLICING IN THE ROMAN EMPIRE

With the development of formalized governments and states, laws became a prerequisite. The Roman Empire, for example, had formal laws and a judiciary system. However, they did not establish formalized police departments as we know them today. It was the responsibility of the Roman legions to maintain order and peace. The Romans did institute the Praetorian Guard, composed of Roman legionnaires, to protect the emperor, and they created the *praefectus urbi,* composed of approximately 500 to 600 legionnaires, to maintain order in the city of Rome and to keep watch for fires, which were a frequent danger. The Roman model of policing was more in line with the traditional policing of the twentieth century than with the community policing philosophy, in that it was more reactive to incidents than proactive.

POLICING IN ENGLAND

The Anglo-Saxon Period

The Roman Empire expanded to the British Isles and maintained order through its legions. But with the collapse of the Roman Empire, Germanic tribes, known as Anglo-Saxons, invaded England. Unlike the Romans, who had written laws and a legislative body, the Germanic tribes had unwritten law based upon tribal customs. Eventually, the Anglo-Saxon culture fused with the Roman culture, and the Anglo-Saxons developed a peacekeeping system for England. The concept of the king's peace evolved, whereby the king assured the people a state of security and peace in return for their allegiance to him. The peacekeeping system revolved around land ownership. The landowner was responsible for all those on his land who violated the king's peace. Ten homesteads formed a *tything,* and ten tythings formed a *hundred,* which was responsible to maintain order.

When a crime was committed, the tythings and hundreds had to be informed, and it was their responsibility to chase and arrest the culprit and bring him or her to the justice of the peace. If the guilty party did not appear in court, the tythings and hundreds were liable for his or her offense. Imprisonment and fines were both used as corrective measures at this time (Melville 1971, 5). As Melville described the Anglo-Saxons' police function,

> The internal peace of the country was held by them to be of the first importance, and every free man had to bear his part in maintaining it; theoretically all men were policemen, and it was only for the sake of convenience that the headborough (or tythingman as he came to be more generally called) answered for those of his neighbors, on whom he had to rely in case of necessity. The word "peace" was used in its widest possible meaning, and a breach of the peace was understood to include all crimes, disorders, and even public nuisances. The principle on which the police system was based was primarily preventive. (8)

This system, referred to as *mutual pledge,* can be traced to King Alfred the Great (870–901 A.D.). Under the mutual pledge system, every male over 12 years of age, unless exempted because of his social status, was accountable for maintaining peace and answerable for his neighbor's behavior. When a male over the age of 12 saw an offense committed, it was his duty to raise the "hue and cry" and give chase with other males after the culprit.

Over the hundreds was the *shire reeve,* the forerunner of the modern-day sheriff. The shire reeve had responsibility under the king for maintaining peace in the shire. To maintain peace, the shire reeve could assemble all the men in the shire in what was known as a *posse comitatus* to keep order. Thus, law and order was considered a community responsibility, as it is in community policing today.

The Norman Invasion

With the Norman invasion of England in 1066, the Anglo-Saxons came under the control of the Normans. Under the Normans, peacekeeping was more important than individual rights. The Normans centralized the peacekeeping system under the crown and revised the Anglo-Saxon system of peacekeeping in what became known as *frankpledge.* The frankpledge system continued the Anglo-Saxon principle of every male's obligation to maintain the king's peace.

By the end of the thirteenth century, another officer for keeping the peace was established under the Normans: the *constable,* a royal officer who had the responsibilities of the tythingman. The constable was elected by the parish, a church district. The Statute of Winchester (1285), which regulated policing between the Norman Conquest and the advent of modern policing in 1829 with the passage of the Metropolitan Police Act, outlined basic principles of policing, which may be summarized as follows:

1. It was the duty of everyone to maintain the king's peace, and it was open to any citizen to arrest an offender.

2. The unpaid, part-time constable had a special duty to maintain the peace, and in the towns he was assisted in this duty by his inferior officer, the *watchman.*
3. If the offender was not caught red-handed, hue and cry was to be raised.
4. Everyone was obliged to keep arms with which to follow the cry when required.
5. The constable had a duty to present the offender at the court *leet* (judge) (Critchley 1972, 7).

The constable system did not survive beyond the latter part of the seventeenth century. Three primary causes are given for its collapse as a peacekeeping mechanism. First, the constables were held in contempt by the people. Second, the justices of peace were held in low esteem because they were known to trade justice for a fee. Third, population and wealth were increasing with the expansion of towns during the initial stages of the Industrial Revolution, and thus the opportunities for crimes were increasing as well. This created an unstable society, with which the constable system was ill equipped to cope (Critchley 1972, 18).

The constable system provides an example of a policing system that failed because it lacked community support. Today, enlightened police administrators recognize that effective policing depends upon the people of the community.

The Seventeenth Century

During the seventeenth century in London, night watches, known as "Charlies" after the king who initiated the system, were paid a small fee to assist the constables in keeping the peace. They were recruited from the poor, old, and weak. They were ineffective in performing their task but would receive their pay whether they were awake or slept on duty (Critchley 1972, 43).

During this time, the English system of maintaining order was not concerned with preventing crime. No reward was provided for prevention. Indeed, peacekeeping efforts encouraged crime: for example, magistrates relied on fees and fines and constables and thief-takers on apprehension and convictions for their income (Critchley 1972, 48). The paid night watches, constables, and thief-takers all exemplify poor policing strategies that failed. Employing incompetent individuals who do not effectively perform good policing creates poor relations with the community.

The Eighteenth Century

In the eighteenth century, the term *police* began to be used, much to the disappointment of many Englishmen. One contributor to policing was Henry Fielding, who was appointed a magistrate to Bow Street in 1848. Fielding conceived the idea of crime prevention as a role for the police. He believed that controlling crime required the cooperation of the public, a strong police force, and the elimination of the causes of crime. He initiated victims' reporting of crimes to his Bow Street office, broadcast the description of robbers in a journal that he published, established a horse patrol, and used the Bow Street office as a clearinghouse for crime information. Also, he established a night horse patrol to guard the roads leading into London. This patrol would eventually become the Bow Street Runners. Upon Henry Fielding's death, his brother John Fielding was appointed to his position. John Fielding continued his brother's work, establishing the Bow Street Runners to investigate crime (Critchley 1972, 32–33; Pringle 1955, 80–83).

Another police reformer of the eighteenth century was a magistrate named Patrick Colquhoun. In a publication titled *A Treatise on the Police of the Metropolis* (1797), he argued that a well-regulated police force should have as its primary goal the prevention of crime, that judicial and police powers should be separated, and that professional police should be established in every parish. He further recommended that the police be paid and be under the control of a central board, that the police have an intelligence unit that would maintain a record of known offenders and classify information about specific crime groups, and that a police gazette be published with the purpose of assisting in detecting crime and educating the general public about crime. Colquhoun's proposals represent a link between the old and new ideas of peacekeeping and order maintenance (Critchley 1972, 38–40).

Both Henry and John Fielding and Patrick Colquhoun contributed to the professionalism of policing. Like the Fieldings and Colquhoun, police administrators today who are committed to the community policing philosophy consider crime prevention an important element of policing.

The Nineteenth Century

In the early nineteenth century, the English Parliament was concerned about poverty, unemployment, and lawlessness. Between 1780 and 1820, there were five parliamentary commissions dealing with the issue of public disorder. Nothing came out of these commissions until Sir Robert Peel was appointed Home Secretary. Peel suggested that an acceptable system would hold each community responsible for maintaining order in its own locale, a principle advocated today as community policing. In addition, Peel recommended to Parliament that London establish a paid police force appointed from the civilian population. Parliament approved Peel's proposal, and in 1829 the London Metropolitan Police was formed. The London police wore blue coats and white pants and carried only a truncheon. They were not provided with firearms (Chapman and St. Johnson 1962, 13–14).

One year later, the Metropolitan Police of London had a force of 3,000 officers. The turnover was high, primarily because of dismissal for drunkenness.

The Peelian Reform principles are as follows:

1. The police must be stable, efficient, and organized along military lines.
2. The police must be under government control.
3. The absence of crime will best prove the efficiency of police.
4. The distribution of crime news is essential.
5. The deployment of police strength by both time and area is essential.
6. No quality is more indispensable to policemen than a perfect command of temper; a quiet, determined manner has more effect than violent actions.
7. Good appearance commands respect.

8. The securing and training of proper persons is at the root of efficiency.
9. Public security demands that every police officer be given a number.
10. Police headquarters should be centrally located and easily accessible to the people.
11. Policemen should be hired on a probationary basis.
12. Police records are necessary for the correct distribution of police strength. (Germann et al. 1973, 60–61)

These 12 principles still hold true today, and supporters of community policing should review them. Police agencies are changing over from traditional policing to community policing to become more effective and efficient. Most police departments adopting community policing are establishing police substations and ministations in order to become more accessible to the community. Modern technology such as computers and crime analysis are being used by community policing officers to make the community aware of the crime problem and to deploy police officers. That the absence of crime is a measure of police competence is as true now as it was in 1829, when Peel published his principles. Many police chiefs in the last decade of the twentieth century have attributed the decrease in crime to community policing.

On July 19, 1829, the Metropolitan Police Act officially became law, and Sir Robert Peel appointed two police commissioners—one an ex-soldier to enforce discipline and the other an efficient attorney. The police headquarters were housed in a building known as Scotland Yard. The metropolitan police department was divided into 17 police divisions, each containing 165 officers. Each division incorporated a superintendent in charge, four inspectors under the superintendent, and 16 sergeants reporting to the inspectors. All sergeants supervised nine patrol officers. Because the English people were not supporters of a formal police organization, considering it a threat to their liberty and a potential suppressive force, it was recognized that public support was necessary. The metropolitan police department was founded on the premise that the police should be attuned to the people and should gain their cooperation. Today's community policing philosophy holds this as a key premise. The intent of the metropolitan police planners was to establish a homogenous democratic body. The police recruits were expected to be literate, of good character, under age 35, in good physical condition, and at least five feet, seven inches tall (Critchley 1972, 51–52).

America traces its culture and legal system to our English heritage. Our founding fathers who created the United States were Englishmen before they became Americans. Although we divorced ourselves from the English with the Revolutionary War, we cannot divorce ourselves from the contributions of our English ancestors—not only our legal system but also, to a large extent, our criminal justice process. Our police system was inherited from the English. The early English settlers brought with them the procedures they knew, many of which have become ingrained into American culture. Although there have been many changes in America since the Revolutionary War, our policing and legal systems have incorporated many English concepts. Most Americans consider these concepts ours, not realizing that they were inherited from our English forebearers.

AMERICAN POLICING

The Metropolitan Police Act of 1829 that established the London police force became the model for American policing. The New York City Police Department is often credited for initiating the modern police department in America in 1845. Before that time, American peacekeeping strategies were similar to those of the English.

Boston founded the first night watch in America in 1634; it consisted of six soldiers and an officer. Two years later, a town watch, consisting of townspeople and not military personnel, was initiated to keep the peace. This system lasted until 1712, when Boston voted to pay their watchmen (Bopp and Schultz 1972, 17–18). Charlestown in 1680 and Philadelphia in 1682 appointed constables who had responsibilities similar to those of their English counterparts. As in England, county governments appointed sheriffs as their primary peace officers, and it was an obligation of citizens to serve as watchmen and constables (Johnson 1981, 5).

The American Revolution

During the American Revolutionary War peacekeeping was maintained by troops, who were English or American, depending upon who occupied the area. After the war, peacekeeping reverted back to civilian control. Providence, in 1797, retained 12 night watchmen to take over the public safety duties performed by the militia. In 1801, Boston became the first city to mandate that permanent night watchmen be employed. That same year, Detroit hired its first civilian police officer. More and more civilian peacekeepers were being employed after the Revolutionary War because of the growth of cities and the increase of crime associated with cities (Bopp and Schultz 1972, 25–28).

Traditional peacekeepers—sheriffs, constables, and night watchmen—were unable to control crime occurring in American cities. The crime during this period consisted of violence, riots, and vice. Modern America still has these problems, but today the responsibilities to solve them have been given to community policing officers.

To regulate Philadelphia's crime problem, Stephen Girard, upon his death, left the city of Philadelphia financial resources for public safety. With the Girard money, Philadelphia established a day force of 24 police officers and a night force of 120 night watchmen. Maintaining separate day and night watches created difficulties in peacekeeping. Recognizing this, the New York legislature unified the day and night watches of the New York City Police Department in 1845. New York City had the first unified police force in the United States, modeled on the London Metropolitan Police Department. The combining of day and night patrols became the model of American police departments (Bopp and Schultz 1972, 33–38).

Modern policing developed in America during a period marked by disorder and violence. Most American cities experienced mob disorder. A variety of factors appear to have contributed: conflicts of native Americans with Irish and German immigrants; racial conflict, with whites attacking abolitionists and free blacks; economical disasters, which led mobs of financially ruined investors to attack banks and employees to destroy the property of their employers; attempts to solve issues of morality, as when brothels were attacked as a means of cleaning up the city; and political conflicts, as in the case of the frequent mob riots on election days in major cities such as New York, Philadelphia, and Baltimore. It was generally left to the militia to suppress riots. The series of major riots in American cities led to the establishment of the "new police," the forerunner of our contemporary police departments (Walker 1977, 4–5).

The New Police

Unlike our police of today, the "new police," when first formed in New York City in 1845, did not wear uniforms or carry firearms. Not until the mid-1850s did New York City require police officers to wear uniforms. Eventually, the requirement that officers wear uniforms was adopted by other cities. Firearms began to be used in the 1840s. Without formal approval, police officers began carrying them to protect themselves from armed thugs. Eventually, the cities gave formal approval for police officers to carry firearms. Today, we expect police officers to be armed and to wear uniforms.

During the nineteenth century, city populations increased substantially. Cities that were hamlets grew into urban centers. Industrialization increased the demand for workers. Cities grew substantially, with immigrants arriving from foreign countries. There also was a migration from the farms to the cities. This influx of population into cities affected policing in three ways:

> First, it increased the geographical size of the cities enormously as new housing and business areas opened to accommodate these people. This meant that the physical effort involved in patrolling a city increased beyond anyone's expectations. No police force could maintain a suitable presence on the street at all time. Budgetary problems complicated the effort to protect cities. Throughout the century cost-conscious politicians kept the number of officers as small as possible. During the frequent economic downturns which occurred, department strength suffered as mayors sought ways to reduce city payrolls. Thus the sheer size of the cities, and their economic problems, jeopardized one of crime prevention's cardinal principles: pervasiveness.

> Second, urbanization encouraged specialized use of land within cities. Expanded businesses needed convenient locations. Prosperity and changing tastes prompted many people to seek more spacious homes. The poor crowded into the inner city neighborhoods where they could afford the rents. In sum, more distinct downtown, suburban, and slum areas emerged. Opportunities for crime varied within these districts. Careless merchants and homeowners made tempting targets in the business and middle class areas. The tensions of slum life created numerous situations where violence often erupted. Crime thus expanded with urban growth.

> Finally, the process of urbanization threw thousands of strangers together. Irishmen, blacks, Germans, Scandinavians, Italians, Poles, and Jews, to

mention just a few groups, . . . lived cheek-by-jowl. The initial contacts between these groups sometimes were anything but friendly. Ethnic rivalries, class distinctions, and different levels of political influence mixed in the urban neighborhoods to produce a fruitful source of crime. These differences created solutions in which uniform law enforcement policies became impossible. Criminals would learn to exploit these situations. (Johnson 1981, 36)

The English Heritage

Historically, America can trace its policing heritage to the English Middle Ages. Throughout most of English history, there was a civilian orientation toward police activities. Traditionally, the English held that male citizens were to function as peacekeepers and public safety officers. They had a strong belief in individual rights and did not readily accept the formalization of the police. They recognized that the military can suppress individual rights and freedom, and their concern was that a formalized police could do so as well. The founding fathers of the London Metropolitan Police Department recognized this concern by establishing several policies to curtail citizens' fear of abusive police practices. For example, the police wore blue uniforms rather than the red uniforms of the British Army. They were recruited from the general civilian population and not the military. The founding fathers paid the police an average salary for that period in order to discourage the upper classes from entering the police field. They wanted to have a police department that reflected the general population of the community. The London police were also not armed, being provided only with a truncheon. The police department was centralized, and officers had to follow rigid procedures that had been instituted to control their behavior.

The English had a tradition of local control of peacekeeping. Unlike the French or Germans, they never established a national peacekeeping force. It was the responsibility of the men in the villages and towns to maintain order.

The English concepts of individual rights and freedoms and of policing as a local matter to be handled by the local government were transposed to America with the early English colonists. Like their English cousins, the Americans were not great supporters of establishing a formalized police agency; they considered it un-American. In response to the seriousness of crime in American cities, they, like England, organized municipal police agencies to maintain law and order.

As in England, policing had a civilian orientation. America, like England, did not recruit police officers from the military but from the community of civilians. American police departments were, at least in theory, highly centralized. As in the English system, rank structure corresponded with authority, so that, for example, the police chief had more authority than the captain, and the captain had more authority than the sergeant. There was a formal chain of command in which lower ranking police officers were accountable to superiors: for example, a patrol officer was supposed to report to the sergeant. This has been called the bureaucratic model of policing. However, in reality, this system was often dysfunctional. Supporters of the community policing philosophy have criticized the bureaucratic model as rigid and as not allowing for creativity and innovation. A number of community policing departments are reorganizing to be more in tune with the community and to allow their line personnel to be decision makers.

Generally, American police, unlike the London police, did not function as professionals. Although municipal police departments technically were centralized, they tended to function as decentralized neighborhood police units. Because police officer positions were obtained by appointments from local politicians, it should not come as a surprise that officers were frequently puppets of politicians. During certain periods of the nineteenth century and in some cities, the person who ran the police department was not the police chief but the precinct captains or ward bosses. Often, the police chief could not even approve the hiring of a police officer for a specific precinct. It was not uncommon for police officers to lose their positions with a change in political parties. Policing today has become much more professional than that of the mid- to late nineteenth century. Generally, to become a police officer today, an individual has to pass written, medical, agility, and psychological examinations. In addition, he or she often must pass a polygraph and undergo an intensive background examination. Once accepted as a recruit, the police candidate must successfully pass training at a police academy, followed by field training under the supervision of a seasoned officer. When training has been successfully completed, the recruit will serve a probationary period that lasts, depending upon the police department, from three months to about one year. At any time during this period, the recruit can be terminated. Once the recruit has successfully completed the probationary period, he or she can lose his or her position only for cause. Unlike the nineteenth-century police officer, he or she does not owe his or her police position to a politician.

Nineteenth-century police officers came from the lower classes and tended to be recently arrived immigrants. They were not well educated, and no major skills were required for the position. Personal qualifications for police recruits were not standard. Robert Fogelson (1977) wrote the following about big-city police in the nineteenth century:

> The police served themselves and the machines a lot better than the citizens. They drew an adequate salary; if they put in a full day, which was unusual, they worked no harder and no longer than other Americans; and they enjoyed greater security. They also supplemented their salaries with payoffs; most patrolmen preferred taking graft to raiding gamblers and most detectives preferred collecting rewards to arresting criminals. (35)

Early police departments were often responsible to perform activities that did not fall under the realm of public safety. For example, the police in New York City cleaned streets, distributed supplies to the poor in Baltimore, assisted the homeless in Philadelphia, and inspected vegetable markets in St. Louis. Nineteenth-century police did more than peacekeeping. Their peacekeeping role was more reactive than preventative. They were expected to respond to complaints or for requests for assistance. They were not expected to curtail crime (Fogelson 1977, 16–17). In contrast, contemporary community police officers are expected to be problem solvers but are not expected to clean streets or inspect vegetable markets.

One of the difficulties police departments have had since the implementation of patrols is that of supervising patrol officers. When on patrol, officers can hide in alleyways or visit friendly merchants in order to avoid their supervisors. In earlier times, the patrol officer, because of the lack of two-way communications, could evade his sergeant and not be found unless he wanted to be found. To a great extent, policing was influenced by politics and the neighborhood culture.

Changes in Policing

Policing in American cities during the nineteenth century was primary decentralized and neighborhood oriented. The major dilemmas of this policing model were corruption, inefficiency, incompetence, and political influence. A number of Americans wanted to see improvements in American policing, and eventually there were attempts to improve the administration of police departments. The first attempt to control city police departments was in New York State in 1857. On the basis of the claim that New York City was too corrupt to govern itself, the state legislature passed a law—the Metropolitan Police Bill, modeled after Sir Robert Peel's London Metropolitan Police Bill—that placed the New York City Police Department under state control by putting it under the supervision of a board appointed by the governor. State control of police departments spread to other states: Baltimore in 1860, St. Louis and Chicago in 1861, Kansas City in 1862, Detroit in 1865, and Cleveland in 1866. There was some opposition to this development: politicians and cities did not want the state to control their local police department. Also, there was a lack of uniformity in controlling police agencies throughout the state (Fogelson 1977, 43–44).

Around the turn of the century, the Progressive movement emerged as a voice to improve city government, including policing departments. This movement reflected upper-middle-class concerns to safeguard honesty of public officials, establish uniform standards, and operate by well-defined regulations and procedures (Johnson 1981, 67). A city police department in which each captain responded to his neighborhood's customs could not expect to possess uniform standards. Disorder and drunkenness were acceptable among some ethnic groups, and the police who recruited from these ethnic groups looked the other way when minor offenses were committed. A tolerance toward brothels, public intoxication, and rowdiness was not acceptable to the upper classes of America's cities. Police critics of this period supported scientific management and the hiring of specialists to solve a city's problems. They were disturbed not only by the lack of uniform police standards but also by the lack of discipline and efficiency and the poor quality of police personnel.

The National Prison Association, founded in 1870, championed civil service, police training, and a nonpartisan police department through its Standing Committee on Police. But changes in policing came slowly. Civil service was adopted by New York City in 1883 and by Chicago in 1895. A major crusader for police change in New York City was the Reverend Charles Parkhurst, president of the Society for the Prevention of Crime. Parkhurst charged that police corruption and vice were rampant in New York City. His campaign led to the New York legislature's establishing the Lexow Commission in 1894 to investigate police activities. The commission found political interference with the police process, police connivance with criminals, and police payoffs for appointments and promotions (Johnson 1981, 66–67).

An outcome of the Lexow Commission was the creation of a four-member board of police commissions for New York City. Theodore Roosevelt was appointed president of the board and served a two-year term. Roosevelt worked to improve the quality of police recruits, raise standards of police performance, and curtail corruption. The crime-fighting image of the police was projected under Roosevelt's tutelage through awards of medals and certificates to patrolmen for specific acts of bravery. Roosevelt's purpose was to motivate the rank-and-file patrolmen to be productive (Walker 1977, 46). Although the image of police as crime fighters was not accurate, it persisted and is still with us today. Even today, there are police chiefs who sell policing to politicians and the public by arguing that the police are crime fighters. This myth is even believed by some rank-and-file police officers. However, the community policing philosophy runs counter to the crime-fighting image. Community policing officers are problem solvers who work in partnership with neighborhood residents to solve crime and disorder problems.

The Crime-Fighting Role

Critics of policing in the nineteenth century advocated police training; the elimination of corruption, police brutality, and inefficiency; and the removal of politics from police affairs. Nineteenth-century police officers, as mentioned earlier, were expected to perform activities not normally associated with policing, including charity work, street cleaning, and boiler inspection. Critics believed that such assignments distracted police from crime-fighting tasks and diminished the image of the police. Their goal was to professionalize the police, and they felt that this could best be accomplished by giving the police the new image and goal of crime fighting. Police chiefs in America readily accepted that police should be primarily crime fighters. The foremost change agent for improving police departments in the first half of the twentieth century was a strong advocate of the crime-fighting mission. August Vollmer, who began his police career as a town marshall in Berkeley, California, in 1905, and ended up being a police chief there until 1932, is credited with being the chief spokesman for the professionalization of the police and their training in the science and technology of all aspects of crime-fighting work (Carte and Carte 1975, 2).

With Vollmer and other police chiefs advocating crime fighting as the most important mission of police, it did not take long before this philosophy became ingrained in the police culture. Crime fighting was sold as the primary function of the police, even though crimes such as homicide, rape, and assaults are usually not preventable by the police and are often not even solved. The philosophy of crime fighting is reflected in those departments that still do "bean counting," or counting the number of arrests that each officer makes. In some police departments, officers are rewarded for what are considered "good arrests," or arrests for which there is sufficient evidence to obtain a conviction.

The police have difficulties in implementing their crime-fighting philosophy when the community refuses to cooperate with them in providing information about drugs, gambling, or prostitution and when they are dealing with juvenile gangs. To be successful in fighting crime, they must decide which crime or crimes their resources should be concentrated upon.

While Vollmer was president of the International Association of Chiefs of Police in 1922, he made a number of recommendations. He suggested that police departments begin employing clerks, typists, photographers, and other professionals such as identification experts. He also supported lateral entry for police officers, claiming that the system of requiring all police officers to work their way up through the promotional process was inadequate. Vollmer had nine additional recommendations for improving policing and the criminal justice process (Carte and Carte 1975, 56–57).

1. Increased use of policewomen, especially in the "vast field of pre-delinquency."
2. Police schools for training purposes, wherein the content "may vary slightly in different communities, but the fundamentals in the police school curriculum should be identical in all departments."
3. Modern equipment, such as signal devices, wireless telephony and telegraphy, automobiles, motorcycles, motorboats, and laboratory apparatus. We must be prepared to meet the criminal with better tools and better brains than he possesses if we hope to command the respect of the community that we serve.
4. Greater emphasis on crime prevention as our principal function.
5. More police contribution toward solving the problem of unnecessary delay and miscarriage of justice in criminal trials.
6. Uniform national and even international laws, uniform classification of crimes, simplified court procedures, better methods of selecting and promoting properly trained jurists and modern requirements.
7. Abandonment of trial-and-error methods of crime solving for more efficient scientific techniques and crime laboratories, enlisting the aid of microscopists, chemicals analysts, medico psychologists, and handwriting experts.
8. Centralized and improved methods of maintaining police records. A bureau of records, if properly organized, is the hub of the police wheel.
9. Petitioning of universities to devote more time to the study of human behavior and its bearing upon political and social problems and to the training of practical criminologists, jurists, prosecutors, policemen, and policewomen.

The crime-fighting model did establish the image of the American police as trained professionals who concentrated on solving crimes and had the skills to control crime in their communities.

Crime fighting is associated with a model of traditional policing that contemporary advocates of community policing consider a failure. But I believe that the founders of the crime-fighting philosophy deserve credit for bringing the police closer to professionalism. Traditional policing is not a failure, because it has provided successful services to the public—services that people have learned to expect from their local police department. Still, crime-fighting tactics have only been partially successful in solving crimes. Generally, police departments solve only 50 to 70 percent of the homicides in their cities—a record that can hardly be considered good. Although projection of a crime-fighting image has led the public to believe that police investigate all crimes and will solve all crimes, police officers know that this is not true. Because police administrators recognize that the crime-fighting model has not been a complete success, they are adding the community policing philosophy to those aspects of the crime fighting model that work.

Crime Prevention

Since the founding of the modern police system in London by Sir Robert Peel, the British police have considered crime prevention as one of their primary responsibilities. But this philosophy did not take root in the United States until much later. The founders of the American police did not consider crime prevention to be a primary objective of the police. The American police were trained to react to the commission of a crime, not to try to prevent the crime. Vollmer's recognition of crime prevention as a crucial police objective was an important step forward. The police and the community, according to Vollmer, should form an alliance to develop strategies to control crime (Carte and Carte 1975, 33–34). This premise is right in line with the philosophy of community policing today.

Raymond Fosdick's book *American Police Systems,* first published in 1920, devoted a chapter to crime prevention. Fosdick asserted that the policeman, like the watchman of an earlier time, had the obligation to forestall and apprehend the law violator and that police work had as its goal the prevention of crime. However, he argued that large numbers of police officials were skeptical about the possibilities of crime prevention, and he reported two of New York City's highest ranking officers as stating that crime prevention had only limited value in crime control. According to Fosdick, many police administrators disclaimed any responsibility for crimes committed in their jurisdiction. They felt that the solution to rising crime was simply to add more policemen, and they lacked vision in that they never considered working with the resources at hand and developing creative strategies to control crime. The emphasis of the typical police department was to function defensively against crime and not to attack the roots of crime by limiting crime opportunities. Law enforcement strategies took precedence over crime prevention strategies.

Fosdick portrayed Commissioner Woods of New York City as a progressive police administrator who supported crime prevention programs. Under Woods' tenure, a system of junior police was established for boys between the ages of 11 and 16. The boys received training in first aid and safety rules and were involved in athletic activities under the supervision of police officers. Woods was instrumental in establishing playgrounds and "play streets" that were closed to traffic for several hours to allow children to play. Also, the New York City police officers went to the schools to talk to children and held Christmas parties for neighborhood children at precinct houses. The goal of these crime prevention programs was to divert youngsters away from trouble before they became involved with the criminal justice process. Other cities followed New York and established crime prevention programs for children (Fosdick 1929, 354–370).

In the first quarter of the twentieth century, women who had been primarily trained as social workers began to enter the field of policing. The involvement of women in policing had its roots in the protection movement of this period, launched by private organizations, to oversee the welfare rights of women and children. It also developed out of the "white slavery scare" of the 1910s and 1920s. In the early 1900s, there was a major concern, fueled by the entertainment and news media, that women were being abducted and sold into prostitution. Movies and newspapers all had stories about innocent girls being kidnapped and sold into prostitution. Reformers pointed to the failure of police to enforce vice laws and to the culpability of men in fostering prostitution (Appier 1992, 5–7, 11–12).

Women's role in policing was considered to be primarily preventive. Although women police officers had the same arrest powers as their male counterparts, neither they nor the male officers considered male and female assignments to be the same. The expectation was that the female police officer would give counseling and guidance while the male police officer would have the duty to make arrests.

Thus, the crime prevention model became, to a large degree, shaped by and associated with women. It had three major principles:

1. The highest form of policing is social work.
2. Crime prevention is the most important function of the police.
3. Women are inherently better than men at preventing crime. (Appier 1992, 5)

Women police officers of this era readily accepted the crime prevention role. Many of them brought the social service philosophy with them when they entered the police field. They worked with children, helped find missing persons, and interviewed juvenile and woman offenders. They also gave advice to parents about mischievous children and to married couples on how to deal with domestic situations. In Detroit, women police officers were involved in preventing girl delinquency. The steps that they outlined for establishing a delinquency prevention program for girls included the following (Connolly 1944, 19):

1. Creation in the police department of a crime prevention bureau that thoroughly investigated cases of delinquency and potential deliquency and referred them for solutions to the right public and private agencies
2. Provision of "clean, proper detention places" (19)
3. Provision of a children's court or the equivalent, with competent probation officers and judges who were not political appointees
4. Financing of public and private agencies so that they could give delinquents and potential delinquents "careful long-term aid" (19) that many of the cases required

(including not only rehabilitating delinquent girls and boys but cleaning up the areas that produced them)
5. Punishment of parents for willful neglect

Such ideas fit in well with today's concepts of community policing. Community police officers are expected to perform the same activities that Commissioner Woods required. They are involved in crime prevention and are expected to work with neighborhood residents to prevent crime.

The 1930s and 1940s were decades of little change for policing. The Great Depression marked the decade of the 1930s whereas World War II consumed the 1940s. During these decades, the influence of the crime prevention model began to wane, and crime control, achieved by making arrests, became the most important function of police departments. This *crime control model* remained dominant until the political upheavals of the 1960s and 1970s began to shake up policing, as well as expose it to closer scrutiny.

The Decade of Turmoil, Violence, Chaos, and Change

The 1960s was a decade of turmoil, violence, and chaos that resulted in changes in society and policing. Although there were many negative aspects to this decade, there was also much that was positive. First, Congress passed the civil rights law, creating opportunities for minorities. The civil rights movement was followed by the movement against the Vietnam war, the feminist movement, the gay rights movement, and the movement for rights of the disabled. The police, because of their positions as law enforcement officers, played a major role during this time and were perhaps more abused than at any other time in American history. They were placed in an awkward position because their duties required them to defend governmental policies and laws that many Americans were against. Police officers are sworn to uphold the U.S. Constitution and the constitution of their state, to enforce the laws of the federal and state government, whether these laws are popular or unpopular, and to protect governmental property and public officials from acts of violence.

The police of the 1960s were placed under more scrutiny than police in any other decade before or since, and the police officers of later generations have reaped the benefits. America might not have had community policing if American policing had not been evaluated and analyzed by various governmental commissions that often found fault with police procedures and behavior. Credit must also be given to progressive-thinking police administrators who were willing to experiment with new strategies of policing. And various experimental strategies of policing instituted in eras before community policing have contributed to the community policing philosophy implemented in the later part of the twentieth century.

The 1960s began with hope, promise, and enthusiasm but ended in darkness and malaise. It began with the election of John F. Kennedy, a young and charismatic president who offered hope for a better life to all Americans, regardless of their race, class, or ethnic background. But by the time it ended, five political and civil rights leaders—John F. Kennedy, his brother Robert Kennedy, Martin Luther King, Jr., Medgar Evers, and Malcolm X—had been assassinated. Americans became cynical and mistrustful toward their fellow citizens and their government. The five assassinations were not the only acts of violence occurring in American communities and streets. Others were the bombing of a black church in Birmingham, Alabama, and the attacks on civil rights workers and demonstrators in the South, including the murder of young civil rights workers in Mississippi.

There were also numerous riots in America's urban centers. The nonviolent civil rights movement espoused by Martin Luther King, Jr., that attracted many people in the first part of the 1960s became unacceptable to a number of civil rights activists for whom "equality" and "desegregation" were occurring too slowly. Some civil rights leaders broke with the nonviolent movement of Reverend King and adopted violence as a solution to obtaining equal opportunity and rights. During the mid-1960s, a number of militant black groups were formed that advocated violence and espoused the ideology that all white people were evil. Leaders such as Stokely Carmichael and H. Rap Brown took to the streets to proclaim that the only way the black man would get his rights would be by taking them through violence. Militant organizations such as the Black Panthers, led by activists like Bobby Seale, had open and at times violent confrontations with the police. Black activists openly advocated disrespecting the police and called them "pigs." This often caused agitation between the police and name callers and at times led to confrontation between the police and the militants. Militants argued that the American government was evil and that the police were the protectors of the American government; therefore, the police were enemies of the people. Black militant acts included arson and shooting down police officers.

Riots in American cities by blacks, similar to the Los Angeles riot of 1995, were common in American cities during the 1960s. Not only large urban centers such as Harlem (New York), Watts (Los Angeles), Newark (New Jersey), and Detroit (Michigan) but also midsized and smaller American cities experienced burning and looting. Often police actions would precipitate a riot because of exaggerated rumors. For example, when the police made a traffic stop and arrested the driver for a crime, rumors might travel through the predominantly black community that the police had unmercifully beaten the driver. Also, riots during this decade seemed to be contagious. For the first time, Americans could watch a riot in progress on the 6:00 p.m. national network news. Individuals prone to rioting observed what rioters were doing in other cities and became copycats in attempting to start a riot in their neighborhoods. Although the rioters burned and looted the stores and businesses in their own neighborhoods, they did so not only to get something for nothing but to get back at the majority of owners, who did not live in the black community. Anti-Semitism against the white business owners played a role as well.

An investigation of these civil disturbances by the National Advisory Commission on Civil Disorders (1970) found that police behavior toward blacks, particularly aggressive police patrol and harassment in urban communities, was a major factor contributing to riots. According to the commission's report,

> In Newark, in Detroit, in Watts, in Harlem—in practically every city that has experienced racial disruption since the summer of 1964—abrasive relationships between police and Negroes and other minority groups have been a major source of grievance, tension and ultimate disorder. (8)

Other commissions of the 1960s reported that at times the police had even created potential rioting situations. The Walker Report on the major confrontation that occurred between peace demonstrators against the Vietnam war and the police during the 1968 Democratic Convention in Chicago called this disturbance a police riot.

The Vietnam war escalated under the presidency of Lyndon Johnson. This led to demonstrations and violence in American streets. A segment of the American population believed that America had no business being involved in Vietnam. Many of the protesters were college students. They often demonstrated on campuses and took over buildings of departments, such as science departments, that had contracts with the Pentagon or took over the administrative offices of the university. The police were frequently called in to retake university buildings and protect university property. At times, violent confrontations took place between the police and university students. At Columbia University, the Ivy League university in New York City, both students and police officers were seriously injured; one student jumped on a police officer's back, paralyzing him. Demonstrations not only by college students but also by other Americans continued until the war came to an end under the Nixon presidency in the 1970s.

The police in the 1960s found themselves attempting to control disturbances not only by minority citizens but also by university and college students who were predominantly white and middle class and confronting Vietnam war protesters who included entertainment celebrities, business leaders, professionals, and average citizens. Serving as the defenders of an unpopular government policy could only tarnish the police image. The police in the 1960s were severely criticized for their handling of public demonstrations as well as for their poor relations with minority neighborhoods.

In the last decades of the twentieth century, the police have been more respected and treated with more courtesy than they were during the 1960s. The attitude of the public has become more positive toward the police, making it possible for community policing to be established in many communities.

The Challenge of Crime in a Free Society

In 1965, President Lyndon Johnson established the President's Commission on Law Enforcement and Administration of Justice, referred to as the President's Crime Commission. Its report *The Challenge of Crime in a Free Society* (President's Commission 1967) contained a number of recommendations to improve policing (Exhibit 1–1). Many of these recommendations were initiated, and some are still in effect. For example, many states have a state agency overseeing police standards.

The 1960s was a decade when crime increased substantially. Between 1960 and 1969, murder increased 62 percent, aggravated assault 102 percent, and robberies 177 percent (Federal Bureau of Investigation [FBI] 1970). The establishment of the commission and the passage of the Law Enforcement Assistance Act in 1965 were the first steps in providing federal grant-in-aid programs for helping state and local governments to reduce crime (National Advisory Commission on Intergovernmental Relations 1970, 8). Since the late 1960s, the federal government has continued to provide

Exhibit 1–1 Recommendations of the Challenge of Crime in a Free Society

1. Police departments should have community-relations machinery consisting of a headquarters unit that plans and supervises the department's community-relations unit.
2. A minority-group neighborhood should have a citizens' advisory committee that meets regularly with police officials to work out solutions to problems of conflict between the police and the community.
3. Departments in communities with a substantial minority population should have as a high priority the recruitment of minority officers and should deploy and promote them fairly.
4. Every jurisdiction should provide adequate procedures for the processing of all citizens' grievances and complaints about the conduct of any police officer.
5. Police departments should develop and enunciate policies that give police personnel specific guidance for the common situations requiring exercise of police discretion.
6. Basic police functions should be divided among three types of officers termed the community service officer, police officer, and police agent.
7. Police departments should recruit more actively at college campuses and inner-city neighborhoods.
8. The ultimate aim of all police departments should be that all personnel with general law enforcement powers have baccalaureate degrees.
9. Police departments should take immediate steps to establish a minimum requirement of a baccalaureate degree for all supervisory and executive positions.
10. Background investigations and personal interviews should be used by all departments to determine moral character and intellectual and emotional fitness of police candidates.
11. Primary emphasis for police recruiting should be on education, background, character, and personality of a candidate for police service.
12. To attract college graduates, salaries should be competitive with other professions that seek the same graduates.
13. Police salaries within local government should stand on their own merit and not be tied to other city departments.
14. Promotion eligibility should stress ability over seniority.
15. Police specialists should be selected for their talents and abilities without regard to their prior service.
16. A national police retirement system should be established to encourage lateral movement of police personnel.
17. All training programs should provide instruction on subjects that prepare recruits to exercise discretion properly, and to understand the community, the role of the police, and what the criminal justice system can and cannot do.
18. Professional educators should teach specialized courses such as law and psychology.
19. Formal police training for recruits should consist of a minimum of 400 hours of classroom work spread over a 4–6 month period so it can be combined with carefully selected and supervised field training.

continues

Exhibit 1–1 continued

20. Entering officers should serve probation periods preferably of 18 months but certainly no less than 1 year.
21. Every police officer should have at least one week of training annually.
22. States should establish a commission on police standards.
23. Police departments should employ a legal advisor.
24. Police departments should implement the organizational principle of central control.
25. Police departments should organize key staff and line personnel into an administrative board similar in function to a corporation's board of directors.
26. Police departments should have a comprehensive program for maintaining integrity and an internal investigative unit only responsible to the chief.
27. Police departments should commence experimentation with a team policing concept that envisions those with patrol and investigative duties combining under unified command with flexible assignments to deal with the crime problems in a defined sector.
28. Police departments should have a firearms policy that firearms can only be used when the officer's life or the life of another person is in imminent danger.
29. States should assume responsibility for assuring that area-wide records and communications needs are provided.
30. Crime laboratories should be provided to police departments by either the city, county, regional area, or state.
31. State agencies should assist smaller agencies in specialized investigations.
32. Consolidation of police services should be considered in metropolitan or county areas to keep down expenses and to provide a service that normally would not be provided.
33. Police standard commissions should be established in every state and be empowered to set mandatory requirements and to give financial aid to governmental units for the implementation of standards.

Source: Reprinted from President's Commission on Law Enforcement and Administration of Justice, *The Challenge of Crime in a Free Society,* pp. 91–123, 1967, Government Printing Office.

police agencies with funding to improve policing. Most police departments initiated their community policing strategy with federal funding. Federal grants were available for the hiring of additional police officers and equipment if the police department agreed to establish community policing. The City of Wichita Police Department, Kansas, and the Sedgwick County Sheriff's Department, Kansas, are two law enforcement agencies that instituted community policing upon receiving federal monies.

Through federal funding, made available by the Law Enforcement Assistance Act through the Justice Department's Office of Law Enforcement Assistance (OLES), 27 states established new criminal justice planning committees or broadened the activities of existing committees, 17 states began police science and college degree programs, 20 states initiated or expanded police standards and training systems, 20 states started planning for statewide integrated in-service correctional training systems, and 33 large cities developed police and community relations programs. The Safe Streets and Crime Control

Act was passed during the Johnson administration in 1967 to implement many of the recommendations of the President's Commission on Law Enforcement and Administration (National Advisory Commission on Intergovernmental Relations 1970, 7–10). Since the passage of the Law Enforcement Assistance Act, the federal government has influenced local policing through its funding process. Obviously, federal monies have played a major part in establishing standards, training, and innovations in police departments. Just as President Johnson wanted to improve policing in the 1960s, President Clinton has supported funding for those police departments initiating the community policing philosophy. In addition, the Clinton administration has created federal agencies to assist local police agencies that need training and professional consultants to establish community policing for their departments.

Police Response to a Decade of Turmoil, Violence, and Chaos

As a result of the turmoil of the 1960s, police officers, responding to what they considered to be antipolice sentiment in the form of unwarranted attacks and unnecessary criticism, began to organize. Benevolent and fraternal organizations evolved into police unions to protect the rank-and-file officer. According to Samuel Walker (1992), the noted police scholar, police officers succeeded in organizing unions in the 1960s because of several factors:

1. *Lagging salaries and benefits.* Officers were angry over the fact that their salaries and benefits had fallen behind those available in other jobs that a police officer might consider.
2. *Poor police management.* Rank-and-file officers were extremely angry and alienated over the way their departments were managed. Police chiefs at that time had almost unlimited power to run their departments, and many operated in an arbitrary and vindictive manner. Officers who criticized the department were often punished with frequent transfers or assignment to low-status jobs. Officers who were disciplined were often treated as badly as some criminal suspects, with long interrogation, no access to an attorney, no right to appeal disciplinary actions, and so on.
3. *Social and political alienation.* During the turmoil of the 1960s, police officers felt they were being attacked from all sides. They resented accusations of discrimination from civil rights groups. They also felt that Supreme Court decisions were limiting their ability to fight crime.
4. *A new generation of officers.* The police union movement was led by a new generation of police officers. They were generally younger and more assertive than the established leaders of police fraternal groups. Ironically, in asserting their own rights, the police union leaders resembled the civil rights activists and student antiwar activists whom they resented.
5. *The law and order model.* Police unions succeeded in part because the opposition to them was weak. In a period of great concern about "law and order," mayors and members of city councils did not want to appear to be hostile to the police. Thus, they were more willing to make concessions to the police unions.

6. *A new legal climate.* Police unions also succeeded because the attitude of the courts had changed. Earlier, courts had held that police and other public employees had no legal right to form their own unions, under the principle that public employment was a privilege and not a right. By the 1960s, the courts had adopted the position that employees did have some right to form unions. A 1965 Michigan Public Employees Law, for example, helped the Detroit police union establish itself. (371)

The police of the latter part of the twentieth century and early twenty-first century owe much to those police officers of the 1960s and 1970s for demanding respect and courtesy. The leaders of these decades helped reshape policing. Today, the police are not socially and politically alienated from their fellow Americans. In fact, community policing wants to bring the police closer to their fellow citizens.

National Advisory Commission on Standards and Goals

In 1971, the head of the Law Enforcement Assistance Administration (LEAA) established a National Advisory Commission on Criminal Justice Standards and Goals to formulate national criminal justice standards and develop goals for crime prevention at the local and state levels. The commission published six reports. The *Report on Police,* published in 1973, made several hundred recommendations to improve policing, including recommendations for recruiting minorities and women and for having all police agencies require a baccalaureate degree as a minimum condition of employment by 1982. Some of these recommendations have been incorporated into policing and some have not. For example, police departments generally do not require a baccalaureate degree as a condition for employment.

The *Report on Police* developed several standards that police administrators and supporters of community policing should review and incorporate into community policing programs. These standards encouraged that a police-community team work to control crime. Specific standards deal with

- local definition of police function
- accountability to the public
- communication with the public
- public understanding of the police role
- developing community resources
- community crime prevention
- police-public workshops and seminars
- responding to personnel complaints (National Advisory Commission 1973, 5)

POLICE PROFESSIONALISM

Throughout much of their history, both the English and the American police have striven toward professionalism. Robert Peel's formation of the London Metropolitan Police in 1829 had as its goal the establishment of a professional body of crime prevention specialists. When the "new police" were formed in New York City in 1845, the goal was to develop a professional body of peacekeepers. Since the mid-nineteenth century, American police periodically reinvented themselves to obtain prestige as professionals. This reinvention of philosophies and strategies, which marks the successive eras of policing, has been, for the most part, a self-serving attempt to promote an "image."

Barbara Raffel Price, in her book *Police Professionalism* (1977), claimed that the police have not achieved professionalism for several reasons. Although her book was published in the late 1970s, her analysis is still accurate today. The police are not professional for three reasons:

1. *A lack of systematic knowledge available for appropriation by the occupation.* The vacuum is apparent and is reflected in the limited evolution of police strategies and formation of experts. The absence of sufficient systematic knowledge for occupational expertise is a serious shortcoming in professionalization. A prime characteristic of all professions is the presence of a coherent body of theory or practice that the occupation can claim as its own and that can be transmitted through a lengthy training process.
2. *The bureaucratic structure of the police.* Centralization was a principal organizing force for police departments in England and subsequently was adopted in the United States as well. Today, an elaborate bureaucratic model has been built around the centralized structure. Among the many bureaucratic features of the police department are rules, specialization of functions, communications patterns, universalism, an elaborate control system of rewards and punishment, and the civil service structure. The development of a centralized command and a strict hierarchy of control was no accident. The Metropolitan Police of London, the historical model of American police, was formed to respond to civil chaos. Military discipline was imposed upon a civilian militia-type organization as a strategy for obtaining performance. It was the judgment of early police leaders in both England and America that without strict supervision there was no guarantee that the patrolman would perform his duties. No presumption existed that highly motivated men of great dedication had been recruited.
3. *The internal leadership of the occupation.* Positions taken by police leaders, those who most often speak for the occupation, have created obstacles to the police's becoming professional. The generally conservative leadership directs more attention to efforts to protect its own position and keep external criticism to a minimum than to promote substantive change. (8–10)

The issue of professionalism is still with us today. Although the police claim to be professional, they have not yet achieved professional status. Most police departments do not require a baccalaureate degree to enter the field. Further, police officers, for the most part, do not believe in lifelong learning and do not feel that they have a responsibility to keep abreast of the knowledge in their field. Many officers stop learning anything new once they leave the police academy. They believe that the police department has the sole responsibility to train officers, and, if they are not paid for attending training classes, they refuse to attend training. In police departments that have unions, the officers are backed by their unions, which hold that they should be paid for training. Unlike members of the traditional professions of medicine, law, and the ministry, many police officers want to be paid for everything they do: not only training but court time and working overtime. Yet traditional professionals do not expect to be paid extra for putting extra time into their chosen profession, and they are expected to keep abreast of knowledge in their field. Police who are not committed to their craft are more like mercenaries, who do their job only for money, than professionals.

Police professionalism is highly controversial. According to James Q. Wilson (1978), the noted police scholar,

> Occupations whose members exercise, as the police do, wide discretion alone and with respect to matters of great importance are typically "professions"—the medical profession, for example. The right to handle emergency situations, to be privy to "guilty information," and to make decisions involving questions of life and death or honor and dishonor is usually, as with a doctor or priest, conferred by an organized profession. The profession certifies that the member has acquired by education certain information and by apprenticeship certain arts and skills that render him competent to perform these functions and that he is willing to subject himself to the code of ethics and sense of duty of his colleagues. . . . Failure to perform his duties properly, if detected, will be dealt with by professional sanctions—primarily, loss of respect. Members of professions tend to govern themselves through collegial bodies, to restrict the authority of their nominal superiors, to take seriously their reputation among fellow professionals, and to encourage some of their kind to devote themselves to adding systematically to the knowledge of the professions through writing and research. The police are not in any sense professionals. They acquire most of their knowledge and skill on the job, not in separate academies; they are emphatically subject to the authority of their superiors; they have no serious professional society, only a union-like bargaining agent; and they do not produce, in systematic written form, new knowledge about their craft. (29–30)

ERAS OF POLICING

The history of policing has been subdivided by police scholars into "eras" of policing. These are not exact time frames, and the number of eras propounded varies depending upon who is presenting the information. The most well-known scheme is that of Kelling and Moore (1988), who described a political era, a reform era, and a community problem-solving era.*

---

*I deviate from several of Kelling and Moore's premises and include concepts contrary to their work in the following paragraphs.

Political Era

Kelling and Moore (1988) denoted the political era of policing as the period from 1840 to the early 1900s. During this period, there was a close relationship between the police and politicians. Cities and towns were growing, and modern policing was in its infancy. Policing came under the control of local governments. In urban centers, politicians oversaw police operations. The police and the politicians worked together to control the neighborhood, often referred to as a *ward*. It was the ward politician who controlled the police in his neighborhood: because job security was not available to ward police officers, they were obligated to the ward politician for their positions. Police tasks included not only crime prevention and order maintenance but also a broad range of social service activities.

Although police departments had a centralized, semimilitary, hierarchical organizational structure, they generally did not function along these lines. Rather, they functioned as decentralized units, with the ward politician running the police department in his ward. Police officers reflected the ethnic makeup of the ward in which they lived and worked. Policing in the ward was conducted primarily by foot patrol. Communication between police officers and a centralized police headquarters was almost nonexistent. When call boxes came into existence, they were used for the patrol officer to maintain contact with supervisors. Some police departments had detectives, but they were not well trained or held in great esteem. Detectives were often used by politicians to obtain information for political purposes. Their caseloads were usually based upon persons and not criminal offenses. The strengths of the political era of policing were:

> First, police were integrated into neighborhoods and enjoyed the support of citizens—at least the support of the dominant economic and political interests of an area.

> Second, and probably as a result of the first, the strategy provided useful services to communities. There is evidence that it helped contain riots. Many citizens believed that police prevented crimes or solved crimes when they occurred. And the police assisted immigrants in establishing themselves in communities and finding jobs. (Kelling and Moore, 1988, 4)

The political strategy also had weaknesses.

> First, intimacy with community, closeness to political leaders, and a decentralized organizational structure, with its inability to provide supervision of officers, gave rise to political corruption. Officers were often required to enforce unpopular laws foisted on immigrant ethnic neighborhoods by crusading reformers who objected to ethnic values. Because of their intimacy with the community, the officers were vulnerable to being bribed in return for nonenforcement or lax enforcement of laws. Moreover, police closeness to politicians created such forms of political corruption as patronage and police interference in elections.

Second, close identification of police with neighborhoods and neighborhood norms often resulted in discrimination against strangers and others who violated those norms, especially minority ethnic and racial groups. Often ruling their beats with the "ends of their nightsticks," police regularly targeted outsiders and strangers for rousting and "curbstone justice."

Finally, the lack of organizational control over officers resulting from decentralization and the political nature of many appointments to police positions caused inefficiencies and disorganization. The image of Keystone Cops—police as clumsy bunglers—was widespread and often descriptive of realism in American policing. (4)

Hubert Williams and Patrick Murphy (1990) have critiqued Kelling and Moore's description of the political era as pertaining primarily to Northeast urban centers and as omitting mention of social and racial conflicts related to police work, such as the work of "slave patrols" up to the time of the Civil War to apprehend runaway slaves, racial oppression by the police, the enforcing of racially biased laws, and the scarcity of black police officers. During this period, blacks had limited legal rights and were discriminated against in the North and eventually in the South after Reconstruction, when slavery was outlawed.

In still another perspective on the political era of policing, Victor Strecher (1991) questioned "whether policing was deviantly political or simply part of an era which was inherently political" (2). During this period, government services were developing not only for policing but also for such areas as education, social service, and sanitation. The development of government services came under the influence of politics, and the police were not an exception. Strecher also pointed out that bestowing the name "political era" on this period could lead to the erroneous interpretation that politics no longer influences policing.

Reform Era

The reform era, according to Kelling and Moore, began in the early 1900s and lasted into the 1970s. It was characterized by a movement to reform policing by freeing it from the control of politicians. The implementation of civil service for police departments decreased and in some cases eliminated political patronage in the hiring and firing of police officers. Political influence was considered as deviant and as indicating poor police administrative leadership. The police emphasized their law enforcement function over their previous social service role during the political era.

Administrators followed a management philosophy of control, unity of command, and a division of labor. Attempts were made to routinize and standardize police patrol. Communications within the police organization flowed downward but not upward. Police administrators showed little interest, if any, in what lower ranking police officers had to recommend. Unlike the political era, which had supported a close citizen-police relationship, the reform era deemphasized this relationship. The goal of police administration was to distance the police from citizens, because during the political era police-citizen closeness had led to corruption and favoritism among neighborhood groups. Some communities decreased foot patrols while increasing automobile patrols. Police in cars definitely put a distance between police officers and citizens. In this era, police took the stance that they were the professionals who would handle crime control and that the general public had no business to be involved because they lacked the expertise and knowledge.

In the 1960s and 1970s, the reform era strategy ran into several obstacles:

First, regardless of how police effectiveness in dealing with crime was measured, police failed to substantially improve their record. During the 1960s crime began to rise. Despite large increases in the size of police departments and in expenditures for new forms of equipment, police failed to meet their own or public expectations about their capacity to control crime or prevent its increase. More research conducted during the 1970s on preventive patrol and rapid response to calls for service suggested that neither preventive patrol and rapid response to calls for service was an effective crime control or apprehension tactic.

Second, fear rose rapidly during this era. The consequences of this fear were dramatic for cities. Citizens abandoned parks, public transportation, neighborhood shopping centers, churches, as well as entire neighborhoods. . . .

Third, despite attempts by police departments to create equitable police allocations systems and to provide impartial policing to all citizens, many minority citizens, especially blacks during the 1960s and 1970s, did not perceive their treatment as equitable or adequate. They protested not only police mistreatment, but lack of treatment—inadequate or insufficient services—as well. . . .

Fourth, the civil rights and anti-war movements challenged police. This challenge took several forms. The legitimacy of police was questioned: students resisted police, minorities rioted against them, and the public, observing police via live television for the first time, questioned their tactics. . . .

Fifth, some of the myths that undergirded reform strategy—police officers use little or no discretion and the primary activity of police is law enforcement—simply proved to be too far from reality to be sustained. . . .

Sixth, although the reform ideology could rally police chiefs and executives, it failed to rally line police officers. During the reform era, police executives had moved to professionalize their ranks. Line officers, however, were managed in ways that were antithetical to professionalization. . . .

Seventh, police lost a significant portion of their financial support, which had been increasing or at least constant over the years, as cities found themselves in fiscal difficulties. . . .

Finally, urban police departments began to acquire competition: private security and the community crime control movement. (Kelling and Moore 1988, 8–9)

Strecher (1991) did not completely agree with Kelling and Moore's reform era model. First, he took issue with the years initiating and ending the reform era and claimed that the reform era began in the 1880s with the establishment of civil service. Second, he argued that a variety of research findings were not included in the reform era paradigm.

My own objection to Kelling and Moore's scheme is that it makes no mention of the crime prevention model, which came into vogue in the 1910s and lasted to the 1940s. As discussed earlier in this chapter, this model emphasized crime prevention and social work as the most important functions of policing and was a boom for women entering policing, many of whom were trained and experienced social workers.

In the 1930s, the crime prevention model began to lose ground to the crime control model discussed earlier in this chapter, which fits in better with Kelling and Moore's description of the reform era in that it discarded social service and preventative roles for police. With the adoption of the crime control model, the role of women in policing began to decrease. The new model emphasized making arrests, and at this time men were considered better suited than women to this task.

The crime control model of the reform era has not completely fallen by the wayside. The reform era ideology has been ingrained into many police organizations, and it is difficult for police agencies to change.

Community Problem-Solving Era

Problem-Oriented Policing

In 1979, Herman Goldstein published an influential article that sought to initiate a new mode of problem solving in police departments. Goldstein maintained that in traditional police practice, a citizen calls the police about a specific problem and the police respond, handle the incident, and return to their patrol duties. Goldstein argued that this approach, which he called *reactive,* was standard operating procedure even though the beat officer might be called to the same incident over and over again without correcting the situation. He recommended that the police instead work proactively and direct their efforts toward solving the persistent problems that generate the calls. Eck and Spelman (1987) described this approach as "a departmentwide strategy aimed at solving persistent community problems. Police identify, analyze, and respond to underlying circumstances that create incidents" (xv).

Although many police department claim that they are doing community policing, they are often doing only problem-oriented policing. Chapter 5 will provide more detailed information about problem-oriented policing.

Community-Oriented Policing

In community-oriented policing, as in problem-oriented policing, police take a proactive role in problem solving, but they do so in partnership with the communities they serve. Community-oriented policing has been defined as "a philosophy of full service personalized policing where the same officer patrols and works in the same area on a permanent basis, from a decentralized place, working in a proactive partnership with citizens to identify and solve problems" (Trojanowicz and Bucqueroux 1994, 3). Community-oriented policing will be reviewed in much greater detail in later chapters of this book.

Community Problem Solving

Kelling and Moore described the era from the late 1970s or early 1980s to the present as the community problem-solving era, in that it combines problem-oriented policing with a community orientation. Community problem solving involves a variety of policing strategies that will be outlined in this book.

The wave of the future appears to be community-oriented policing. For several years community-oriented policing has had the support of the federal government. The federal government has provided funding for additional police officers to police departments for implementing community-oriented policing. In addition, grants have been made available to police agencies incorporating the community-oriented policing strategy.

As Strecher (1991) has noted, community-oriented policing is a slippery concept that academics have defined differently for different purposes, and it has stirred up many controversies (6). The purpose of this book is to address some of the questions that Strecher raised. Only by reviewing contemporary police innovations can we ascertain the strategies that make up problem-oriented and community-oriented policing and assess their value.

SUMMARY

Community policing has been advanced as the new policing system for the twenty-first century. How can police administrators, political leaders, or students of policing support or even understand the value of and need for community policing in our contemporary world if they have no knowledge of where the police have been? How can effective decisions and policies be implemented if politicians and administrators making decisions lack knowledge or understanding of police history? Politicians, police administrators, community policing supporters, and students of policing should understand the history of policing, including its various strategies and methods, if they hope to have an impact on the field.

In ancient times, individuals were expected to follow the rules of the majority, and policing was conducted informally. The tribe, clan, and family had the responsibility to control individuals' behavior and to enforce the informal rules or customs established by the tribe or clan. Eventually, rules became laws, and rule breaking was formalized as an act committed against the state.

With the development of formalized governments and states, laws became a prerequisite. The Roman Empire formalized laws when it formalized its government structure and gave the responsibility of enforcing the laws to its legions. When the Romans conquered the British Isles, they brought their legal system with them. But with the collapse of the Roman Empire, Germanic tribes invaded England, bringing with them their tribal customs and unwritten laws. These unwritten laws became integrated into the English legal system. During this period, the concept of king's peace evolved, whereby the king assured the people a state of security and peace in return for their allegiance to him. This peacekeeping system revolved around land owner-ship in that the landowner was responsible for all those on his land who violated the king's peace.

During the seventeenth century in London, night watches were paid a small fee to keep the peace. They were recruited from the poor, the old, and the weak and were ineffective in performing their task. During the eighteenth century, the term *police* began to be used. The idea of police involvement in crime prevention emerged during the early eighteenth century, along with that of citizens reporting crimes to the police. In the early nineteenth century, Sir Robert Peel, the Home Secretary, recommended to Parliament that a paid police force be appointed from the civilian population to main-tain order in London. Parliament approved Peel's proposal, and the London Metro-politan Police was formed.

America traces its legal and police system to England. Americans used the London model to form American police departments. In 1845, New York City had the first "unified" police department in the United States: that is, a department combining the night and day watches in one police agency. Generally, American police, unlike the London police, did not function as professionals. Although municipal police depart-ments were theoretically centralized, they tended to function as decentralized neigh-borhood units. Early police departments were often responsible to perform activities that had little to do with maintaining public order. For example, the police cleaned streets in New York City, distributed supplies to the poor in Baltimore, assisted the homeless in Philadelphia, and inspected vegetables in St. Louis.

In the latter part of the nineteenth century the police came under criticism for being under the influence of politicians, corruption, and inefficiency. Police critics advo-cated police training; eliminating corruption, police brutality, and inefficiency; and removing politics from police affairs. The goal of police critics was to professionalize the police. The police were given a new image; they were to become crime fighters. The expectation was that this would be the avenue to professionalized police.

The 1960s was a decade which began with hope, promise, and enthusiasm but ended in a state of disorder, disruption, and chaos. This decade saw the assassination of political and civil rights leaders, riots in the streets, demonstrations against the Vietnam war, and anti-government disdain by certain segments of the American population. The police rep-resented the government and the status quo; therefore, they received the abuse directed toward the government. Government commissions recommended that police officers re-ceive training and eventually obtain a college degree.

This chapter also reviewed the eras of policing, which are not exact time frames. Police scholars have conceived of at least three eras, and not all scholars agree. George Kelling and Mark Moore outline three eras of policing. They are: political era, reform era, and community problem-solving era. Hubert Williams and Patrick Murphy criticize Kelling and Moore because they completely eliminate minorities from their review of the eras of policing. Victor Strecher finds fault with a variety of premises advocated by Kelling and Moore; for example, he suggests that the police were not unique in their political involvement but a part of an inherently political involvement that included all government agencies.

KEY TERMS

August Vollmer	James Q. Wilson	new police
Bow Street Runners	John Fielding	Patrick Colquhoun
constable	king's peace	President's Crime
crime fighter	Metropolitan Police Act	Commission
crime prevention	mutual pledge	Robert Peel
frankpledge	National Advisory Commission on	shire reeve
Henry Fielding	Criminal Justice Standards and Goals	

REVIEW QUESTIONS

1. What value does an understanding of police history have for a student of policing?
2. Why is the establishment of the London Metropolitan Police important to American policing?
3. How important is the history of English policing to American policing, and in what ways?
4. Describe changes in American policing in the latter part of the nineteenth century.
5. What are the major differences between the crime control model and the crime prevention model?
6. Why is the 1960s considered a decade of violence?
7. Describe the various eras of policing.

REFERENCES

Appier, J. 1992. Preventing justice: The campaign for women police, 1910–40. *Women and Criminal Justice* 4, no. 1 (1992): 5.

Bopp, W.J., and D.O. Schultz. 1972. *A short history of American law enforcement,* Springfield, IL: Charles C Thomas.

Carte, G.E., and E.H. Carte. 1975. *Police reform in the United States: The era of August Vollmer, 1905–1932.* Berkeley: University of California Press.

Chapman, S.G., and T.E. St. Johnson. 1962. *The police heritage in England and America.* East Lansing: Michigan State University.

Connolly, V. 1944. Job for a lady. *Collier's,* June 10, p. 19.

Critchley, T.A. 1972. *A history of police in England and Wales.* 2nd ed. Montclair, NJ: Patterson Smith.

Eck, J.E., and W. Spelman. 1987. *Problem-solving: Problem-oriented policing in Newport News.* Washington, DC: Police Executive Research Forum.

Federal Bureau of Investigation. 1970. *Uniform crime reports 1969.* Washington, DC: Government Printing Office.

Fogelson, R.M. 1977. *Big-city police.* Cambridge, MA: Harvard University Press.

Fosdick, R.B. 1929. *American police systems.* New York: Century.

Germann, A.C., et al. 1973. *Introduction to law enforcement and criminal justice.* Springfield, IL: Charles C Thomas.

Goldstein, H. 1979. Improving policing: A problem-oriented approach. *Crime and Delinquency* 25: 236–258.

Johnson, D.R. 1981. *American law enforcement: A history.* Saint Louis, MO: Forum Press.

Kelling, G.L., and M.H. Moore. 1988. The evolving strategy of policing. *Perspectives on Policing,* no. 4, 1–15.

Melville, L. 1971. *A history of policing in England.* Montclair, NJ: Patterson Smith.

National Advisory Commission on Civil Disorders. 1970. *Report of the National Advisory Commission on Civil Disorders.* 90th Congress, June. Washington, DC: Government Printing Office.

National Advisory Commission on Criminal Justice Standards and Goals. 1973. *Task force report: The police.* Washington, DC: Government Printing Office.

National Advisory Commission on Intergovernmental Relations. 1970. *Making the Safe Streets Act work: An intergovernmental challenge.* Washington, DC: Government Printing Office .

President's Commission on Law Enforcement and Administration of Justice. 1967. *The challenge of crime in a free society.* Washington, DC: Government Printing Office.

Price, B.R. 1977. *Police professionalism.* Lexington, MA: Lexington Books.

Pringle, P. 1955. *Hue and cry: The story of Henry and John Fielding and their Bow Street Runners.* Suffolk, England: William Morrow.

Strecher, V.G. 1991. Histories and future of policing: Readings and misreadings of a pivotal present. *Police Forum* 1, no. 1: 2.

Trojanowicz, R., and B. Bucqueroux. 1994. *Community policing: How to get started.* Cincinnati, OH: Anderson.

Walker, S. 1977. *A critical history of police reform.* Lexington, MA: Lexington.

Walker, S. 1992. *The police in America.* 2nd ed. New York: McGraw-Hill.

Williams, H., and P.V. Murphy. 1990. The evolving strategy of policing: A minority view. *Perspectives on Policing,* no. 13, 1–15.

Wilson, J.Q. 1978. *Varieties of police behavior: The management of law and order in eight communities.* Cambridge, MA: Harvard University Press.

Chapter 2

Understanding Police Culture

CHAPTER OBJECTIVES

1. Understand the importance of police culture to community policing.
2. Understand the meaning of *culture.*
3. Understand the various aspects of the police culture.
4. Be familiar with what makes police culture unique.
5. Be familiar with the various myths about policing.

INTRODUCTION

This chapter will concentrate on the value system reflected in the police culture. An examination of the police culture will provide a better understanding of why community policing is necessary. If the goal of community policing is to change the way the police do business, then an understanding of how police culture works and what its values are is a crucial prerequisite. How can any changes be made if the police culture is not taken into consideration? The culture of policing must be scrutinized to determine what effect community policing or any new policing philosophy or strategy will have on the police. Before a concept can be put in place, innovators need to be aware of the occupational culture they hope to influence or change. The philosophy of community policing, if completely adopted and integrated into policing, will change the culture of policing. The big question that remains is, can the police culture change to fit the philosophy of community policing?

POLICE CULTURE

The police culture cannot be overlooked if policing hopes to change its philosophy. But no student can be expected to understand the meaning of police culture without first having a general understanding of culture. What does the term *culture* mean? What is the culture of American society? What is meant by the culture of an occupational field?

Culture is the foundation upon which a social group functions. It consists of knowledge, beliefs, morals, laws, customs, and practices acquired by the group. Culture can be defined as "the total way of life shared by members of a society. It includes not only language, values, and symbolic meanings but also technology and material objects" (Brinkerhoff et al. 1997, 59). The concept of culture also encompasses our thinking, feelings, attitudes, and communications with each other. We learn culture from the people around us. This includes speaking, facial expressions, body language, and beliefs. Cultures are constantly being altered; people are exposed to new ideas, and new forms of technology are developed. For example, feminism and the computer have been added to our culture. Because police officers are members of American society, they operate within that culture's norms, customs, morals, and value system. In America, our culture includes the American dream, competitiveness, equal opportunity for everyone to succeed, and an attitude of independence ("No one can tell me what to do"). Culture not only changes constantly, at least in America, but endures while humans die. Our American culture has been passed down from generation to generation through parents, relatives, and acquaintances. In contemporary society, culture is also being passed down through books, movies, music, museums, television, and entertainment parks.

Culture can be divided into *material culture* and *nonmaterial culture*. Material culture includes tools, streets, television, VCRs, forks, spoons, knives, houses, and automobiles, to name only a few items. Nonmaterial culture includes beliefs, attitudes, education, language, values, rules, norms, mores, folkways, laws, knowledge, and ways of dealing with similar problems and experiences, such as birth, death, raising children, and marriage.

A common language often reflects a common culture. The American version of English binds us culturally together as a people. Our English language symbolizes unity and gives us an identity as Americans. Values, after language, are central to culture. They guide our feelings about events in our lives and govern our sense of right and wrong. Different cultures have different values, and some social scientists have attempted to promote "dominant value profiles" for specific cultures. The American society has been characterized as a *drive society,* "one oriented towards achievement, competition, mobility, and an ever-higher standard of living" (McNall and McNall 1992, 60). McNall and McNall described the major American value system as follows:

1. *Achievement/success.* This is often measured by quantity and is dominated by the belief that only one's own effort will obtain success. American heroes have often been self-made people: Abraham Lincoln split rails, studied by firelight, saved his money, and eventually became president. . . .
2. *Activity/work.* In colonial America and on the frontier, work had survival value. Idleness was dangerous not only from a practical point of view, but from a religious point of view as well. To be busy, useful, and active had intrinsic value besides the tangible results these qualities yielded. The link between work and morality (a good person works, a bad person does not) was fostered by America's Puritan heritage. . . .

3. *Morality.* The Pilgrims, the frontiersmen, and the prairie settlers were all strangers who migrated to a hostile land and struggled to master it. They also sought to recreate the civilization they had left behind. . . . A strong link was forged between morality and civilization, since morality in America was associated with Protestant Christianity. This heritage contributed to a tendency to judge, measure, and weigh, which often meant comparing oneself with others: those who did not work hard, who drifted, who did not save their money, or who drank. A sober, hard-working, God-fearing person was moral and civilized, and those who fell outside of these categories were pitied or despised. . . .

4. *Efficiency/practicality.* Americans value useful activity. Our emphasis on practicality and efficiency makes us tend to accept rational activity and science without question. Efficiency is the standard against which we judge nearly all activities. . . .

5. *Progress.* No idea has more greatly dominated or characterized the American approach to life than progress. The belief in progress is a conviction that the present is better than the past, and that the future will be better still. . . .

6. *Humanitarianism.* Americans believe in helping others, but this value, coupled with morality, has sometimes led to the belief that we must save other people from their shortcomings. A crusading vision has been characteristic of much of American life. The tendency to engage against drinking or against "communism" can have negative consequences because it can result in the suppression of other people's civil rights. . . .

7. *Materialism.* Eighteenth and nineteenth century Americans certainly hoped for a good life. They strove to obtain the material goods that would fulfill such basic needs as food, shelter, and clothing, and some wanted, and got, luxuries like pianos, books, and large houses. Still for them, consumption was not an end in itself. Most Americans' lives were conducted in accordance with a set of values that stressed respectability, hard work, discipline, and helping others. These values have been eroded in our time by a consumer-oriented society. . . .

8. *Democracy.* Americans believe in freedom and equality. (61–63)

Do American police officers share these values? Are these values brought with them when they enter the police field? We as Americans reflect the value system of our nation, neighborhoods, family, relatives, and acquaintances that we copied and adopted as our own. It has traditionally been accepted that police officers should be recruited from the community and should reflect the value system of the community. This unit will discuss in more detail the value system of police officers as it relates to their occupational culture. Other elements pertaining to culture that should briefly be reviewed include norms, folkways, mores, and laws.

Norms are rules of conduct. They designate what people should or should not do, think, or feel. The importance of norms can vary substantially in importance to individuals and society. For example, some people place no importance on their apparel, whereas other people consider it to be of extreme importance. There are two types of informal norms: folkways and mores. Folkways are customary, normal, and habitual ways of doing things. An example of a folkway could be setting off fireworks on July 4. Folkways have no right or wrong assigned to them. However, a person could be considered weird if he or she had green hair. Mores reflect a strong awareness of right or wrong. A violation of mores could be abusing a child. Someone who abuses a child, either physically or sexually, has violated not only mores but also a third type of norm known as a law. Laws, simply, are rules that are enforced by governmental authorities (Brinkerhoff et al. 1997, 67).

Folkways, mores, and laws are forms of *social control.* All cultures have checks on individual, group, and crowd behavior. The control of individual or group behavior may be either informal or formal. Individuals who do not follow the norms or mores of the culture will find themselves ostracized. However, if they break the rules, or more specifically the laws, of government, a more severe penalty can be imposed, such as a fine, confinement to a specific geographical location, or imprisonment. Governmental agencies involved in controlling behavior are criminal justice personnel: judges, prosecuting and defense attorneys, corrections personnel, and police officers. The vanguard in maintaining internal social control in American society are the police. The police are given the legal responsibility to assert social control and the option to use various means to achieve this goal; including arrests and the use of physical force—necessary, deadly force, or force with the potential of causing death. This book discusses the various philosophies, strategies, and methods of social control used by the police. Community policing is the latest social control philosophy to be adopted by the police.

Occupational Culture

Various groups of workers develop a culture unique to their specific occupation. Policing, like other occupations, has developed a unique culture. Before we examine the police culture in detail, an understanding of occupational culture will be helpful. To understand occupational culture, we need to understand how it fits into an organization and to answer several questions: How does occupational culture serve the organization and its members? How does occupational culture evolve? Where does an organizational culture come from?

For starters, we know that the external social environment of an organization exerts an influence. Our American culture influences occupational culture—managers and workers bring the dominant culture to the workplace. The personality of the head of a specific organization may influence the culture of that organization. So may the management style of an organization, whether authoritarian or democratic.

Researchers of organizational cultures study organizational stories; language, including jargon and slang; jokes; rituals; ceremonies; myths; heroes; values; organizational structure; social roles; and standard operating procedures. They also study the intentions of organizational members, what members of the organization think is necessary to move ahead within the organization (Ford 1988, 346) and "the degree of trust, communications, and supportiveness that exists in the organization" (Hellriegel et al. 1983, 590).

It should be understandable that when people work together for any length of time, they will generate particular patterns of thinking, behaving, and feeling. The development of a unique understanding of the world and shared values, symbols, distinctive language, and behavioral norms separates occupational members from nonmembers and legitimately characterizes *subcultures*. Occupational cultures reflect a cooperative transformation to social and physical working conditions. People who perform similar tasks are exposed to similar experiences and are confronted with similar problems. They also interact with similar people and may receive comparable rewards. Workers who share experiences and interact create a feeling of cohesion. They develop knowledge unique to their occupations, a common language, and informal rules for dealing with problems. Also, work clothing may become a symbol of membership in the group (Rothman 1987, 40–41).

Subcultural Occupations

Subcultural occupations tend to isolate individuals from outsiders and from the society at large. Occupations exist, such as the police, where segregation extends beyond the workplace to include leisure and social activities. Workers of specific occupations can form closed communities, associating only with people from their same occupation to the exclusion of individuals from other fields. The people they work with are the people they drink with, golf with, and entertain at home. Attitudes and behaviors that are distinctive and shared by a specific occupational group are instrumental in accomplishing the technical and social goals of the job. Shared work patterns, found in many occupations, ensure predictability and reliability. Occupational standards allow workers to anticipate how other workers will react in a specific situation. Career fields, through their subculture, provide protection from a social or physical environment that appears to be threatening. Symbols of subcultures integrate members into a feeling of solidarity. Uniforms, traditions, and ceremonials help create a cohesive group. Self-image, social status, and social placement are also achieved through an occupational subculture (Rothman 1987, 42–43). As a general concept, subculture encompasses a number of elements:

- Language
  1. argot
  2. gestures
  3. posture
- Artifacts
  1. tools
  2. grooming
  3. uniforms
- Beliefs
  1. knowledge
  2. myths
  3. stereotypes

- Values
- Norms
  1. technical norms
  2. interpersonal norms
- Rituals
  1. magic
  2. rites of passage
  3. naming
  4. expulsion ritual (Rothman 1987, 43)

To consider these elements one by one: the police do have their own language, which includes argot—a distinctive police vocabulary. They have their own tools, such as firearms, nightsticks, mace, and two-way radios, and their own special knowledge. Police often hold certain beliefs in common, including myths about the police and stereotyping of outsiders. The police value system includes a commitment to the role of being society's protectors from "bad guys": "thin blue line." The norms of policing include "backing up a fellow officer" in need of assistance. The police have their own rituals, which may include the drinking with colleagues after duty. Thus, the police do have their own subculture, which this section will discuss in some detail.

What sort of young person enters the police field? Studies indicate that young people entering policing generally are recruited from lower-middle-class backgrounds. John Van Maanen (1978) gave several reasons why young people enter policing:

1. The literature notes that police work seems to attract local, family-oriented, working-class whites interested primarily in the security and salary aspects of the occupation. . . . The available research supports the contention that the police occupation is viewed by the recruits as simply one job of many and considered roughly along the same dimensions as any job choice. . . .
2. The out-of-doors and presumably adventurous qualities of police work (as reflected in the popular culture) were perceived by the recruits as among the influential factors attracting them to the job. With few exceptions, the novice policemen had worked several jobs since completing high school and were particularly apt to stress the benefits of working a nonroutine job.
3. The screening factor associated with police selection is a dominating aspect of the socialization process. From the filling out of the application blank at City Hall to the telephone call which informs a potential recruit of his acceptance into the department, the individual passes through a series of events which serve to impress an aspiring policeman with a sense of being accepted into an elite organization. . . .
4. As in most organizations, the police department is depicted to individuals who have yet to take an oath of office in its most favorable light. A potential recruit is made to feel as if he were important and valued by the organization. Since virtually all recruitment occurs via generational or friendship networks involving police officers and prospective recruits, the individual receives personalized encouragement and support which helps sustain his interest during the arduous screening procedure. Such links begin to attach the would-be policeman to the organization long before he actually joins. (295–296)

In surveys of graduates from the New York City Police Academy, it was found that 80 percent of the fathers of police officers were employed either as service workers or as laborers. Most police recruits have not had high-status employment positions. Many of them have drifted from job to job. This is especially true when police agencies are in a hiring mode. It appears that police departments are not as selective about whom they employ when they are hiring a large number of recruits. The various police scandals of police corruption during the 1980s reflect this. Miami, Houston, New York City, and Washington, D.C., all had police scandals due to poor recruiting. The general requirement for most police agencies is either a GED or a high school diploma. There are police departments that require a baccalaureate degree, but this is more the exception than the rule. The New York Police Department, the nation's largest municipal police agency, adopted in 1997 a policy requiring all police recruits to have two years of college or to substitute two years of military service for the college. Although more police officers are now entering police work with college degrees, many of them having a major in criminal justice, they are still a minority. Some police departments will give extra points on promotional examinations to police officers who have college degrees. Other police departments, such as the Wichita (Kansas) Police Department, require a college degree for promotion.

Since the late 1960s and early 1970s, police departments have concentrated on increasing the numbers of African Americans, Hispanics, Asians, and women in policing. The goal of police agencies is to have a department reflect the racial, ethnic, and gender makeup of the community. The numbers of minorities and women have increased substantially since the 1970s but are still small when compared against the composition of many communities.

Occupational prestige ratings place police officers in the middle, slightly below average. Police officers rank lower in prestige than electricians, bank tellers, or athletes (Davis and Smith 1983). Surveys of police officers concerning prestige found that police officers considered their job better than that of a furniture mover, auto mechanic, or bus driver but not as good as that of a high school teacher, business executive, or druggist. This could be attributed to police officers' working-class background and, for most officers, limited educational achievements. A New Orleans survey of police officers discovered that most officers viewed themselves as extroverted and dependable. These officers also ranked themselves as being low on ambition, intellect, and sophistication. Officers' projection of low self-esteem could have a bearing on their work performance (Hahn 1974, 17–18). The New Orleans survey, however, may not be an accurate reflection of the nation's police officers: for decades the New Orleans Police Department has had a reputation for corruption, inefficiency, and brutality.*

---

*As a former police officer and a researcher of police departments, I believe that the New Orleans Police Department is the exception rather than the rule and that police scandals are rarities rather than common occurrences. Because police scandals such as those that have occurred over the years in New Orleans receive national media play, one can get the false impression that this is the way police do business, but this would not be a true assessment of policing in America. On the whole, police departments are staffed by people of integrity and honesty who are committed to serving their community. One police department that has been free from scandal is the Wichita Police Department. This department, like most of America's police departments, takes immediate steps if a police officer violates a departmental policy or law. When they occur, police scandals are tragedies that unfairly tarnish the image of honest committed police officers.

Identification of Symbolic Assailants

Jerome Skolnick (1994) considered a continually watchful stance toward what he termed *symbolic assailants* to be an integral part of the police culture (51–63). Because of the danger associated with police work, according to Skolnick, the police develop intuitive techniques to identify undesirables or potential law violators. These indicators could include a person's appearance, such as the clothing or hairstyle or, more recently, tattoos and body piercing. For example, gangs are known to use tattoos as symbols that other gang members would recognize. Characteristics of the symbolic assailant as described by Skolnick also include language, gestures, attitude, and body language. Police are trained to be suspicious (though suspicion should be controlled and not too obvious) and to look for the unusual: persons who do not "belong" where they are observed, automobiles that do not "look right," and businesses opened at odd hours, or not according to routine or custom. One article in a journal gives the following partial list of subjects who should be given field interrogations:

1. Suspicious persons known to the officer from previous arrests, field interrogations, and observations
2. Emaciated appearing alcoholics and narcotics users who invariably turn to crime to pay the cost of habit
3. Person who fits descriptions of wanted suspect as described by radio, teletype, daily bulletin
4. Any person observed in the immediate vicinity of a crime very recently committed or reported as "in progress"
5. Known trouble-makers near large gatherings
6. Persons who attempt to avoid or evade the officer
7. [Persons displaying] exaggerated uncertainty over contact with the officer
8. [Persons] visibly "rattled" when near the policeman
9. Unescorted women or young girls in public places, particularly at night in such places as cafes, bars, bus and train depots, or street corners
10. "Lovers" in an industrial area (make good lookouts)
11. Persons who loiter about places where children play
12. Solicitors or peddlers in a residential neighborhood
13. Loiterers around public rest rooms
14. Lone male sitting in car adjacent to school ground with newspaper or book in lap
15. Lone male sitting in car near shopping center who pays unusual amount of attention to women, sometimes continuously manipulating rearview mirror to avoid direct eye contact
16. Hitchhikers
17. Persons wearing coat on hot days
18. Car with mismatched hub caps, or dirty car with clean license plate (or vice versa)
19. Uniformed "deliverymen" with no merchandise or truck (Adams 1963, 28)

The training that recruits receive while attending the police academy contributes to the identification of symbolic assailants. The police are trained to be suspicious. In fact, police officers believe that a suspicious nature is a key ingredient of being a good police officer. Police officers know the environment they patrol and often can detect the unusual. I myself know police officers who are familiar with the residents who live in their patrol areas and who know which automobiles go with which houses. Some officers are so good that they know what garage doors should be opened or closed and whether a car visor should be down or up. Good patrol officers who know how to work their beat are familiar with the small details of their patrol area. Police officers who develop the attitude of suspicion in their work are working hard to make arrests and to control crime.

Police officers who start working the street quickly learn to direct the "policeman's stare" toward the symbolic assailant. In our American culture, staring at an individual is considered rude or abnormal, but police officers consider staring as a means to control a situation. They are encouraged when patrolling in vehicles to do so with their windows open in order to pick up sounds or unique smells. They should use all their senses to determine if anything out of the ordinary, such as the smell of a gas main break, is occurring on their beat, and the law recognizes that they can use the evidence of their senses—seeing, hearing, smelling, touch, and/or taste—to make an investigation without a warrant.

Secrecy

"The concept of ethos encompasses the distinguishing character, sentiments, and guiding beliefs of a person or institution" (Kappeler et al. 1998, 99). A major ethos of policing is secrecy. As early as 1930, August Vollmer was critiquing the "blue code of secrecy" that he considered characteristic of many police departments.

Kappeler et al. (1998) listed the specific injunctions of the code of secrecy as follows:

- *Don't give up another cop.* Perhaps one of the most important factors contributing to secrecy and to a sense of solidarity, this injunction warns officers never, regardless of the seriousness or nature of a case, to provide information to either superiors or nonpolice that would cause harm to fellow officers.
- *Watch out for your partner first and then the rest of the guys working that tour.* This injunction tells police officers they have an obligation first to their partners, and then to other officers working the same shift. "Watching out," in this context, means that officers have a duty not only to protect a fellow officer from physical harm but also to watch out for his or her interests in other matters. If, for example, an officer learns that another member of his or her squad is under investigation by an internal affairs unit, the officer is obligated to inform the fellow officer of this fact.
- *If you get caught off base, don't implicate anybody else.* Being caught off base can involve a number of activities, ranging from being out of one's assigned

sector to engaging in prohibited activities. This injunction teaches officers that if they are discovered in proscribed activities, they should accept the punishment and not implicate others.

- *Hold up your end of the work; don't leave work for the next tour.* This injunction tells officers that if they neglect their work responsibilities, two results are likely to occur. First, other officers must cover for those who shirk their responsibilities. Second, malingerers call attention to everyone on a shift. Thus, there are pressures for all officers to carry their own weight to avoid being detected for deviance. If, however, an officer fails to follow this edict, other officers are expected to "cover" for him or her and to deflect attention from the group.

- *Don't look for favors just for yourself.* This injunction tells officers not to "suck up" to superiors. In essence, it tells them that their primary responsibilities are to their peers and that attempts to curry favors with superiors will meet with severe disapproval. It prevents line officers from developing relationships with superiors that might threaten the safety of the work group. (103–105)

Although the "code of secrecy" is not always as rigid as it is made out to be, secrecy does exist in all police organizations to some extent.

Several factors contribute to the development of a code of secrecy in policing. First, policing is fraught with the potential for mistakes, and officers often observe the code of secrecy to protect their fellow officers from being accused of making inappropriate decisions, as they would wish to be protected themselves. They feel that they are often called upon to make split-second decisions that can be unfairly second-guessed by people not directly involved in policing. To some extent this is true, even though the "split-second syndrome" is sometimes used by the police as a rationalization "to provide after-the-fact justification for unnecessary police violence" (Fyfe 1997, 540). John Crank (1998) has argued along these lines, observing that

> the veil of secrecy emerges from the practice of police work, from the way in which everyday events conspire against officers. The veil has no remedy. It may be desirable to penetrate the veil, but it is not reasonable. To look at secrecy in terms of good or bad, right or wrong, is to miss the point. It is a cultural product, formed by the environmental context that holds in high regard issues of democratic process and police lawfulness, and that seeks to punish its cops for errors they make. Secrecy is a set of working tenets that loosely couple the police to accountability, that allow them to do the work they do without interfering oversight. As long as police conduct law enforcement under a mantle of due process and accountability in the United States, police culture will be characterized by secrecy. (226)

The trust that must exist between police officers who work together also provides a basis for secrecy. Police culture recognizes that officers can only function successfully when they work with colleagues whom they trust. Officers know the most about their fellow officers' performance, but if they revealed this information to superiors, they would jeopardize the trust on which they must rely.

The traditional or professional model of policing has reinforced an ethos of secrecy. Under this model, as discussed in Chapter 1, police are the experts in fighting crime, and the public has no business to know much about their work or to intervene. All nonpolice are outsiders whose knowledge of police activities could only harm the department or its personnel:

> Policemen are under explicit orders not to talk about police work with anyone outside the department. There is much in the nature of a secret society about the police; and past experience has indicated that to talk is to invite trouble from the press, the public, the administration and their colleagues. (Wesley 1951, 30)

To maintain the integrity and reputation of a police department, the professional model of policing underscores the importance of policies and rules and procedures that police officers must follow. Violating policies and rules can be a cause for suspension or even dismissal. Many departments use large looseleaf notebooks to hold the departmental policies, procedures, and rules and expect all police officers to know them all. This is an impossible task for most people, but lack of knowledge about a departmental policy, rule, or procedure cannot be used as an excuse for violating it. Under such a system, officers feel so vulnerable to discipline for rule breaking that they are especially motivated to cover up their own and others' violations.

Further, many police officers have traditionally believed that they have the right, at times, to step out of legal and ethical bounds and to violate department policies and procedures while performing their duties, particularly in the interests of their own survival or job tenure. Lawrence Sherman (1982) has described these beliefs as follows:

1. *Discretion A:* Decisions about whether to enforce the law, in any but the most serious cases, should be guided by both what the law says and who the suspect is.
2. *Discretion B:* Disrespect for police authority is a serious offense that should always be punished with an arrest or the use of force.
3. *Force:* Police officers should never hesitate to use physical or deadly force against people who "deserve it," or where it can be an effective way of solving a crime.
4. *Due Process:* Due process is only a means of protecting criminals at the expense of the law-abiding and should be ignored whenever it is safe to do so.
5. *Truth:* Lying and deception are an essential part of the police job, and even perjury should be used if it is necessary to protect yourself or get a conviction on a "bad guy."
6. *Time:* You cannot go fast enough to chase a car thief or traffic violator, nor slow enough to get to a "garbage" call; and when there are no calls for service, your time is your own.
7. *Rewards:* Police do very dangerous work for low wages, so it is proper to take any extra rewards the public wants to give them, like free meals, Christmas gifts, or even regular monthly payments (in some cities) for special treatment.

8. *Loyalty:* The paramount duty is to protect your fellow officers at all costs, as they would protect you, even though you may have to risk your own career or your own life to do it. (14–15)

Obviously, such beliefs, if acted upon, provide an excellent reason to keep many police activities secret. Because of the nature of police work, secrecy will always be a component of police work, and some aspects of policing, such as details of ongoing investigations and police personnel decisions that are required by law to be kept confidential, will always need to be kept secret and known only to a few individuals. But the code of secrecy is losing its dominance as police departments transfer from the professional model to the community policing model. The philosophy of community policing requires that the police be more open with the information they have and gives them new roles as cooperative problem solvers and decision makers in the communities they serve. Unnecessary secrecy is an impediment to the police-community partnership, a major premise of the community policing philosophy.

Isolation

A by-product of police secrecy is social isolation. "Isolation is an emotional and physical condition that makes it difficult for members of one social group to have relationships and interact with members of another group" (Kappeler et al. 1995, 252). The police subculture isolates police from the rest of society. Often the police place restrictions on themselves that lead to limited interactions with the neighborhoods and people they serve. The "stay-out-of-trouble" mentality that some officers have contributes to the social isolation associated with the police field. Jerome Skolnick (1994) has cited James Baldwin's (1962) description, from the early 1960s, of the isolation of police officers from the people in black ghettoes of the 1950s and 1960s. Baldwin claimed that the only way the police could control the ghetto was through oppressive tactics, the police could not understand the people, and their very presence was an insult. The police represented the white world and were there to control black people—in other words, to keep them in their place. Even when a police officer had done nothing to be hated he was serving as an occupying soldier of the group oppressing them, so people in the ghetto would wish to see him dead. Because police officers worked in a hostile environment, they had to patrol their walking beats in twos and threes (65–67). Baldwin's description may be more accurate for the 1960s and 1970s and for policing under the professional model than it is today, when police agencies throughout America are making the transition to community policing and trying to develop cooperative relationships between communities and the police.

A number of forces contribute to the isolation of the police. First, individuals who have been subject to restrictive actions by the police resent the invasion of their privacy. Second, problems that have recurred throughout policing history, such as police incompetence, brutality, corruption, and pursuit of private interests, have created feelings of mistrust, fear, and disdain between the police and the public. Third, police operations are set apart from society because of their nature. Police work involves dealing with the seamy side of human behavior.

In addition, however, various governmental studies have claimed that since the police have been placed in patrol cars, they have lost contact with the community. They maintain that this chasm has been wider between the police and the public in some locales than others but that it exists everywhere. One of the reasons for adopting the community policing philosophy has been to bridge the gap existing between the police and the community. Police officers who are working the street (beat patrol officers) develop a set of beliefs that regulates the pace, style, and direction of their "on-the-street" behavior. These beliefs are produced by patrol officers' unique social role and their "outsider" position in the community, as well as the nature of police responsibilities and the requirements of personal survival while performing police work. The professional model of policing discouraged interaction with the community. The patrol officer had to be available to answer radio service calls. Police officers were not encouraged to get out of their cars to interact with the public. In fact, communication with the public was not encouraged. For example, as a young police officer, I worked for a midsized city police department that assigned officers a different patrol sector each day or night an officer worked. The purpose of this was to prevent police officers from becoming too familiar with the people on their beat. Police administrators who used this model were attempting to maintain the integrity of the police officers and the department by preventing overfamiliarity between police officers and people on their beat because it was believed that such overfamiliarity could lead to corruption. It should be kept in mind that the professional model of policing was instituted in reaction to the political model of policing which was associated with numerous police corruption scandals. However, community policing is taking the opposite approach from the professional model in that it encourages police officers to get to know the people on their beats so that they can solve problems of concern to the neighborhood. Today, police agencies that have adopted community policing have concentrated their efforts on ending the alienation that has existed for decades between the police and the community. Community members are being included in police policies that affect their communities and neighborhoods, and community-police cooperative efforts are breaking down the old barriers of hostility and mistrust. The more ingrained the community policing philosophy is in an organization, the less isolated the police will be.

Solidarity

Occupational groups have a sense of identification and inclusiveness. People working together and performing the same or similar tasks and facing the same problems are most likely to bond and develop a sense of solidarity. Although all occupations have solidarity, this seems to be especially true of the police. The term *solidarity* implies consensus, integration, friendship, personal intimacy, emotional depth, moral commitment, and continuity in time (Durkheim 1966, 47–48). The sense of solidarity may change with a police officer's rank and age. According to Kappeler et al. (1998),

> Police solidarity . . . may be said to be an effect of the socialization process inherent to the subculture and police work. New members are heavily socialized to increase their solidarity with the group, and those who move away from the subculture, either through age or promotion, are gradually denied ties of solidarity. This cohesion is based in part on the "sameness" of roles, perceptions, and self-imagery of the members of the police subculture. (102)

Three sources of solidarity are defensiveness, professionalization, and depersonalization. Defensiveness reflects in-group/out-group tensions and is an adaptation to an external social world perceived as hostile or critical. Police officers know that other officers will support them against outsiders, and this gives them the moral support to perform their duty. The sense of professionalism instilled in police recruit training further reinforces solidarity and the perception that outsiders can never know what policing is really like and what it demands. Finally, solidarity is strengthened by the depersonalization of outsiders. Police officers who have this attitude often treat citizens inhumanely and may even abuse them. Possible contributors to depersonalization are the need for police efficiency, timed response to calls, and the requirement that the police be impartial (Harris 1973, 163–165).

The alienation from the public that is so often the flip side of police solidarity is partly based on the belief, held by some police officers, that the public does not support them, takes them for granted, does not respect them, and expects them to do a dangerous, low-paying job alone without any help from the community. Although the police have the responsibility to maintain order, they feel that the public should play a part. Also, they may feel as depersonalized by ordinary citizens as citizens feel depersonalized by them.

The danger involved in policing alienates the police officer not only from the criminal offender but also from the law-abiding citizen (Skolnick 1994, 52–53). A similarity may exist between police and military solidarity; Janowitz (1964) has argued that "any profession which is continually preoccupied with the threat of danger requires a strong sense of solidarity if it is to operate effectively. Detailed regulation of the military style of life is expected to enhance group cohesion, professional loyalty, and maintain the martial spirit" (175).

The professional policing model reinforces this kind of solidarity among police officers. But with the implementation of community policing and a restructuring of the police organization to allow community policing strategies to be successful, there may be less need for solidarity among police officers because, as police officers increase their interaction with the community, there should be greater rapport and trust between the police and the community.

Cops' Rules

"Cops' rules" are an important component of police culture (Rubinstein 1973, 267–337). Police officers develop specific skills that help them perform their job successfully. Unlike many other career fields, policing is dangerous. Officers must be concerned not only about their own personal safety but about the safety of their fellow officers and citizens. When they confront citizens who want to harm themselves or a fellow citizen, they must be good manipulators who know when to cajole and when to threaten an individual or group to defuse a violent situation and keep the peace. The police officer on the street would say you have to be a good "bullshitter" to be a good police officer. Experienced police officers usually do not want to confront a disturbed, violent person or suspect; they want the person to voluntarily follow their directions. They do not want the reputation of being a "hothead" and do not want their fellow officers to think that every time they confront a citizen, no matter how trivial the matter, they must make an arrest or get into a scuffle.

Due to the nature of their work, officers develop an intimate knowledge of the street they patrol and become more familiar with the area than most of its residents. Police officers' knowledge about the territory they work during a tour of duty is extremely important to their safety and the safety of others. All police officers assigned to a beat or sector are expected to know the houses, businesses, automobiles, adults, and children. They should know all the names of the streets, and the alleyways, one-way streets, parks, and playgrounds. Most police officers become familiar with what times businesses close, what lights are left on in the evening hours, and what days stores close. All this is expected of beat patrol officers in the police culture.

At some point in their careers, police officers learn to read people and dangerous situations. To survive on the street, or at least not to be considered incompetent by their peers, they must be able to read threats and the possibility of resistance or flight. They must be familiar with all the tools available to them, including verbal and physical skills and weapons.

The most deadly tool at police officers' disposal is their service revolver. In many police departments, police officers in patrol vehicles have shotguns in a gun rack available to them. Today, police departments have a variety of sophisticated firearms at their disposal, but, for practical purposes, officers depend primarily on their service revolver for safety. Because firearms are deadly weapons, officers are trained to use them only as a last resort. They are also taught that the service revolver should not be used to threaten, but should only be pulled from the holster if one is willing to use it.

Another weapon of importance to the police officer is the nightstick or baton. Although the baton can be a deadly weapon, it should be used only as a defensive tool. It generally is used to get a suspect or violent person under control. The baton can do physical damage, but it generally does not cause serious physical injuries. Blows or raps to the shins, knees, elbows, or ankles can take the fight out of many people. The baton helps the police officer gain control of a potentially dangerous situation without causing excessive physical damage to the officer or to the offender.

A deadly weapon that some police officers carry is the slapjack or blackjack. The slapjack, a flat weapon, has a hard surface and thick edges and can cause serious cuts. The blackjack, a round weapon with a spring handle, has a whipping action that increases the force of a blow when used with a rapid motion. Both the slapjack and blackjack are lead pieces encased in leather. Because of the serious damage they can cause, many police departments have banned them.

Mace and pepper mace are used by police officers to stop a suspect. Mace has been used since the 1960s, but pepper mace has become more commonly used in the 1990s because it is generally more effective. Both mace and pepper mace are chemical irritants that act on the central nervous system and induce pain by activating receptor cells within the brain. Pepper mace causes a swelling of the eyes and breathing passage. An intense burning occurs in the eyes, throat, and the skin area sprayed. When pepper spray has been inhaled, breathing will be restricted, and the respiratory tract will be inflamed. The physical effects can include choking, shaking, involuntary closing of the eyes, nausea, lack of coordination, and loss of upper body strength. Disorientation and fear can also occur (Onnen 1993, 1–2).

Other weapons available to the police are the taser and the stun gun. The taser is a low-powered (5-watt) gadget that operates on a 7.2-volt battery and shoots barbs attached to 15 feet of wire. The taser causes "charley horse" spasms that cause subjects to lose control of their bodily movements. The stun gun generates an electrical current when it is placed against a suspect's body for several seconds. The taser and stun gun are both intended to keep police officers from using deadly force.

It is expected within police culture that cops control the space around them. The control of space enables officers to control a situation and make the decisions that will keep them safe. Police officers do not like to go through buildings looking for a suspect or prowler because they do not have complete control of the situation: the suspect or the prowler may be able to see them without their seeing the suspect or prowler.

Controlling people goes along with controlling space. Police officers need to be in charge. In a potentially dangerous situation, whether it is a domestic disturbance, a shooting, or a traffic accident, they cannot allow a suspect, citizen, or criminal to seize the control; if they do, then for all practical purposes they cease to be police officers. Police officers control people by giving them commands, searching them, or arresting them. Police officers are not afraid to use their authority to control people.

An important tool possessed by police officers to control people is force. Although most police officers would prefer not to use force, they are almost required to use it against a rowdy or violent individual to gain control of the situation and to ensure their own safety and the safety of others. As Rubinstein (1973) has described cops' attitudes about using force,

> Very few policemen use physical force gratuitously. A man will cajole, joke, advise, threaten, and counsel rather than hit, but once his right to act is questioned, once his autonomy is threatened, he is prepared to respond with whatever force is necessary. And once his power is contested, he can do no wrong. The legitimacy of this authority allows him to do whatever he must to preserve it. An attack on him is treated as an attack on the state. A man resisting a policeman is suddenly an alleged criminal, and although he has not been convicted, the policeman knows he can treat him in a manner in which he cannot treat others. (327)

Cops' rules often change very little, if at all, under the community policing philosophy. Even under community policing, police officers are expected to be law enforcement officers. That means they investigate crimes, make arrests, interrogate suspects, and place themselves in dangerous situations. Community policing does not mean that police officers discontinue traditional policing. They are still expected to respond to radio service calls, stop traffic violators late at night, chase criminals down dark alleys, and go into darkened buildings searching for burglary suspects, so they need the skills and tools to protect themselves and other innocent people from harm. It would be unrealistic to assume that under community policing police officers would never be in a situation where they had to use force.

Divisions within the Police Culture

Although police culture may be similar across the country, each police department's culture and value system is to some extent unique. Further, police culture can be subdivided according to rank (police officers vs. supervisors and administrators), function (patrol vs. administration), and selection of head administrator (elected vs. appointed) (Hunt and Magenau 1993, 71).

Elizabeth Reuss-Ianni (1983) claimed that there are two police cultures: "street cop" culture and "management cop" culture. The street cop and the management cop do not have similar value systems and do not perceive policing in the same ways:

> Now there are two cultures which confront each other in the department: a street cop culture of the good old days, working class in origin and temperament, whose members see themselves as career cops; opposed to this is a management cop culture, more middle class, whose members' education and mobility have made them eligible for jobs totally outside of policing, which makes them less dependent on, and less loyal to the street cop culture. In a sense, the management cop culture represents those police who have decided that the old way of running a police department is finished (for a variety of external reasons, such as social pressures, economic reality of the city, increased visibility, minority recruitment, and growth in size that cannot be managed easily in the informal fashion of the old days) and they are "going to get on the ground floor of something new." They do not, like the street cops, regard community relations, for example, as "Mickey Mouse bullshit," but as something that must be done for politically expedient reasons if not for social ones. . . . The street cops who are still in the old ways of doing things are confused and often enraged at the apparent change in the rules of the system. So they fight back in the only way they have at their disposal, foot dragging, absenteeism and a host of similar coping mechanism and defensive techniques. Nor is all of this likely to change soon; the old and the new will continue to coexist for some time because the attitudes, values, and ways of doing things have not changed throughout the system. (121–122)

Both street police officers and management police officials recognize that a schism exists between the two groups. Lower ranking street officers cannot influence the management officer, because they lack the authority and power to do so. Management police officials constantly try to influence street police officers by intimidating them into conforming to policies, rules, and procedures. They use the police bureaucracy and the authority and power that come with it to force street officers to toe the line. Street police officers consider the police managers as being "out of touch with police work": because police managers sit behind a desk and do not work the street, they have forgotten what police work is about.

Reuss-Ianni described street cop culture in terms of a cop's code that contains the following maxims, some of which have already been discussed earlier under the heading of "Secrecy":

- Watch out for your partner first and then the rest of the guys working the tour.
- Don't give up another cop.
- Show balls.
- Be aggressive when you have to, but don't be too eager.
- Don't get involved in anything in another guy's sector.
- Hold up your end of the work.
- If you get caught off base, don't implicate anyone else.
- Make sure the other guys know if another cop is dangerous or "crazy."
- Don't trust a new guy until you have checked him out.
- Don't tell anybody else more than they have to know; it could be bad for you and could be bad for them.
- Don't talk too much or too little.
- Don't leave work for the next tour.
- Protect your ass.
- Don't make waves.
- Don't give them to activity.
- Keep out of the way of any boss from outside your precinct.
- Don't look for favors just for yourself.
- Don't take on the patrol sergeant by yourself.
- Know your bosses.
- Don't trust bosses to look out for your interests. (13–16)

This is a code of survival for street police officers. Over a period of time, they are expected to become familiar with it and to live by it, and, if they do not, they become outsiders. They are ostracized by their fellow officers—not spoken to, recognized, or acknowledged. Even rookies are expected to ostracize a police officer who has failed to follow the street officer code.

The code of management cop culture may be considered antithetical to the street cop code. Reuss-Ianni described it as follows:

- Management cop culture demands a watchdog system that can uncover and expose any internal wrongdoing or corruption and so protect the public image of the department.

- Management cop culture expects supervisors to act on clearly defined rules and procedures.
- Management cop culture tries to do away with the old reward system so that the old rules for getting ahead no longer work.
- Management cop culture sees in job stress the potential for police action that can lead to public relations problems. The public needs to be protected from cops who are out of control.
- Management cop culture has imposed departmentwide techniques and practices for controlling activities by officers that are not in keeping with departmental rules and regulations.
- Management cop culture has to produce numbers to prove its accomplishments.
- Management cop culture attempts to reconcile the demand for greater productivity, efficiency, and responsiveness to the community with the demand for improved employee relations and working conditions.
- Management cop culture often implements new management techniques, such as a system in which numerical indicators of performance are set as the basis for reward or punishment.
- Management cop culture pays little attention to actual police practice. It designs new programs and procedures on the basis of seemingly rational and logical factors, while ignoring street-level practice that has more impact on day-to-day operations and the practical outcome of the intended programs.
- Management cop culture has bureaucratized police work to make it more easily managed. (50–116)

After reading the description of management cop practice and comparing it with the cop's code, we can easily recognize that a schism exists between the street cop culture and the management cop culture. However, as community policing becomes more ingrained in the police culture, the gap between the management cop and street cop cultures may decrease. The community policing philosophy will put management and street cops together to solve problems and develop partnerships with the community. Police administrators attempting to implement community policing should recognize that two police cultures exist. For community policing to be successful, police managers need to work within the street cop culture—perhaps winning over street cops by using participatory management. Also, management cops need to recognize street cops' practical knowledge of the neighborhoods they patrol. Only by respecting the line officer can the management cop hope to successfully implement policing philosophies such as community policing.

MYTHS ABOUT POLICING

There are a large number of myths pertaining to policing. Some have been encouraged by police, administrators, officers, and unions. The entertainment and news media have also contributed to myths about policing by portraying police as superheroes. In real life, the police cannot realistically perform the acts or achieve the success that they do on television or in the movies. The news media frequently create a "crime wave" by leading the nightly news with the murder of the day. The media policy of "if it bleeds, it leads" often creates fear in the public that may be far more serious than the actual crime situation. The public can have unrealistic expectations of the police. The police are not all powerful, all knowing, or omnipresent. Students of policing know that police success in solving crime depends on individuals or groups coming forward to assist the police by providing information. Peter Manning (1997) defines the police myth as "Uncertainty, a basic informational matter, structures police work and is the basis from which the defining myth that grounds the work grows. The dramaturgical nature of policing is not an accident but is a function of information levels possessed by the police when considered in the context of their claimed place in a political-moral system." He further states that "if we consider mythological formulations to be products of historical periods, then a change in social structure can produce a change in beliefs, and previous beliefs and myths can become problematic" (281). Manning also delineates a number of functions on how the police myth serves the public and the police:

1. The function of any myth is to integrate themes that are in reality unreconcilable.
2. Myth also obviates the concern with the special interests that might be served by a given explanation; it removes a matter from everyday discourse and places it in the realm of the nebulous and the mystical—that which stands to serve all in a removed and fair, almost dispassionate, fashion.
3. A myth alleviates societal crises by providing a verbal explanation for causes, meanings, and consequences of events that might otherwise be considered inexplicable. Because the actual probability that police action will occur to prevent, punish, or obviate the threat of crime is low, the myth must be publicly maintained.
4. The myth provides, in repetitive form, the axiom that police force will be applied systematically to isolated and thoroughly evil persons in a predictable and routine fashion. The contradictory values involved, of procedural guarantees for all and swift and certain punishment of the guilty (about whom there is likely to be little consensus), are met in the myth, while daily practice falls short of this ideal.
5. The police myth sets apart the actors in the drama of crime, gives them faces and names, and makes them subject to predictable scenarios with beginnings, middles, and ends. In these scenarios, largely a police/media construction, government action is seen as both relevant and effective.
6. Myths of police action concentrate public attention on their force and conserving potential, even in times of rapid change. They are reassuring.
7. The police myth freezes the organization in time and space, giving it a reified authority over the thing it opposes, and establishes, in a timeless dynamic, a tension between the two poles of social life—"good" and "evil." Law enforcement is no longer seen as mere work, involving decision, discretion, boredom, and unpleasantries, it becomes a "creed."

8. Internally, the myth provides, as do all myths, a sense of solidarity, of common purpose, of collective unity in the face of a threatening environment. (278–280)

David Bayley, in his book *Police for the Future* (1994) similarly argued that the idea that the police prevent crime is the biggest police myth of all (3–12). The inability of the police to prevent crime has been a well-kept secret of modern times. Bayley asserted that the police have created an image of themselves as the public's best defense against crime and have managed to sell this myth to politicians and the general public, who have responded by giving them more resources, especially additional police officers. Yet, as Bayley pointed out, repeated analysis has consistently failed to find any connection between the number of police officers and crime rates, and the primary strategies adopted by modern police have been shown to have little or no effect on crime (3).

Bayley described three strategies of modern policing for which there is no evidence of effectiveness in preventing crime: patrolling of city streets by uniformed patrol officers, rapid response to emergency calls, and investigation of crime by detectives. Although the police maintain that these strategies are important for public safety, there is no support for their claims. Foot patrol makes people feel safer but does not prevent crime. A quick police response to a call has no bearing on preventing crime and has value only when the offender is still at the scene and can be apprehended when the crime is in progress. A predictable response time to a call appears to be more valuable than a rapid response. Criminal investigations by detectives have no major effect on public safety.

Bayley concluded that "police actions cannot be shown to reduce the amount of crime" (9) and that the police should "put up or shut up" (10). If the findings on police strategies are accurate, the police should change their tactics. According to Bayley, three fundamental issues need to be addressed if we really want the police to prevent crime more effectively:

1. Why haven't the police been effective in preventing crime?
2. What might the police do to become effective crime deterrents?
3. Considering the requirements, do we want the police to be effective? (11–12)

Realistically speaking, the police cannot prevent or solve all crime. The police are recognizing that their success in preventing crime depends upon the average citizen. Only by giving the citizen a role to play in preventing crime can we expect a long-term decrease in the crime rate.

Police officers in my classes who read Bayley's theory strongly disagreed with him, claiming that the police do prevent crime. Police officers, former police officers, and those who feel the police can do no wrong may be angry about Bayley's thesis. But what is needed here is not anger but open-mindedness. A person can make a comment that the police don't like but still not be antipolice. Police philosophies and strategies should be open to evaluation, and, if they are not working or can be changed to work, then appropriate action should be taken. Although I myself do not completely agree with Bayley, I feel that he raises an important issue: whether the police should have the sole responsibility in crime prevention and what role, if any, the community, the neighborhood, and the individual have in crime prevention. In the community policing philosophy, the community, the neighborhood, and the individual must assume the *primary* role in crime prevention if they want to keep crime under control, and the police should provide them with training and assistance. A closer police-community partnership and a crime prevention orientation are emphasized.

Another myth sold to the public by the police is the "crime fighter" image. Studies indicate that police spend 80 to 90 percent of their time performing service activities. Further, if police are supported to be crime fighters, they are failures, because their performance record for solving crime is not impressive. For example, over the last several years, the police nationally solved approximately 65 percent of the homicides committed, about 52 percent of the forcible rape crimes, and 24 percent of the robbery crimes. It may not be the fault of the police that they have not solved more crimes. However, they should stop selling the myth that they are "crime fighters" if they cannot deliver on their promise. The "crime-fighting" image of the professional model of policing should be considered as one of this model's weaknesses. If police agencies keep what works from the professional model of policing and incorporate the strategies of community policing as well, they should be more successful in solving crime. The police should not have the sole responsibility to solve crime. The community has a responsibility not only for preventing crime but also for solving crimes. Community policing attempts to provide citizens with a sense of security by making them partners with the police in solving crime.

In all police departments, there are officers who joined up because they wanted to do the "real" police work of crime fighting. John P. Crank (1998) had the following to say about "real police work":

> Real police work engages the vital self, invokes the warrior's dream to make a difference in the battle against crime. Officers have the opportunity to use the special skills that derive from the unique experience and enforcement training—the powerful themes of danger, officer safety, weapon training, and protecting other officers come into play with real police work. Real police work is the "symbolic rites of search, chase and capture." Such work gives the police a sense of self-worth absent in the give-and-take of normal everyday routines. (117–118)

Many police officers consider "real police work" to be catching bad guys: making arrests, conducting high-speed chases, catching burglars, stopping bar fights, and making the big bust. For them, real police work does not include such mundane tasks as doing paperwork and providing services. But in reality, except in a few busy patrol areas, paperwork and service activities, such as rescuing an elderly gentleman who has fallen on the floor and putting him back into bed, usually make up 80 to 90 percent of a patrol officer's day. Some patrol officers feel that they are primarily report takers: they make a call, write up a report, and go back on patrol to take another call and make another report. They are expected to be in service to respond to citizens' calls, no matter how trivial these may be. They are not expected to solve a crime; if they do, it's a big event. If a crime has been committed, it normally will be turned over to the detectives to investigate. Such patrol operations do not necessarily make for happy patrol officers. Since most patrol officers would prefer performing "real police work," it is understandable that they are not necessarily big supporters of community policing, which they often consider "touchy feely."

Police work, with its image of action, offers individuals who want to work outdoors, have some autonomy, and put bad people away an opportunity that few professions offer. But community policing just expands the concept of real police work. Community policing can eliminate the dead time that officers complain about when not responding to radio calls for service.

SUMMARY

The police culture cannot be overlooked if policing hopes to change its philosophy. The foundation upon which social groups function is referred to as culture. Culture consists of knowledge, beliefs, morals, customs, and practices acquired by a group. It also encompasses our thinking, feelings, attitudes, and communications with each other. We learn culture from the people around us. Because police officers are members of American society, they reflect American customs, morals, and values.

Elements pertaining to culture include norms, folkways, mores, and laws. *Norms* are rules of conduct. They designate what people should or should not do, think, or feel. *Folkways* are customary, normal, and habitual ways of doing things. *Mores* reflect a strong awareness of right or wrong. *Laws* are rules that are enforced by governmental authorities. All of these are forms of social control.

All cultures have checks on individual, group, and crowd behavior. Governmental agencies involved in controlling behavior are criminal justice personnel. The police are given the legal responsibility to assert social control and the option to use various means to achieve this goal, including arrests and the use of physical force—if necessary, deadly force, or force with the potential of causing death.

Various groups of workers develop a subculture unique to their specific occupation. Occupational subcultures tend to isolate individuals from outsiders and from the society at large. Workers of specific occupations can become closed communities, associating only with people from their same occupation to the exclusion of individuals from other fields. Policing is one such occupation.

Because of the dangers associated with police work, officers develop intuitive techniques to identify undesirable or potential law violators by their appearance, language, body language, gestures, and attitudes. They develop an alert and watchful stance toward what Skolnick (1994) has called the "symbolic assailant."

Another major feature of police culture is the ethos of secrecy. Police officers quickly learn that to survive in policing they must keep much of what they and their fellow officers do secret from their superiors and the public at large. The traditional police bureaucracy and the professional model of policing have supported the culture of secrecy in that police administrators wish to keep from outsiders any information that would reveal the inefficiencies and failures of the police department. Also, the perception of all nonpolice personnel as outsiders who should not be involved in police business has encouraged secrecy.

The term *solidarity* means consensus, integration, friendship, personal intimacy, emotional depth, moral commitment, and continuity in time. People working together and performing similar tasks are most likely to bond and develop a sense of solidarity. Policing is characterized by a high degree of solidarity, not only because of the dangers of the work but because of defensiveness toward what is believed to be a critical and nonsupportive public, shared professionalism, and depersonalization of outsiders.

"Cops' rules" are another important component of the police subculture. These are specific skills that help police officers to perform their job successfully. Police officers must be concerned about their own safety and the safety of fellow officers and citizens. They must be good communicators and be able to defuse volatile situations. An important tool possessed by the police officer to control people is force, including weapons. Although most police officers would prefer not to use force, they are almost required to use it against a rowdy or violent individual to gain control of the situation.

The policing culture may be similar in some ways all across the country, but each police organization's culture has aspects that are unique. Also, Elizabeth Reuss-Ianni has described two policing cultures: "street cop" culture and "management cop" culture. The management cop and the street cop do not perceive policing in the same ways. The street cop emphasizes survival techniques, whereas the management cop emphasizes community relations. Today, community policing would be stressed by the management cop.

Many myths about policing have become a part of the police culture. These myths have been encouraged by police leaders, administrators, officers, and union officials. Some of them are the crime-fighting role of the police and the ability of the police to prevent crime.

KEY TERMS

anomie	material culture	out-groups
cops' rules	mores	"real police work"
cynicism	myths	secrecy
folkways	nonmaterial culture	solidarity
isolation	norms	street cop
management cop	occupational culture	symbolic assailant

REVIEW QUESTIONS

1. Define *culture.*
2. Define *occupational culture.*
3. Explain how the identification of symbolic assailants relates to the police culture.
4. Explain how secrecy relates to the police culture.
5. Discuss isolation and how it relates to the police culture.
6. Discuss solidarity and how it relates to the police culture.
7. Discuss cops' rules and how they relate to the police culture.
8. Discuss the division between street cop culture and management cop culture.

REFERENCES

Adams, T.F. 1963. Field interrogation. *Police* (March–April): 28.

Baldwin, J. 1962. *Nobody knows my name.* New York: Dell.

Bayley, D.H. 1994. *Police for the future.* New York: Oxford University Press.

Brinkerhoff, D.B., et al. 1997. *Sociology.* 4th ed. Albany, NY: Wadsworth.

Crank, J.P. 1998. *Understanding police culture.* Cincinnati, OH: Anderson.

Davis, J.A., and T.W. Smith. 1983. *National Data Program for the Social Sciences: General social survey, cumulative file, 1972–1982.* Ann Arbor, MI: Inter-University Consortium for Political and Social Research.

Durkheim, E. 1966. *The sociological tradition,* trans. R.A. Nisbet. New York: Basic Books.

Ford, R.L. 1988. *Work, organization, and power: Introduction to industrial sociology.* Needham Heights, MA: Allyn & Bacon.

Fyfe, J.J. 1997. The split-second syndrome and other determinants of police violence. In *Critical issues in policing: Contemporary readings,* 3rd ed., ed. R. Durham and G.P. Albert. Prospect Heights, IL: Waveland.

Hahn, H. 1974. A profile of urban police. In *The police community: Dimensions of an occupational subculture,* ed. J. Goldsmith and S.S. Goldsmith. Pacific Palisades, CA: Palisades.

Harris, R.H. 1973. *The police academy: An inside view.* New York: John Wiley.

Hellriegel, D., et al. 1983. *Organizational behavior.* 3rd ed. St. Paul, MN: West.

Hunt, R.G., and J.M. Magenau. 1993. *Power and the police chief: An institutional and organizational analysis.* Newbury Park, CA: Sage.

Janowitz, M. 1964. *The professional soldier: A social and political portrait.* New York: Free Press of Glencoe.

Kappeler, V.E., et al. 1995. Breeding deviant conformity: Police ideology and culture. In *The police and society: Touchstone readings,* ed. V.E. Kappeler. Prospect Heights, IL: Waveland.

Kappeler, V.E., et al. 1998. *Forces of deviance.* 2nd ed. Prospect Heights, IL: Waveland.

Manning, P.K. 1997. *Police work: The social organization of policing.* 2nd ed. Prospect Heights, IL: Waveland.

McNall, S., and S. McNall. 1992. *Sociology.* Englewood Cliffs, NJ: Prentice Hall.

Onnen, J. 1993. Oleoresin capsicum. *Police Chief* 60, no. 5, 1–2.

Reuss-Ianni, E. 1983. *Two cultures of policing: Street cops and management cops.* New Brunswick, NJ: Transaction.

Rothman, R.A. 1987. *Working: Sociological perspectives.* Englewood Cliffs, NJ: Prentice Hall.

Rubinstein, J. 1973. *City police.* New York: Farrar Straus Giroux.

Sherman, L. 1982. Learning police ethics. *Criminal Justice Ethics* 1, no. 1: 14–15.

Skolnick, J. 1994. *Justice without trial: Law enforcement in democratic society.* 3rd ed. New York: Macmillan.

Van Maanen, J. 1978. Observations on the making of policemen. In *Policing: A view from the street,* ed. P.K. Manning and J. Van Maanen. Santa Monica, CA: Goodyear.

Wesley, W.A. 1951. The police: A sociological study of law, custom and morality. Unpublished doctoral dissertation, University of Chicago.

Chapter 3

Police Discretion, Police Misconduct, and Mechanisms To Control Police Misconduct

CHAPTER OBJECTIVES

1. Have an understanding of police discretion.
2. Have an understanding of what constitutes police misconduct.
3. Have an understanding of the various mechanisms to control police misconduct.
4. Have an understanding of the value of community policing in controlling police discretion and police misconduct.

INTRODUCTION

Police discretion—the power to take action or not to take action or to arrest or not to arrest—is an important component of police powers and makes the police officer in our society an extremely powerful official. How officers use the discretion allowed them usually determines whether justice is done and whether the law has been applied fairly.

Before the mid-1950s, our politicians and police administrators constantly told us that they enforced all laws and investigated all crimes equally. Today, we recognize that it would be impossible for the police to do this.

The major issues concerning the use of police discretion are the abuses that can occur with unchecked police discretion. Police abuses, often referred to as *police misconduct,* have been a major problem with many police departments in the 1990s.

All vocations and professions have their outlaws, and policing is no exception. The police field has its share of officers who violate not only departmental policies and procedures but also the very laws they are to enforce. When a police officer is involved in misconduct, a scandal is created that tarnishes not only the reputation of the police officer but the reputation of the department that employs him or her and, even worse, the image of all police officers.

Since some police officers are involved in misconduct, mechanisms must be put in place to check police misconduct. These mechanisms investigate accusations of police misconduct and attempt to eliminate any police improbity when it is found. They are important not only for police personnel but for the community as well. Police officers following the rules of their department and doing their jobs strongly support mechanisms that punish and, if needed, remove police officers from the force. The community wants to have confidence in its police officers. Community policing cannot be successful without this trust.

POLICE DISCRETION

America is a nation of laws. Our laws are passed by the legislative branch of government and enforced by the executive branch. From a broader sociological perspective, law can be viewed as the "various agencies and procedures by which rules are made, applied, and enforced, including that law which is at the end of a policeman's nightstick" (F.J. Davis 1975, 19).

Laws are passed to keep the peace and maintain order. Political scientists often claim that the purpose of laws is to "keep the underclass under control" for the benefit of the middle and upper classes. These groups could not remain affluent if chaos and disorder became rampant. Therefore, laws are passed and are enforced, not only for public safety purposes, but also to keep our society functioning well economically. A former Bronx borough commander of the New York City Police Department expressed this philosophy in commenting on the political implications of his policing role:

> To the degree that I succeed in keeping the ghetto cool, to the degree that I can be effective—to that degree, fundamentally, am I deflecting America's attention from discovering this concern? . . . The fact of the matter is that we are manufacturing criminals.
>
> . . . We are very efficiently creating a very volatile and dangerous subelement of our society. And we are doing it simply because we don't want to face the burdens and the problems and the responsibilities that their existence imposes on any society with conscience.
>
> . . . And I am very well paid, almost to be the commander of an army of occupation in the ghetto. And this is a great tragedy. (Brown 1988, xiv)

Individuals in our society, including those responsible for enforcing the law, do not always specifically follow the letter of the law. In addition, legal norms are constantly changing (Aaronson et al. 1984, 4). For example, a few years ago it was acceptable to discriminate against homosexuals, but today this discrimination would be a violation of the law. Also, several decades ago it could be appropriate behavior to smack a youngster who talked back to a police officer and send him home to his parents, where he could expect to get smacked again. Today, this would be considered misconduct on the police officer's part.

Police discretion is the authority of police to decide how much effort they will apply to enforcing specific laws and which laws they will or will not enforce. According to Reiss (1984), the concept implies several things: "One is quite simply that an officer who has discretion has the power to choose among alternatives. A second and closely related aspect is that a choice made by an officer in the exercise of his discretion must be permitted by law. A third is that an officer who has discretion exercises individual judgment in making a choice" (89–90). Although the police are sworn to enforce all the laws all the time, they enforce only the laws they choose to enforce, and even when they choose to enforce specific laws, they often will enforce these laws only some of the time. In many situations, they will warn, reprimand, or release an individual rather than make an arrest. Often they decide not to arrest for crimes they do not consider to be violent or a serious threat, such as traffic violations, petty gambling offenses, and simple assaults between acquaintances. When the police decide not to arrest someone who qualifies to be arrested, it can be said that they are using discretion.

Joseph Goldstein (1960) argued that full enforcement would be difficult, if not impossible, to achieve. Police officers take an oath on the day they officially are sworn in as police officers that they will uphold and defend the U.S. Constitution, their state constitution, and federal and state laws. They are expected to respect individual rights of all people as outlined with the Bill of Rights of the U.S. Constitution, including the rights of those suspected of committing a criminal offense; to know and follow all the various court decisions establishing legal police procedures; and to protect the due process rights, as guaranteed by the Fifth and Fourteenth Amendments, of everyone that they arrest, search, or interrogate. Given the ambiguities in criminal law and due process boundaries, the conflicts between requirements, the time constraints, and the limitations of personnel and investigative techniques, full enforcement is simply not feasible. Therefore, actual enforcement of the law is inevitably selective enforcement (554–561).

Selective enforcement emphasizes the "spirit of the law" rather than the letter of the law. The unwritten selective enforcement policy of police departments enables police to enforce only the most serious laws and the laws the community wants enforced. Although legislation is written as if full enforcement is expected, lawmakers do not really expect it, and police agencies do not really expect it either. Kenneth Culp Davis (1975) argued that the police should not be expected to enforce all laws when prosecutors do not prosecute all cases. Prosecutors pick and choose what to prosecute, and police officers who observe them soon know what it takes for a prosecutor to take a case to trial. Further, insufficient funds allocated to police departments allow only for selective enforcement. Finally, lawmakers frequently pass laws that are difficult to obtain evidence to enforce (79–97).

Herman Goldstein (1963) discussed several factors influencing police exercise of discretion:

1. Both state statutes and city ordinances may be explicit in defining conduct to be considered criminal, but there may be little expectation on the part of those who enacted the laws that they be enforced to the letter.
2. The problem does not always stem from a double standard of matters of morality. Often it stems from mere obsolescence.
3. Another major factor which forces the exercise of discretion is the limitation of manpower and other resources. Few police agencies have the number of personnel that would be required to detect the total amount of criminality which exists in a community and to prosecute all offenders. Rarely is consideration given to

the relationship between the volume of what can be termed criminal acts and the resources available to deal with them.

4. Since there are no established priorities for the enforcement of law prohibiting one type of conduct as against another, the police official must determine the manner in which available manpower and equipment will be used. The daily assignment of manpower is, therefore, perhaps the most easily identifiable exercise of discretion on the part of the police.

5. In establishing priorities of enforcement, greater attention is ordinarily given to more serious crimes. A determination not to arrest is more common at the level of petty offender—and especially if the offender is an otherwise law-abiding citizen. Policies—albeit unwritten—begin to evolve. Just as social gamblers may be arrested only if their activities become organized and move into public places, so drunkards may be arrested only if they are belligerent and homeless as distinct from those who are cooperative and long established residents.

6. Discretion may be exercised on the basis of a police officer's particular assignment. Many police agencies have officers assigned to specific types of investigations, such as those relating to homicide, burglaries, or narcotics. Officers so assigned understandably consider their respective specialized function as being of greater importance to the department. The generalization can be made that police officers frequently refrain from invoking the criminal process for conduct which is considered of less seriousness than that which they are primarily responsible for investigating.

7. Where the volume of criminal activity is high, it is common to observe police policies which result in the dropping of charges against minor assailants when the victim is unwilling to testify. Without a complainant, the case cannot usually be prosecuted successfully.

8. Discretion is often exercised by the police in a sincere effort to accomplish some social good. This is a sort of humanitarian gesture in which the police achieve the desired objective without full imposition of the coldness and harshness of the criminal process. (142–143)

The recognition of police discretion can be traced back to the early twentieth century, when the Supreme Court of Michigan, in *Gowan v. Smith,* 122 N.W. 286 (1909), asserted that

the police commissioner is bound to use the discretion with which he is clothed. . . . To enable him to perform the duties imposed upon him by law, he is supplied with certain limited means. It is entirely obvious that he must exercise a sound discretion as to how those means shall be applied for the good of the community.

Although police discretion has been recognized by the courts since the early 1900s, it was not until the late 1950s that the police began to officially claim that officers possess discretion in performing their policing duties. The legitimation of police discretion was aided by an American Bar Association survey on the administration of justice in the mid-1950s. The survey found that criminal laws were frequently used for a variety of purposes other than their specified purposes—for example, to deal with numerous social problems, such as family problems, mental illness, and alcohol abuse. Thus, it was acceptable practice for police officers to use their own judgment about when to arrest and for what purpose (Walker 1993, 10). The 1987 U.S. Supreme Court ruling in *McClesky v. Kemp,* 411 crL 4107, further upheld the use of police discretion. Although police administrations under the professional model of policing have held onto the claim that the police fully enforce all the laws, progressive police administrators and personnel know that selective enforcement is the norm.

Discretion in when and whether to arrest or to use force against an individual places police officers in an extremely significant position in our society. Michael Brown (1988) went so far as to claim that "police discretion—the day to day decisions of policemen—is tantamount to political decision making, for the role of the police is based upon the legitimate use of coercion" (3–4). Brown further argued that "the police have crucial policy-making powers by virtue of their power to decide which laws will be enforced and when. At issue is not simply the legality of these decisions, but the routine use of the legitimate means of coercion in society. The day-to-day choices of policemen affecting the meaning of law, order, and justice within American society" (5–6).

Fairness issues necessarily arise when race or class biases influence who is arrested or apprehended and who is not. Kenneth Culp Davis (1971) provides a realistic example:

> A statute, an ordinance, and a police manual all provide that a policeman "shall" arrest all known violators of the law. A policeman lectures a boy from a middle-class neighborhood, but he arrests a boy from the slums, although he knows that both are equally guilty of violating the same statute. Because the evidence against both boys is clear, the policeman's decision is the only one that counts, for the release of the first boy is permanent and the conviction of the second follows almost automatically. (218)

Further, police discretion can be abused. One common form of abuse is the dispensing of *street justice.* Street justice occurs when the police believe that someone should be punished for what he or she did or is about to do. The police usually have no qualms about using street justice on an individual who defies their authority or insults them. They feel that if they are to function effectively, they must be in control, and that if they are insulted and allow an individual to get away with it, then they have lost control of the situation. The Rodney King beating by the Los Angeles police officers could be considered an example of street justice. Rodney King gave the police a high-speed chase. He refused to stop when the police were in pursuit. According to police culture, when the police want you to do something, you do it. If you refuse to follow police orders, you may have to pay the price. Rodney King was beaten to teach him a lesson—and also to teach a lesson to the passengers in the vehicle King was driving.

Many police officers see nothing wrong with street justice. They consider it to be a part of their right and duty to teach an individual a lesson. However, when police officers dispense street justice, the individual has been denied due process and judicial review. By using street justice, a police officer has violated the law. Assaulting an individual is a crime, and police officers should not be immune from criminal charges if they have violated the criminal law or denied someone his or her constitutional rights.

The nature of the role assigned to the police by the community may be a factor contributing to the use of street justice:

> Studies of police discretion largely reinforce the view that street decisions to intervene, arrest, use force or issue traffic citations are primarily a function of situational and organizational factors that reflect interpretations of community needs and expectations by police. "Street justice" in this sense is a response to a community mandate that something be done about situations where formal institutions cannot or will not respond for a variety of reasons. (Sykes 1986, 498)

Sykes argued that

> the liberal reformers fear police repression and overlook the functional role of "street justice" in creating a sense of community. They choose to emphasize justice defined as due process over the community-based idea of justice as righting a wrong and protecting the physically weak from the powerful. Police accountability remains a problem, and order maintenance continues to create the potential for abuse of power. However, the answer provided by the professional model, which emphasizes the enforcement role, leaves many citizens without the means to civilize their community. The trade-off must be recognized and provides an on-going moral dilemma in liberal society. (507)

The dilemma here is that street justice is used by police officers to enforce the moral code of the community they police but that the use of street justice pushes justice aside.

According to Carl Klockars (1986), the police, for practical reasons, do not use street justice all that frequently. If an individual given street justice has been beaten severely, the police officer has to transport the individual to the emergency room and file a report explaining how the individual received his injuries. Because of the threat of departmental charges and the reactions of the community, which may not know the context of the officer's use of force, street justice is kept to a minimum (515–516)— although this may be hard to believe at a time when publicity has surrounded the Rodney King fiasco and the case of Brooklyn police officers accused of shoving a broom handle up a suspect's rectum. The next section will discuss the various forms of police misconduct in detail.

As Joseph Goldstein (1960) pointed out, police, unlike the other actors within the criminal justice process, generally operate without being seen, so their actions are not reviewed. If police officers assigned to investigate a disturbing event report the complaint as unfounded, in most cases their report will be accepted and not visible to the public. Even the records maintained by the police are usually too incomplete to be used to evaluate nonenforcement decisions (543–553).

Because of concerns over fairness issues in police discretion, police abuses of discretionary powers, and the public invisibility of much police practice, many commissions and agencies have recommended that police departments establish rules to guide police discretion. In 1967, the President's Commission on Law Enforcement and Administration of Justice recommended such guidelines, stating that "in view of the importance, complexity, and delicacy of police work, it is curious that police administrators have seldom attempted to develop and articulate clear policies aimed at guiding or governing the way policemen exercise their discretion on the street" (103). In 1973, the National Advisory Commission on Criminal Justice Standards and Goals reiterated this theme. It recommended that

> every police agency should acknowledge the existence of the broad range of administrative and operational discretion that is exercised by all police agencies and individual officers. That acknowledgment should take the form of comprehensive policy statements that publicly establish the limits of discretion, that provide guidelines for its exercise within those limits, and that eliminate discriminatory enforcement of the law.

> 1. Every police chief executive should have the authority to establish his agency's fundamental objectives and priorities and to implement them through discretionary allocation and control of agency resources. . . .
> 2. Every police chief executive should establish policy that guides the exercise of discretion by police personnel in using arrest alternatives. . . .
> 3. Every police chief executive should establish policy that limits the exercise of discretion by police personnel in conducting investigations, and that provides guidelines for the exercise of discretion within those limits. . . .
> 4. Every police chief executive should establish policy that governs the exercise of discretion by police personnel in providing routine peacekeeping and other police services that, because of their frequent recurrence, lend themselves to the development of a uniform agency response.
> 5. Every police chief executive should formalize procedures for developing and implementing the foregoing written agency policy.
> 6. Every police chief executive immediately should adopt inspection and control procedures to insure that officers exercise their discretion in a manner consistent with agency policy. (21–22)

In 1974, the American Bar Association argued that

> police discretion can best be structured and controlled through the process of administrative rule-making by police agencies. Police administrators should, therefore, give the highest priority to the formulation of administrative rules governing the exercise of selective enforcement, investigative techniques, and enforcement methods. (8)

Internal rule making gives police agencies a means of limiting the control of external agencies that can regulate police decision making. The courts, state legislatures, and city councils, and county commissioners can all influence police policies and discretion. The numerous United States Supreme Court decisions have dictated police policies and have limited police discretion in many situations. State legislatures and city councils have directed the police to arrest the abuser during a domestic violence call. When the police observe or believe that a person has been assaulted by a spouse or live-in companion, they are legally required to make an arrest. This requirement to arrest during a domestic dispute is an example of how the legislative branch has taken away the power of the police to exercise discretionary decision making in certain situations. Rule making exists to increase predictability, fairness, and efficiency in daily policing activities. It should also contribute to more effective and responsive policing by minimizing procedural errors, improving police-community relations, establishing uniform policy, and centralizing accountability for decision making (Aaronson et al. 1984). Kenneth Culp Davis (1975), a strong advocate of rule making, provides 14 reasons why rule making is needed for controlling police discretion:

1. The quality of enforcement policy will be improved because it will be made by top officers instead of patrolmen. The top officers obviously have skills and broad understanding that patrolmen typically lack. . . .
2. The quality of enforcement policy will be improved because the preparation of rules will lead to appropriate investigations and studies by qualified personnel, including specialists with suitable professional training. No longer will it be made primarily by the offhand guess work of patrolmen.
3. The quality of enforcement policy will be improved because it will be made by officers who are addressing their minds to the problems of what the policies should be and why. . . .
4. The quality of enforcement policy will be improved by openness, for the police will further develop those practices that the public approves and will minimize or eliminate [others].
5. The quality of enforcement policy will be improved by suggestions and criticisms that come from the public. Even the best of administrators in federal agencies usually find that written comments on proposed rules call to their attention effects that even the most careful studies have failed to uncover.
6. Policy formulated through rulemaking procedure is more likely to carry out community desires. Today the police usually make guesses about what the public wants. . . .
7. A great gain from use of rulemaking procedure will be the education of the public in the reality that the police make vital policy. The public are now inclined to assume, as even the police do much of the time, that the police merely enforce the law and have little or nothing to do with policy making. . . .
8. Bringing enforcement policy out into the open will increase the fairness to those affected by the policy. Fairness requires opportunity to know not only the formality of the law (statutes and judicial opinions) but also the reality of the law (what is enforced). When excessive criminal statutes are cut back to nonenforcement, one who wants to act contrary to the statute but in accordance with what the enforcement policy permits should be entitled to know the enforcement policy. . . .

9. Open rulemaking based on the realities of the policy problems that confront the police should demonstrate to legislative bodies the need for reworking criminal legislation to bring it in accord with what is practicable from the standpoint of enforcement. . . .

10. Police rulemaking can and should gradually take the place of the somewhat unsatisfactory exclusionary rule, elaborately fabricated by the courts, now governing two or three percent of police activities. . . .

11. Police rulemaking can gradually ease the judicial burden of fabricating and administering the exclusionary rule, a task the federal courts should never have undertaken. . . .

12. Police rulemaking will mean that police enforcement policy will no longer be almost completely exempt from judicial control, as it has been from the beginning of American government. A limited judicial review of the kind that is customary with respect to other administrative action is clearly desirable. . . .

13. Open rulemaking will promote equal justice by reducing policy differences from one officer to another. The present system of allowing most enforcement policy to be made by the patrolman handling each case causes unnecessary disparity. . . .

14. Possibly most important of all is the idea that rulemaking can reduce injustice by cutting out unnecessary discretion, which is one of the primary sources of injustice. Necessary discretion must be preserved, including especially the needed individualization—the adapting of rules to the unique facts of each case. (113–119)

Although rule making has been advocated as a means of controlling police discretion and police behavior, most police departments have not followed the recommendations to implement rule making for their police officers. For the most part, departments initiate rules for their officers as a reaction to predicaments or in response to lawsuits, political pressure, or some other crisis. Rule making by police departments reflects a reactive approach rather than a proactive approach (Walker 1986, 362–363). But when police departments have failed to implement rules to control police discretion, the judicial branch has taken over this function. Enlightened police leaders have realized that when police activities have intruded on the liberty of citizens, policies and procedures are required to give police officers guidance on

1. the use of physical and deadly force
2. arrest and alternatives to arrest
3. stopping, questioning, and frisking citizens
4. the handling of disorders and/or minor nuisances short of arrest
5. the use of all intrusive investigative techniques including informants, infiltrators, and electronic surveillance (Walker 1986, 388–389)

How does police discretion fit into community policing? The community policing philosophy emphasizes a closer police-community working relationship. Traditional, or professional-model, policing requires that police respond to citizen complaints that are telephoned in to a dispatcher. When patrol officers have free time and are not responding to citizen calls or investigating a traffic accident, they are expected to look for crime on their beat—for example, by patrolling areas that are known to have a high rate of burglaries or a high rate of automobile accidents. Under the community policing philosophy, all police officers are community policing officers. If this concept becomes operationally successful, the use of discretion by patrol officers may be more carefully thought out. As the beat patrol officer and the neighborhood develop a closer working relationship, breaking down barriers and sharing a stake in crime prevention and control, police discretion may become more attuned to community needs, and police misconduct may decrease.

POLICE MISCONDUCT

According to organizational scholars, most occupations provide their members with the opportunities for behavioral misconduct, and police departments are no exception. Three elements of occupational misconduct are: "(1) opportunity structure and its accompanying techniques of rule violations, (2) socialization through occupational experiences, and (3) reinforcement and encouragement from the occupational peer group, i.e. group support for certain rule violations" (Barker 1977, 356).

The quality of work performance varies considerably from one police department to another. Most police departments employ police officers who generally function in a fair and impartial manner. However, some police officers become involved in police misconduct. Despite the publicity that police misconduct receives, police departments or even state or federal agencies keep few records of misconduct. It is not uncommon for departments to allow officers involved in police misconduct to resign their position rather than to be fired or prosecuted so that both the officers and the department can avoid publicity. In some states, policy agencies that do this may be violating state training standards that certify police officers. By these standards, an officer involved in police misconduct can lose his certification so that he is no longer employable as a police officer. But when a department allows a police officer to resign, this individual can obtain employment in another police department.

Kenneth Culp Davis (1974) outlined five basic facts about police policy:

1. Much of it is illegal or of doubtful legality.
2. Subordinates at or near the bottom of the organization, not top officers, make it.
3. Most of it is kept secret from those who are affected by it.
4. Police policy is characteristically based on superficial guesswork and hardly at all on systematic studies by staffs of qualified specialists or on investigations like those conducted by our best administrative agencies and legislative committees.
5. It is almost completely exempt from the kind of limited judicial review deemed necessary for almost all other administrative agencies. (703–704)

When the police are involved in misconduct, or even when the community believes that the police have committed acts of misconduct, the community loses confidence in the police. To obtain and maintain the community's backing, the police must refrain from wrongdoing. The community expects its police officers to be beyond reproach. When the police lose the community's respect and trust, the department cannot be an effective tool in criminal apprehension and in controlling crime. Without the community's support, victims and witnesses will not come forward. This is crucial because victims and witnesses willing to participate in the legal process contribute to the arrest and prosecution of offenders.

Forms of Police Misconduct

Police misconduct may be either a criminal violation or a violation of departmental rules. It may be unethical or amoral without being criminal, and it does not have to result in personal gain, such as monetary rewards. It may be a reflection of the community, the police department, or an individual police officer.

Sleeping on Duty, Drinking on Duty, and Drug Abuse

Sleeping on duty usually occurs on the midnight shift during slow periods when there are few calls for service. Although this violates police policies, there are officers who attempt to take catnaps during working hours. Some police officers have second jobs and work during the day and need to rest during their midnight tour of duty.

Most police departments have police officers who are alcoholics who will drink on duty. Police officers with drinking problems are very creative in finding ways to drink on duty. These officers will not refrain from drinking simply because it violates a departmental policy.

In the modern era of policing, the use of drugs by police officers can be added to the police misconduct list. Police officers who were exposed to drugs growing up and who in all probability experimented with drugs do not necessarily see anything wrong in using drugs. With drugs easily available in our society, it can be expected that some police officers will use illegal drugs. Police officers have been known to confiscate drugs from users and dealers to use for themselves. Drug abuse has become such a serious problem in our society that police departments require all police recruits to take a drug test before being employed as a police officer.

Police Brutality

Police brutality is an ambiguous term that has had different meanings depending upon who uses it. Does police brutality include verbal abuse? To some people, it does. A number of police actions, including profane and abusive speech, commands to move or go home, field stops and searches, threats, prodding with a nightstick or approaching with a pistol, and the use of "physical force," may be called police brutality (Barker 1986, 71). Most of these actions may or may not be police misconduct depending on the situation.

However, police officers should never use profanity or abusive language toward any citizen, including those suspected of a crime, and police officers threatening any person should be considered to be violating good police procedures and open to police misconduct charges.

Citizens should recognize that police have the right to direct citizens' behavior. For example, if a person is double parked and blocking traffic on a city street, the police officer has the responsibility to require that person to move his or her vehicle. Also, if a group of teenagers are blocking access into a store and intimidating adults from entering the store, the police officer has the right to tell the teenagers to move. In these situations, the police officer should be civil and courteous but should expect the citizen or teenagers to comply with his or her request.

Many citizens are unaware that when police are patrolling their beat they are expected to conduct field stops and searches. The stops are to determine if the person or group stopped has legitimate business in the area. Generally, stops will take place during the evening hours. Citizens not versed in the law may not know that police officers have the legal right to conduct searches to protect themselves or others from potential harm. Whether police misconduct occurs during a field stop or search depends upon the specific situation and the police officer's behavior.

The use of physical force by police officers is not illegal or a violation of a departmental policy if it is justifiable. Every situation where physical force has been used must be judged on its own merit.

Generally, acts of police brutality are known only to the police officer and a few witnesses, normally fellow police officers, who observe excessive force being used. Segments of the American population, especially minority members, who have the largest number of brutality complaints against the police, consider police brutality to be more of a serious problem than politicians or police administrators care to admit. Police brutality is usually discovered when the victim of police brutality reports it to police authorities or to other governmental agencies. Most states provide for a state investigative agency to investigate police brutality by local police officers. On the federal level, the Federal Bureau of Investigation has the authority to investigate police misconduct that can be in violation of federal laws.

Today, with the prevalence of video cameras, cases of police brutality can be videotaped and shown later. The videotaping of acts of police brutality is evidence that a specific officer or officers were involved in the use of excessive force. One case in which an act of police brutality was brought to the world by a citizen's videotaping was the beating of Rodney King on March 3, 1991, by Los Angeles police officers. George Holliday, the citizen who videotaped the Rodney King beating, brought the tape to the Los Angeles Police Department, who refused it. Upon the department's refusal, Holliday brought the tape to a local Los Angeles television station, which aired it. Eventually, CNN received a copy of the videotaped beating of Rodney King and gave it national and international coverage.

Another well-publicized case of police brutality took place on August 9, 1997, when a Haitian immigrant, Abner Louima, was taken into custody by police while sitting in a Brooklyn music club. The police took him to the 70th Precinct station house, where they strip-searched him and beat him. Two police officers allegedly shoved a toilet plunger into his rectum and then forced the toilet plunger into his mouth, breaking his front tooth. Louima asserted that while this was occurring, the officers taunted him, saying, "That's your shit, nigger," "We're going to teach niggers to respect police officers," and "This is Giuliani time, not Dinkins time" (Farley 1997, 38). Later, Louima recanted these comments. Eventually, Louima was hospitalized after a one and one-half hour ordeal in the station house. His injuries included a ripped bladder and a punctured lower intestine. The New York City police commissioner Howard Safir called the incident a "horrific crime" and stated, "We are going to make sure the perpetrators . . . go to jail." He also pointed out that a police officer who had been present had come forward to implicate his colleagues (Farley 1997, 38). One police officer was arrested and charged with sexual assault and sexual abuse, and 12 police officers were either transferred, suspended, or demoted to desk duty.

When the public becomes aware of acts of police brutality in incidents such as the Rodney King or Abner Louima cases, they cannot have much confidence that the police will refrain from using excessive force. As acts of police brutality keep coming to the attention of the public before they can forget the last act of brutality brought to their attention, the police forfeit the community's trust more and more. The gap between police and minority communities grows wider or at least cannot be lessened while such inhumane acts are being committed.

Police brutality cannot be condoned. For community policing to be successful, police brutality must be eliminated. Why would citizens want to cooperate with the police when police are abusing community residents? The negative publicity from the Rodney King beating and the Abner Louima abuses will take decades for the police to overcome.

Throughout the 1990s, there have been numerous documented incidents of police brutality in American police departments of all sizes and in all parts of the country, rural and urban. But very little statistical information is available to document the overall extent of police brutality. Obtaining reliable data on police brutality is extremely difficult. To correct this deficiency, the U.S. Congress passed the Violent Crime Control and Law Enforcement Act in 1994, requiring the Attorney General to collect data on the use of excessive force by the police and to publish an annual report of the findings. The President's Commission on Law Enforcement and Administration of Justice (1967) reported in its task force report that physical abuse by police officers was not a serious problem but did exist. Commission researchers observed patrol officers in a number of cities and did observe instances where excessive force was used when none was necessary. Those abused were poor, drunks, sexual deviates, or youngsters considered hoodlums. The key ingredient discovered by the task force members was that those abused had verbally challenged the police officer's authority (181–182). If police officers are willing to use excessive use of force in the presence of the president's task force, what can the community expect when officers are not in the presence of witnesses?

Corruption

Police corruption is the abuse of police authority to obtain personal gain. This gain may be either monetary or nonmonetary. In committing acts of corruption, police officers gain financially by failing to perform services they are required to perform or by giving services that they should not be giving. Barker and Wells (1981) characterized acts of corruption in three ways: "(1) They are forbidden, by some norm, regulation or law, (2) they involve the misuse of the officer's position, and (3) they involve a material gain, no matter how insignificant" (4). Stoddard (1995) has listed 10 identifiable forms of corruption:

1. *Mooching:* An act of receiving free coffee, cigarettes, meals, liquor, groceries, or other items either as a consequence of being in an underpaid, undercompensating profession or for the possible future acts of favoritism which might be received by the donor.

2. *Chiseling:* An activity involving police demands for free admission to entertainment whether connected to police duty or not, price discount, etc.

3. *Favoritism:* The practice of using license tabs, window stickers, or courtesy cards to gain immunity from traffic arrest or citation (sometimes extended to wives, families, and friends of recipient).

4. *Prejudice:* Situations in which minority groups receive less than impartial, neutral, objective attention, especially those who are less likely to have "influence" in city hall to cause the arresting officer trouble.

5. *Shopping:* The practice of picking up small items such as candy bars, gum, or cigarettes at a store where the door has been accidentally unlocked after business hours.

6. *Extortion:* The demands made for advertisements in police magazines or purchases of tickets to police functions, or the "street courts" where minor traffic tickets can be avoided by the payment of cash bail to the arresting officer with no receipt for payment.

7. *Bribery:* The payment of cash or "gifts" for past or future assistance to avoid prosecution; such reciprocity might be made in terms of being unable to make a positive identification of a criminal, or being in the wrong place at a given time when a crime is to occur, both of which might be excused as carelessness but with no proof as to deliberate miscarriage of justice. Differs from mooching in the higher value of a gift and in the mutual understanding regarding services to be performed upon the acceptance of the gift.

8. *Shakedown:* The practice of appropriating expensive items for personal use and attributing them to criminal activity when investigating a break-in, burglary, or an unlocked door.

9. *Perjury:* The sanction of the "code," which demands that fellow officers lie to provide an alibi for fellow officers apprehended in unlawful activity covered by the "code."

10. *Premeditated Theft:* Planned burglary, involving the use of tools, keys, etc., to gain forced entry or pre-arranged plan of unlawful acquisition of property which cannot be explained as a "spur of the moment" theft. (190–191)

Police corruption, according to veteran police officers, is always around and comes to life whenever the police department's vigilance subsides. For example, New York City police officers have been charged with selling protection to drug dealers and sometimes trafficking in drugs themselves. A report by the Mollen Commission investigating corruption in the New York City Police Department asserted that the department's internal affairs unit routinely buried corruption cases. The report also stated that police officers routinely made false arrests, tampered with evidence, and perjured themselves in court ("Policing the Police," 1994, 16).

Stoddard (1995) found that the police culture has an informal code that protects police misconduct and tolerates corruption. Recruits are indoctrinated into the police culture and soon come to the realization that they need the support of their fellow police officers, because their personal safety may depend upon a fellow officer being there when they need assistance. Further, recruits quickly learn that they themselves are bound to make mistakes, which may result in violations of departmental policy or criminal law, so that it may be best "not to see anything" if they observe misconduct by other officers.

Corruption has been found in many police departments—among them, Atlanta, Miami, Houston, New Orleans, Philadelphia, Savannah (Georgia), and Los Angeles. Further, suburban and rural areas are not free from corruption. For example, a number of rural sheriffs from Georgia have been sent to prison for protecting drug dealers.

Scandals involving police officers in corruption cannot benefit the police-community relationship. Police departments must check corruption if they hope to develop a successful community policing program.

Commission of Criminal Acts

A police officer lying in a court of law to obtain a conviction of a suspect has committed a crime. Lying in court has been termed *perjury* and is considered a criminal offense. Other violations of criminal law by police officers include murder, bank robbery, burglary, domestic violence, child abuse, and rape. In New Orleans, an off-duty police officer committing a robbery killed her partner, who was moonlighting as a security guard. Four Detroit police officers were charged with the murder of a black motorist who was beaten to death. In Galesburg, Illinois, a police officer was charged with bank robbery, and in Chatham County, Georgia, a police officer was caught robbing a convenience store on his own beat. In Wichita, Kansas, a police officer was charged with child molestation. A Savannah, Georgia, police officer placed his own child in scalding hot water.

Although few police officers commit criminal acts, the few that do create a negative image for all police officers. Police departments must recruit people of integrity and investigate all complaints of improbity. Also, they must be proactive in their investigations and emphasize eliminating not only police crime but all police misconduct.

MECHANISMS TO CONTROL POLICE MISCONDUCT

For police misconduct to be controlled, mechanisms must be in place to control police behavior. These mechanisms can be either internal or external to the police agency. There are strong arguments for both approaches. Groups and individuals that distrust the police claim that the police cannot clean their own house, so that outside mechanisms are necessary to control police misconduct. On the other hand, police personnel find it offensive that "outsider" civilians who do not understand police work will be reviewing their behavior. This controversy is a hot topic in policing. If community policing is to be accepted by both the community and the police, this dispute will eventually have to be settled to the satisfaction of both.

Internal Affairs Units

Since the 1950s, midsized and large police departments have established internal affairs units to investigate citizen complaints of police misconduct. The size of an internal affairs unit can range from one officer to several hundred officers. The norm for staffing internal affairs is to have a ranking officer, captain or lieutenant, involved in investigating the conduct of police officers.

The ideal would be for every complaint against a police officer to be investigated impartially. When the internal affairs unit conducts an investigation, it interviews all witnesses and examines all pertinent evidence. The internal affairs unit commander reports directly to the chief of police. A report is made of the unit's findings to the chief, who then makes a decision for the appropriate action to be taken. This decision made by the chief could be any of the following:

- *Unfounded*—the investigation found evidence that the act did not occur.
- *Exonerated*—the investigation found that the act was justified, lawful, and proper or that the act complained about never took place.
- *Not sustained*—the investigation did not find sufficient evidence that could either prove or disprove the allegation.
- *Sustained*—the investigation found sufficient evidence to prove the allegation.

Internal affairs units can be assigned to investigate all forms of police misconduct, from sleeping on duty to acts of corruption. One major area the unit investigates is the use of force by police officers. There are many complaints against police officers using excessive force against suspects and citizens. The policy of the Savannah, Georgia, Police Department provides for immunity to police officers being questioned. Giving police officers immunity means that any information supplied by an officer cannot be used against him or her in a criminal trial. Most departments require that the police officer give a written response to the allegations made against him or her. The internal affairs procedures established by the Savannah Police Department serve several purposes, which could be applied to any police department that has an internal affairs unit:

- The procedure permitted citizens to seek redress of their legitimate grievances against officers when the citizens felt they were subjected to improper treatment by an officer.
- The procedure provided the chief of police with an opportunity to monitor employee compliance with departmental procedures and rules. When violations

were established, appropriate discipline, training and direction were applied as needed in order to correct the problem.
- The procedure of investigating all citizen complaints, including anonymous complaints, helped perpetuate a positive image and ensure the integrity of the police department.
- The procedure also helped protect the rights of the department employee. (McLaughlin 1992, 96)

Civilian Review Boards

The notion of a civilian review board can be traced to the 1930s. A civilian review board is "an independent tribunal of carefully selected outstanding citizens from the community at large" (Bopp and Schultz 1972, 146). Today, civilian review boards are often referred to as *civilian oversight of the police.* Although there may have been a name change, the two terms have similar meanings.

Every time there is a situation where the police are suspected of police misconduct, we hear the call for implementing a civilian review board if a civilian review board does not already exist. Many of the same arguments made for a civilian review board in the late 1990s are similar to the arguments made by supporters for civilian review boards in the 1960s and 1970s.

Civilian review boards review citizens' grievances against the police and are made up of only civilians, without any police representative. In 1967, the President's Commission on Law Enforcement and Administration of Justice recommended the adoption of such an external grievance system, stating that "the primary need is for the development of methods of external control which will serve as an inducement for police to articulate important law enforcement policies and to be willing to have them known, discussed and changed if change is needed" (32).

Police acceptance of such an external grievance mechanism is a way of showing that their operations are open for inspection. The police can only create doubt and suspicion by rejecting an external grievance mechanism. In keeping with the community policing philosophy, with its emphasis on community-police interaction, it is a friendly gesture on the part of the police to accept civilian review boards, or civilian oversight bodies, as they are referred to today.

Civilian oversight bodies have several characteristics in common. First, civilian oversight has been legally approved, either as an amendment to the city charter or by city ordinance. Second, civilian oversight bodies function independent of the police agency (Terrill 1990, 80).

Those opposed to civilian review boards claim that

1. Civilian review boards ignore other legal resources that citizens have for registering legal complaints, i.e. state attorney's office, federal EOC, civil suits, FBI civil rights investigation, and so forth.
2. It is difficult for citizens to understand operations of law enforcement agencies and have a thorough understanding of laws, ordinances and procedures which law enforcement officers must uphold and operate within.

3. Civilian review boards have a destructive effect upon internal morale.
4. Civilian review boards invite abdication of authority by line supervisors and lower-level management. These are the levels of supervision that should exercise maximum control.
5. Civilian review boards weaken the ability of upper-level management to achieve conformity through discipline.
6. The creation of a civilian review board is tantamount to admitting that the police cannot "police themselves" (Hensley 1988, 45)

Conversely, the supporters of civilian review boards claim that

1. A lack of communication and trust exists between the law enforcement and minority group communities.
2. The lack of trust is accentuated by the belief that law enforcement agencies fail to discipline their own employees who are guilty of misconduct.
3. Civilian review would theoretically provide an independent evaluation of citizens' complaints.
4. Civilian review would ensure that justice is done and actual misconduct is punished.
5. Civilian review would improve public trust in law enforcement.
6. Civilian review provides for better representation of the entire community. (Hensley 1988, 45)

Many of America's largest cities have some type of civilian oversight of police actions. In addition, many of America's midsized cities have adopted civilian oversight boards (Snow 1992, 51). The implementation of civilian oversight bodies does not necessarily mean that the police officer will always be found guilty. In all probability, the police officer will be exonerated. Civilian oversight boards are not in conflict with the community policing philosophy. In fact, civilian oversight boards indicate to the community the seriousness of police intentions to develop alliances and a cooperative spirit. The City of Wichita, Kansas, Police Department, which has a civilian oversight board, rarely has a recommendation of its internal affairs unit overturned by the civilian oversight board.

Civil Liabilities

For the past three decades, the United States Supreme Court has utilized Title 42, Section 1983 of the U.S. Code to control police misconduct. Section 1983 prohibits government employees from using any state law, ordinance, regulation, or custom as a technique to deprive citizens of their constitutional rights. The Court's purpose in establishing Section 1983 has been described as follows:

> As a result of the new structure of law that emerged in the post–Civil War era—and especially of the Fourteenth Amendment, which was its center-

piece—the role of the federal government as a guarantor of basic federal rights against state power was clearly established. Section 1983 opened the federal courts to private citizens, offering a uniquely federal remedy against incursions under the claimed authority of state law upon rights secured by the Constitution and the laws of the nation. . . .

The very purpose of Section 1983 was to impose the federal courts between the states and the people, as guardians of the people's federal rights—to protect from unconstitutional action under color of state law, "whether that action be executive, legislative, or judicial." (Nahmond 1986, 4)

Section 1983 of Title 42 of the U.S. Code pertains to procedural remedies rather than substantial remedies. It provides an avenue of redress for violating the federal constitutional rights addressed in the Bill of Rights. In the 1960s, Section 1983 was expanded by the United States Supreme Court to afford the petitioner the right to take action against any person who—*under color of law* (state law, custom, or usage)—subjects another person to the deprivation of any rights, privileges, or immunities guaranteed by the U.S. Constitution. This means that the Fourth Amendment guarantees against unreasonable searches and seizures are equally applicable to the states via their judicial incorporation into the due process clause of the Fourteenth Amendment.

The Supreme Court of the United States has the authority to enforce the Fourteenth Amendment's provisions against state agents. This includes police officers who in some way fail to act according to their authority or who abuse their authority. The abused citizen, under Section 1983, can file a suit of equality, or other proceedings for redress, for being denied his or her constitutional rights, privileges, and immunities by an official's abuse of his or her position.

The United States Supreme Court concluded that local government officials under Section 1983 could be sued in their official positions for damages and retrospective declaratory and injunctive relief, even though local government may pay. A local government body can be held liable for traditions and customs that are unwritten and practiced by governmental departments or their employees.

Section 1983 plaintiff suing a local government (or, indeed, any defendant) has to show both duty and a cause in fact relationship between breach of duty and plaintiff's constitutional deprivation. If only one of these is shown, plaintiff loses and defendant wins because a prima facie Section 1983 cause of action has been started. (*Jane Monell et al. v. Department of Social Services of the City of New York et al.* 1978)

To establish a prima facie case, sufficient evidence, physical or testimonial, must be presented during the legal process. Police officers, their police departments, and their cities are being sued for their actions. Although some suits may be frivolous, many are legitimate and pertain to police misconduct. All police officers are considered to be agents of the government, and the governmental agency that employs their services is held accountable for their behavior. Primarily, county and city governments are sued because they have "deep pockets." These governments have a tax base and can raise money to pay plaintiffs the amount awarded. Suing a police officer does not generally result in obtaining a large amount of money. Police officers are not affluent, and a large cash settlement cannot be obtained from them.

Mechanisms should be in place to investigate, control, and attempt to eliminate police misconduct. The awesome power of police officers must be checked. Not one but several various mechanisms should be employed to control police misconduct. In an era where police administrators are attempting to improve their relationship with the community that employs them, it behooves governmental bodies to have in place mechanisms to control police misconduct.

When community members become stakeholders in the mission of police officers and have a major interest in crime prevention and control of disorder, a closer relationship between the police and the community can develop. With a police-community partnership, trust and confidence can grow, and fewer complaints will be made concerning police actions.

SUMMARY

Police administrators used to claim that the police fully enforced all laws and investigated all crimes equally. But this has never been true. Full enforcement of the law would be impossible. In reality, selective enforcement is the norm. Police officers assigned to investigate a disturbing event can report the event as unfounded, and in most cases their report will be accepted and not visible to the public. Even the records maintained by the police are usually so incomplete that they cannot be used to evaluate nonenforcement decisions.

Although the police are sworn to enforce all the laws all the time, they enforce only the laws they choose to enforce and enforce even these laws only some of the time. In many instances, they warn, reprimand, or release an individual rather than make an arrest. Often, they decide not to arrest for crimes that they do not consider to be violent or a serious threat, such as traffic violations, petty gambling offenses, and simple assaults between acquaintances. In making such decisions, police are using discretion. Police discretion is the authority of police to decide how much effort they will apply to enforcing specific laws and which laws they will or will not enforce.

It would be ridiculous to eliminate police discretion. Discretion is crucial to policing. But because it can be abused or applied unfairly, many commissions and agencies have recommended that police departments institute rules and procedures to guide and control the use of police discretion.

The quality of police work performance varies considerably from department to department. Most police departments employ police officers who generally function in a fair and impartial manner, but some police officers become involved in misconduct. Police misconduct is wrongdoing committed by a police officer. It can be either a criminal violation or a violation of departmental rules. It may be unethical or amoral without being criminal. Incidents of police misconduct include commission of criminal acts, corruption, excessive use of force, racism, and denying someone his or her civil rights. Police misconduct has taken place throughout history and still takes place today.

For police misconduct to be controlled, mechanisms must be put in place to control police behavior. These mechanisms can be either internal or external to the police agency. There are strong arguments for both approaches. Groups and individuals that distrust the police claim that the police cannot clean their own house and advocate outside mechanisms to control police misconduct. But some police personnel find it offensive that "outsiders"—civilians who do not understand police work—will be reviewing their behavior.

Since the 1950s, midsized and large urban police departments have established internal affairs units to investigate citizen complaints of police misconduct. The size of an internal affairs unit can range from one officer to several hundred officers. The norm for staffing internal affairs is to have a ranking officer, such as a captain or lieutenant, involved in investigating the conduct of police officers.

One external mechanism for controlling police misconduct is civilian review boards. A civilian review board is an independent tribunal of carefully selected outstanding citizens from the community at large who review citizens' grievances against the police. Supporters of civilian review boards recommend that boards be composed of only civilians without any police representative. An additional remedy to check police misconduct is Title 42, Section 1983 of the U.S. Code. Section 1983 prohibits government employees from using any state law, ordinance, regulation, or custom to deprive citizens of their constitutional rights. The United States Supreme Court concluded that local government officials under Section 1983 could be sued in their official positions for damages and retrospective declaratory and injunctive relief, even though local government may pay. A local government can be held liable for traditions and customs that are unwritten and practiced by government departments or their employees.

KEY TERMS

Abner Louima	full enforcement	Rodney King
actual enforcement	*Gowan v. Smith*	rule making
civil liability	internal affairs units	selective enforcement
civilian review board	police brutality	street justice
corruption	police misconduct	

REVIEW QUESTIONS

1. Explain police discretion.
2. What did the American Bar Association study discover about police discretion?
3. Provide an example of police discretion.
4. What do the terms *full* and *selective enforcement* mean?

5. Describe rule making as it pertains to police discretion.
6. Describe the meaning of *street justice.*
7. What does the term *police misconduct* mean?
8. Describe the purpose of an internal affairs unit.
9. What are the functions of a civilian review board?

REFERENCES

Aaronson, D.E., et al. 1984. *Public policy and police discretion: Processes of decriminalization.* New York: Clark Boardman.

American Bar Association. 1974. *The urban police function.* Gaithersburg, MD: International Association of Chiefs of Police.

Barker, T. 1977. Peer group support for police occupational deviance. *Criminology* 15, no. 2: 356.

Barker, T. 1986. An empirical study of police deviance other than corruption. In *Police deviance,* ed. T. Barker and D.L. Carter. Cincinnati, OH: Anderson.

Barker, T., and R.O. Wells. 1981. Police administrators' attitudes toward the definition and control of police deviance. Paper presented at the Academy of Criminal Justice Sciences, Philadelphia.

Bopp, W.J., and D.O. Schultz. 1972. *A short history of American law enforcement.* Springfield, IL: Charles C Thomas.

Brown, M.K. 1988. *Working the streets: Police discretion and the dilemmas of reform.* New York: Russell Sage Foundation.

Davis, F.J. 1975. Law as a type of social control. In *Law and control in society,* ed. R. Akers and R. Hawkings. Englewood Cliffs, NJ: Prentice Hall.

Davis, K.C. 1971. *Discretionary justice: A preliminary inquiry.* Urbana: University of Illinois Press.

Davis, K.C. 1974. An approach to legal control of the police. *Texas Law Review* 52: 703–704.

Davis, K.C. 1975. *Police discretion.* St. Paul, MN: West.

Farley, C.J. 1997. A beating in Brooklyn. *Time* August 22, p. 38.

Goldstein, H. 1963. Police discretion: The ideal versus the real. *Public Administration Review* 23: 142–143.

Goldstein, J. 1960. Police discretion not to invoke the criminal process: Low-visibility decisions in the administration of justice. *Yale Journal* 69: 543–561.

Hensley, T. 1988. Civilian review boards: A means to police accountability. *Police Chief* 55, no. 9: 45.

Jane Monell et al. v. Department of Social Services of the City of New York et al., 436 U.S. 658, 56 Ed. 611, 98 S. Ct. (1978).

Klockars, C.B. 1986. Street justice: Some micro-moral reservations: Comments on Sykes. *Justice Quarterly* 3, no. 4: 515–516.

McLaughlin, V. 1992. *Police and the use of force: The Savannah study.* Westport, CT: Praeger.

Nahmond, S.H. 1986. *Civil rights and civil liberties litigation.* New York: McGraw-Hill.

National Advisory Commission on Criminal Justice Standards and Goals. 1973. *Police.* Washington, DC: Government Printing Office.

Policing the police [editorial] 1994. *New York Times,* May 1, section 4, p. 16.

President's Commission on Law Enforcement and Administration of Justice. 1967. *The challenge of crime in a free society.* Washington, DC: Government Printing Office.

Reiss, A.J. 1984. Consequences of compliance and deterrence models of law enforcement for the exercise of police discretion. *Law and Contemporary Problems* 47 (Autumn): 89.

Snow, R. 1992. Civilian oversight: Plus or minus. *Law and Order* 40, no. 12: 51.

Stoddard, E.R. 1995. The informal "code" of police deviancy: A group approach to "blue-coat crime." In *The police and society: Touchstone readings,* ed. V.K. Kappeler. Prospects Heights, IL: Waveland.

Sykes, G. 1986. Street justice: A moral defense of order maintenance policing. *Justice Quarterly* 3: 498–507.

Terrill, R.J. 1990. Alternative perceptions of independence in civilian oversight. *Journal of Police Science and Administration* 17, no. 2: 80.

Walker, S. 1986. Controlling the cops: A legislative approach to police rulemaking. *University of Detroit Law Review* 63: 362–389.

Walker, S. 1993. *Taming the system: The control of discretion in criminal justice 1950–1990.* New York: Oxford University Press.

Chapter 4

Crime Prevention and Community Policing

CHAPTER OBJECTIVES

1. Understand what is meant by crime prevention.
2. Understand what the police role is in crime prevention.
3. Understand the relationship between crime prevention and community policing.
4. Be familiar with the various crime prevention strategies.
5. Be familiar with situational crime prevention.
6. Be familiar with community crime prevention.
7. Be familiar with crime prevention media campaigns.

INTRODUCTION

The police sell themselves as the best defense against crime and argue that with more resources they would do a better job in protecting the community from crime. But according to David Bayley (1994), the noted police scholar, "The plain fact is that police action cannot be shown to reduce the amount of crime" (9). Differences in crime rates cannot be attributed to variations in the number of police, and additional police officers do not slow, even temporarily, rates of increase in crime. Although there are probably critical thresholds beyond which changing the number of police would affect crime, no one knows what these thresholds are. There is no evidence that patrolling by uniformed officers, rapid response to emergency calls, and expert investigation of crime by detectives will prevent crime. Police foot patrols make people feel better, but they do not prevent crime; and research has consistently failed to show that the intensity of random motorized patrolling by uniformed officers has any effect on crime rates, victimization, or even public satisfaction. Finally, the success of police in criminal investigations has no appreciable effect on public security (4–7).

Other police researchers have also found that many police strategies are not effective in preventing crime (Bouza 1990; Kelling et al. 1974; Walker 1989). The *reactive approach* to crime has been especially criticized. The use of 911 by citizens to call the police after a crime has been committed does not prevent crime. In many cases, the police are working merely as report takers—taking a report after the fact. Although they need to respond to emergencies and *crime-in-progress* calls to arrest an offender or come to the aid of a citizen who requires assistance, they probably should not respond to every 911 call. Many nonemergency calls can be taken over the telephone, or the citizen can come to police headquarters to file a report. Some police departments even take some criminal reports over the telephone: for example, the Wichita Police Department does this with stolen car reports.

The alternative strategy to reactive policing is *proactive* policing. Proactive policing involves taking an active part in going after criminals. Sting operations are one example: the police establish a fencing operation, getting the word out to the street criminals that they are purchasing stolen merchandise, then purchase stolen goods from the thieves and then arrest them.

The community policing philosophy is a proactive approach to policing that emphasizes crime prevention through working with the community. If crime prevention is not incorporated into the community policy philosophy, the strategy of community policing is substantially weakened, and in all probability community policing will not succeed.

Certain police strategies, such as directed patrol and proactive arrests, which will be discussed in Chapter 5 in more detail, have shown tangible evidence of preventing crime. The police *can* prevent robbery, disorder, and domestic violence, but only by using specific methods under certain conditions.

When Sir Robert Peel first established the Metropolitan London Police in 1829, he listed crime prevention as a major goal of policing. Today the police need to return to this goal. All officers should be knowledgeable in crime prevention techniques and strategies. The purpose of this chapter is to provide police officers and potential police officers with a foundation of knowledge about crime prevention. In departments adopting the community policing philosophy, an understanding of crime prevention is essential.

WHAT IS CRIME PREVENTION?

Several definitions of crime prevention are used, depending upon the program or research conducted. Lab (1997) stated that "crime prevention entails any action designed to reduce the actual level of crime and/or the perceived fear of crime" (10). Another definition, from the National Crime Prevention Institute (1986), described crime prevention as "any kind of effort aimed at controlling criminal behavior" (2). The institute classified crime prevention efforts into *direct controls* of crime—those that "reduce environmental opportunities for crime"—and *indirect controls,* which would "include all other measures, such as job training, remedial education, police surveillance, police apprehension, court action, imprisonment, probation and parole" (2). Crime prevention advocates believe in "preventing specific kinds of crimes; mobilizing residents for prevention efforts; and developing physical and social environments inhospitable to crime" ("Crime Prevention" 1996a, 4). They assume that people can take measures to avoid becoming victims of crime and that neighborhoods can reduce crime by becoming more vigilant with regard to checking both potential serious crimes and public nuisance offenses (vandalism, graffiti, public drunkenness) that lead to deterioration of a neighborhood's quality of life.

GUIDELINES AND GOALS FOR CRIME PREVENTION PROGRAMS

The National Crime Prevention Institute (1986) listed 10 guidelines for crime prevention programs:

1. Potential crime victims or those responsible for them must be helped to take action which reduces their vulnerability to crimes and which reduces their likelihood of injury or loss should a crime occur.
2. At the same time, it must be recognized that potential victims (and those responsible for them) are limited in the action they can take by the limits of their control over the environments.
3. The environment to be controlled is that of the potential victim, not of the potential criminal.
4. Direct control over the victim's environment can nevertheless affect criminal motivation, in that reduced criminal opportunity means less temptation to commit offenses and learn criminal behavior and, consequently, fewer offenders. In this sense, crime prevention is a practical rather than a moralistic approach to reducing criminal motivation. The intent is to discourage the offender.
5. The traditional approach used by the criminal justice system (such as punishment and rehabilitation capabilities of courts and prisons and the investigative and apprehension functions of police) can increase the risk perceived by the criminal and thus have a significant (but secondary) role in criminal opportunity.
6. Law enforcement agencies have a primary role in crime prevention to the extent that they are effective in providing opportunity reduction education, information, and guidance to the public and to various organizations, institutions, and agencies in the community.
7. Many skills and interests groups need to operate in an active and coordinated fashion if crime prevention is to be effective in a community-wide sense.
8. Crime prevention can be both a cause and an effect of efforts to revitalize urban and rural communities.
9. The knowledge of crime prevention is interdisciplinary and is in a continual process of discovery, as well as discarding misinformation. There must be a continual sifting and integration of discoveries as well as a constant sharing of new knowledge and practices.
10. Crime prevention strategies and techniques must remain flexible and specific. What will work for one crime in one place may not work for the same crime in another place. Crime prevention is a "thinking person's" practice, and countermeasures must be taken after a thorough analysis of the problem, not before. (20–21)

Jefferies (1977) has similarly listed several key elements of a crime prevention program:

1. It will be set in motion *before* the crime is committed, not after.
2. It will focus on *direct* controls over behavior, and not on indirect controls.
3. It will focus on the *environment* in which crimes are committed, and on the organism within this environment, and not on the individual behavior.

4. It will be an *interdisciplinary* effort, based on all disciplines dealing with human behavior.
5. It will be *less costly* and *more effective* than punishment or treatment. This means that crime prevention is a more just and moral system than the system currently in control. (37)

The success of a crime prevention program can be measured in several ways:

- *Is crime reduced?* Are there fewer robberies, assaults, vandalism, burglaries, than last month? Than this time last year?
- *Is fear of crime reduced?* Do residents see crime as being reduced? Do they act in ways that show they are less afraid to move about their neighborhoods?
- *Are attitudes changed?* Do citizens have more confidence in their community and its institutions? Are they more convinced that their actions can improve the community? Are they more involved in civic activity?
- *Are the needs of the community and the residents met?* Are teenagers finding recreation, employment, and leadership opportunities? Do children have access to crime prevention and child protection instruction, reliable adults with whom to speak if scared or threatened? Can people and businesses function free from intimidation? Are senior citizens comfortable walking and driving in their community? Are all citizens provided with the chance to lend their skills to community betterment? (488)

RESPONSIBILITY FOR CRIME PREVENTION

There are many views about who has the responsibility for crime prevention. One is that the individual bears the sole responsibility of avoiding becoming a victim of a crime. Another is that the neighborhood or community bears this responsibility. Still another is that the police bear sole responsibility. A fourth approach is consonant with the philosophy of community policing: obviously, the police have an obligation to concentrate on crime prevention, but they cannot do it alone; they need the cooperation of the neighborhood and of the citizens who live in the neighborhood. Crime prevention cannot be successful unless it involves the participation of all formal and informal structures, the cooperation of all neighborhood residents, and a heightened awareness, among individual residents, of the dangers of crime and how they may be avoided.

Role of the Police in Crime Prevention

As Sir Robert Peel stated when he first established the London police, the police have the responsibility of preventing and controlling crime. No other component of criminal justice has the vantage point that the police possess in working with the community. The police are the first to know if a crime has been committed. They have to investigate crimes, solve crimes, arrest criminal offenders, and recover stolen property. By the nature of their positions, they also know *how* offenders commit crime—knowledge that can be used to prevent crime from occurring.

In 1971, the National Advisory Commission on Criminal Justice Standards and Goals was appointed by the Law Enforcement Assistance Administration (LEAA). The purpose of the commission was to establish national criminal justice standards and goals for reducing and preventing crime at the municipal, county, and state levels. In 1973, the commission completed its task and published its standards for crime prevention in the report *Police.* The standards for crime prevention as it applies to the police are as follows:

> Every police agency should immediately establish programs that encourage members of the public to take an active role in preventing crime, that provide information leading to the arrest and conviction of criminal offenders, that facilitate the identification and recovery of stolen property, and that increase liaisons with private industry in security efforts. (66)

The report *Police* delineated the following goals that the police should strive toward in crime prevention:

1. Every police agency should assist actively in the establishment of volunteer neighborhood security programs that involve the public in neighborhood crime prevention and reductions.
   a. The police agency should provide the community with information and assistance regarding means to avoid being victimized by crime and should make every effort to inform neighborhoods of developing crime trends that may affect their area.
   b. The police agency should instruct neighborhood volunteers to telephone the police concerning suspicious situations and to identify themselves as volunteers and provide necessary information.
   c. Participating volunteers should not take enforcement action themselves.
   d. Police units should respond directly to the incident rather than to the reporting volunteer.
   e. If further information is required from the volunteer, the police agency should contact him by telephone.
   f. If an arrest results from the volunteer's information, the police agency should contact him by telephone.
   g. The police agency should acknowledge, through personal contact, telephone call, or letter, every person who provides information.
2. Every police agency should establish or assist programs that involve trade, business, industry, and community participation in preventing and reducing commercial crimes.
3. Every police agency should seek the enactment of local ordinances that establish minimum security standards for all new construction and for existing commercial structures. Once regulated buildings are constructed, ordinances should be enforced through inspection by operational police personnel.
4. Every police agency should conduct, upon request, security inspections of businesses and residences and recommend measures to avoid being victimized by crime. (66)

Police style should be considered as important as substance. Both the style and substance of police practices must be considered reasonable from the public's viewpoint. Scientific evidence supports the conclusion that the less respectful the police are toward suspects and citizens, the less people will comply with the law (Sherman et al. 1997, 1).

Traditionally, police departments have established crime prevention units or at a minimum have had an officer who does security checks for residents and businesses. But departments that have adopted the community policing philosophy have gone further by incorporating crime prevention into community policing. They attempt to obtain the cooperation of neighborhood and business associations in implementing crime prevention strategies, and they provide education, training, and consulting services to the neighborhood through programs on preventing such crimes as rape, mugging, shoplifting, and juvenile delinquency.

Role of the Community in Crime Prevention

The term *community* has multiple definitions. It can mean an entire city, for example the community of Wichita, or merely a section of a city, such as the north side or the east side. Crime prevention strategies or plans function more efficiently when they are geared to a small, specific geographical area, or *neighborhood.* According to Bursik and Grasmick's (1993) definition,

> First, and most basically, a neighborhood is a small physical area embedded within a larger area in which people inhabit dwellings. Thus, it is a geographical and social subset of a larger unit. Second, there is collective life that emerges from the social networks that have arisen among the residents and the sets of institutional arrangements that overlap their networks. That is, the neighborhood is inhabited by people who perceive themselves to have a common interest in that area and to whom a common life is available. Finally, the neighborhood has some tradition of identity and continuity over time. (6)

Different neighborhoods in a city will have different crime problems. The biggest crime problems of poorer neighborhoods may be drive-by shootings and drug dealings on street corners, whereas the upper-middle-class neighborhood's biggest crime problem may be burglaries.

Police departments are well aware of where violent crimes, and specific crimes such as robberies, burglaries, auto theft, and drug dealing, usually take place. To keep these crimes under control, police need the cooperation of citizens who live in the places where the crimes occur.

A study sponsored by the National Institute of Justice on Crime Prevention in the late 1990s concluded that

> communities are the central institution for crime prevention, the stage on which all other institutions perform. Families, schools, labor markets, retail

establishments, police and corrections must all confront the consequences of community life. Much of the success or failure of these other institutions is affected by the community context in which they operate. Our nation's ability to prevent serious violent crime may depend heavily on our ability to help reshape community life, at least in our most troubled communities. (Sherman 1997, 1)

According to the National Crime Prevention Council (1989), a neighborhood's crime problems begin with physical decay and public nuisance offenses such as vandalism. Crime prevention and improvements of the neighborhood's quality of life are tied together. Neighborhoods must not only target crime directly but also address the underlying causes of their crime problem. A wide variety of programs can be implemented: they can include improved housing, education, and employment opportunities, to name only a few.

It has been estimated that approximately 10 to 20 percent of neighborhood residents may, at any one time, be involved in crime prevention activities. Obtaining their involvement may require a great deal of effort. Further, even when crime has been decreased, crime prevention participation by citizens needs to be continued so that crime will not take an upward turn. Most citizens can muster enough energy for the short term but long-term involvement is needed to keep crime under control.

CRIME PREVENTION AND COMMUNITY POLICING

Crime prevention and community policing have six major features in common:

1. *Each deals with the health of the community.* Both community policing and crime prevention acknowledge the many interrelated issues that generate crime. They look to build health as much as to cure pathological conditions.
2. *Each seeks to address underlying causes and problems.* Although short-term and reactive measures (e.g., personal security, response to calls for service) are necessary, they are insufficient if crime is to be significantly reduced. Looking behind symptoms to treat the causes of community problems is a strategy that both, at their best, share in the full measure.
3. *Each deals with the combination of physical and social issues that are at the heart of many community problems.* Community policing and crime prevention both acknowledge that crime-causing situations can arise out of physical as well as social problems in the community. An abandoned building may attract drug addicts; area burglars may be unsupervised, bored teens. Both approaches examine the broadest possible range of causes and solutions.
4. *Each requires active involvement by community residents.* Crime prevention practitioners—law enforcement and civilians alike—have acknowledged that their chief task is to enable people—children, teens, adults, senior citizens— to make themselves and their communities safer by helping them gain appropriate knowledge, develop helpful attitudes, and take useful actions. The very essence of community policing requires overt participation of residents in what has been termed the "coproduction of public safety."

5. *Each requires partnerships beyond law enforcement to be effective.* Crime prevention efforts involve schools, community centers, civic organizations, religious groups, social service agencies, public works agencies, and other elements of the community. Experience in community policing documents the need for similar partnerships both to reach people and to solve problems.

6. *Each is an approach or philosophy, rather than a program.* Neither community policing nor crime prevention is a "program"—that is, a fixed model for delivery of specific services. Each is, rather, a way of doing business. Each involves the development of an institutional mind set that holds community paramount and values preventive and problem-solving efforts in all of the organization's business. Each can involve a wide range of programs and other initiatives. ("Crime Prevention" 1996a, 3–4)

There are five basic arguments for linking crime prevention and community policing:

1. Crime prevention and community policing have a common purpose—making the public safer and making communities healthier.

2. Crime prevention offers information and skills that are essential to community policing.

3. Crime prevention and community policing have great potential for enriching each other.

4. Crime prevention responsibilities may be repositioned within the department as it moves to community policing, but successful departments have found a need for a clear focus of responsibility for crime prevention and a driving necessity for the capacity to apply and teach crime prevention knowledge and skills.

5. Thoughtful, planned action that carefully nurtures a core of crime prevention expertise while making the skills and knowledge available to all officers, especially those working at the street level, can substantially benefit the transition to community policing as well as its practice. ("Crime Prevention" 1996a, 4–8)

The state of Oregon's board of public safety and training believes so strongly that crime prevention is a key ingredient of community policing that it provides an 80-hour course in crime prevention specifically designed for community policing officers. The course was designed because state administrators realized that crime prevention officers had been doing the work described by the community policing philosophy, including problem solving, citizen empowerment, and forming a partnership with the community. Under the Oregon program, crime prevention officers function as mentors, resources, catalysts, and troubleshooters for community policing officers who are responsible for the day-to-day community policing activities. The crime prevention officers instruct community organizations in conflict resolution, volunteer management, and program development ("Crime Prevention" 1996b, 3).

The integration of crime prevention and community policing operations can only strengthen the security and safety of neighborhoods and the community as a whole. Crime prevention programs have discovered that neighborhood associations and individuals can play a successful part in ensuring their own security and safety. Programs such as block watches, Neighborhood Watch, business and security surveys, and security training have all proven useful in involving citizens in crime prevention. These programs have often involved the citizens in problem solving and the initiation of a police-citizen partnership. They also have given citizens a greater sense of empowerment with respect to their own safety. They are discussed in depth in the next section of this chapter.

CRIME PREVENTION STRATEGIES

According to the National Crime Prevention Institute (1986), setting up a crime prevention program involves three main tasks: "designing the organization; defining the crime problems and priorities; and developing crime prevention program objectives" (150). As Paul Whisenand (1977) has argued, all these tasks require community input and commitment if a program is to be successful. Therefore, police crime prevention specialists or community policing officers must work with the community to develop and implement the crime prevention program. The National Crime Prevention Institute (1986) has outlined the following steps for this process:

1. *Developing and organizing community background data* such as crime and loss patterns, police patrol districts, census tracts, natural geographical boundaries, and other socio-economic and demographic patterns. The purpose of the background data is to provide a reasonably complete picture of crime patterns and related socio-economical conditions, and to help determine the variety and cohesiveness of the neighborhood population.
2. *Using collected data to select target areas.* High priority areas are identified along with control areas so that the true impact of the project may be assessed. A control area is a comparable neighborhood that does not participate.
3. *Establishing criteria for levels, kinds, and distribution of participation.* The practitioner should make determinations prior to attempting to organize a neighborhood as to what a workable participatory model for that neighborhood should look like. For example, what percentage of residents should be involved? How should the residents be distributed? Should residents remain anonymous for surveillance and reporting purposes? What residents should take the organizing lead? Should police present a low profile at first? What kind of project activities are likely to be best received at first? What level of acceptance is likely? And so on.
4. *Approaching neighborhood leaders.* People who have been identified as having significant influence in the area should be approached first. After they have been informed as to the possible nature of a project and its potential value to the area, they may be asked to invite friends and other potential group members to participate in preliminary, exploratory meetings. It is important to remember that local leaders serve many important group functions, such as information dissemination, recruitment, and stimulation of group interest. They can also provide the practitioner with valuable feedback on the progress and interest in their areas.

5. *Providing education and training.* Group members and related police personnel should become acquainted with their respective roles in the crime prevention efforts. The goal is to build the basic mechanism necessary for citizens and police working together. Initial citizen education and training would probably include crime reporting procedures, guidance on what to report, and the basic security and personal safety tips. As interest is generated and people begin to increase interaction with police, the police must encourage the actions of the group members and provide guidance for further contacts. It is imperative that not only the beat officer, but other police personnel, such as the patrol commander and radio dispatcher, be aware of the group and lend their cooperation.

6. *Providing feedback to police and citizens.* If a citizen's call results in a good arrest, he should be notified and recognized. . . . If the call resulted in an officer being dispatched in a situation where police responses were inappropriate, he should also be notified and courteously advised on what should and should not be notified to the police. Citizens should also provide feedback to the police department on police response. Was it timely, courteous, accurate?

7. *Formulating crime-specific tactics.* When project performance reaches a level that reflects a capacity for the project to function as an efficient, unified, and directed entity, it is time to look back to the crime data files to determine the most serious crime problems facing project areas and devise crime specific tactics to address the problems. Times, places, and methods of criminal attack must be considered to identify what specific things a citizen can do to reduce the chances of criminal victimization.

8. *Implementing crime-specific tactics throughout the organization as they are developed.* This may be accomplished by having the practitioner or local police patrol attend the periodic group meeting to provide specific training, or training sessions for group leaders, who, in turn, train the groups and supervise the process of implementation. Crime-specific tactics should be implemented comprehensively throughout the project to achieve maximum effectiveness and to avoid displacement.

9. *Assessing the performance of the organization.* Performance assessment should come in the form of formal and informal feedback throughout the organizational period. At some point, however, the determination must be made that the project is sufficiently organized, educated, and trained to become essentially self-sustaining and regenerative with only logistical support from the police.

10. *Evaluating the impact on crime in the project area.* The project goal of reducing crime can only be reliably and validly assessed reactive to crime-specific tactics. Also, it is important that the tactics be quickly assessed so that they can be revised as necessary. When refined tactics have been implemented, the practitioner can legitimately establish that specific actions by citizens are affecting the rate of crime in the neighborhood.

11. *Encouraging the group to take on other needed changes.* Once the mechanism for community action has been established and has been proven effective, it can take on various community improvement projects. Widening the scope of the activities of the organization can help sustain the crime prevention effort by

offering participants a diversity of activities to meet their interests. Many of these projects will probably overlap, reinforcing each other and increasing the total chance of success. (146–147)

In some areas, like Knoxville, Tennessee, police have established community advisory boards to help draw up the area's crime prevention plan. In Knoxville, board members included representatives from neighborhood associations, Neighborhood Watch groups, and business associations, as well as numerous civic leaders ("Crime Prevention" 1996c, 4–8). The section later in this chapter on Seattle's Community Crime Prevention Program provides more detail on this kind of planning.

Crime Analysis

Crime analysis is a process of collecting, analyzing, and disseminating information so that it can be put to use by the crime prevention specialist and community police officer. Crime analysis has always been conducted informally by police officers and investigators, but, in the last several decades, some larger police departments have established formal crime analysis units to assist crime prevention specialists in preventing and controlling crime. The crime analysis unit uses information collected from reported criminal offenders to prevent crime and apprehend and suppress criminal offenders.

The criminal analysis unit receives information from the patrol, detective, communications, administration, and special units. The primary source of information is police reports. The information collected from within the police department consists of preliminary investigative reports, supplementary reports, arrest reports, field contact reports, statistical data, and various other department reports. Usually, police reports include the type of incident, the location of the incident, information about the victim, the suspect's description, and any unique characteristics of the incident—the type of weapon used, verbal expressions, facial features, and so forth. The analysis unit also collects information from outside agencies such as corrections, probation, state agencies, private organizations, and other law enforcement agencies. Information collected from external agencies includes status and records of known offenders and crime problems in other law enforcement agencies.

The purpose of crime analysis is to provide practical information about crime patterns, crime trends, and possible leads of possible suspects. For crime analysis to be useful, information about suspects should be forwarded to crime prevention specialists, community policing officers, and beat patrol officers, who generally have the primary responsibility for crime prevention.

A study by the International Association of Chiefs of Police (Chang et al. 1979) identified seven crime analysis functions as being important for effective crime prevention:

1. *Crime pattern detection*: monitoring and discovering crime occurrences that share common attributes or characteristics such as (1) geographical location or occurrence, (2) time of occurrence, (3) modus operandi (MO) information

2. *Crime suspect correlation:* providing information to operations personnel on possible suspects

3. *Target profiling:* analyzing victimized persons and/or premises by specific geographical areas to specify the nature of the objects that might be attached

4. *Forecasting crime potential:* predicting the exact time and location of future crime events for short-range purposes

5. *Monitoring exceptional crime trends:* periodically monitoring at fixed intervals (i.e., daily, weekly, or monthly) the occurrence of crime incidents city-wide or by geographical area to identify "out-of-control" points based on crime thresholds

6. *Forecasting crime trends:* analyzing crime incidents by type, area, and/or time to identify trends in support of long-term police action. Forecasting crime trends involves the prediction of crime volume in time domain, i.e., forecasting by time of day, day of week, based on the historical crime data by using statistical methods and techniques

7. *Resource allocation*: allocating available manpower for (1) patrol operation and (2) criminal investigation for the sake of achieving some departmental operational goals and objectives. (xvi–xviii)

Contemporary crime analysis units are using *computerized mapping* to solve crimes and to aid in the investigation of crime. According to the National Institute of Justice (Rich 1988),

> Mapping software has many crime control and prevention applications. In addition to the location of crime, geographic data that can be helpful in crime control and in efforts to apprehend a perpetrator include the perpetrator's last known address, the location of the person who reported the crime, the location of the recovered stolen property, and the locations of persons known or contacted by the perpetrator. Geographical information valuable in planning, conducting, and evaluating crime prevention programs includes the locations of crimes committed during the past month; the locations of abandoned houses, stripped cars, and other "broken windows" conditions in a neighborhood; and the locations where persons who could benefit from crime prevention and other social programs actually live. (15)

The potential uses for computerized mapping software are numerous. Computer-aided dispatch (CAD) and record-management systems that store and maintain service calls, incidents, arrests, and mapping data are commonly used in large police departments. Geocoding, a CAD feature and a records management system, can verify addresses and can associate other geographic information, including police reporting areas, beats, and districts. The crime analysis unit can use mapping software to prepare crime bulletins and to assist crime prevention specialists by informing them of the geographical areas in the community that should be targeted.

The use of crime analysis and computerized mapping in crime prevention assists in developing victim profiles that can be used in crime prevention programs. Crime analysis can provide specific modus operandi information (offenders' methods of committing crimes) that can be used to prevent crime. The crime prevention specialist or community policing officer can be expected to provide strategies and inform residents and businesses on actions that can keep them from becoming victims of crime.

Neighborhood Watch

Neighborhood Watch programs have received a great deal of national media attention, and students studying community policing should be familiar with them. Neighborhood Watch programs are instituted in neighborhoods to reduce crime as well as the fear of crime. They are based on the concept that citizens of a neighborhood where crimes occur play an important part in controlling crime in their neighborhood. Watch members can be either homeowners or apartment dwellers who band together to establish a sense of security and safety. The association works to create a genuine neighborhood and to exchange information among members on crime prevention techniques. The community police officer or crime prevention specialist should function only as a supportive advisor to the Neighborhood Watch. If the community policing officer runs the meeting, residents will feel that they should not be actively involved in the association. The success of Neighborhood Watch depends on residents' active involvement to make their neighborhood safe and secure.

Community policing officers should be vigorous in looking for ways to improve the security and safety of residents and businesses in their assigned neighborhood, particularly by eliminating criminal opportunities. They should look for poor or inadequate locks and recommend that owners replace them. Owners who leave their establishment open or unlocked need to be informed of the dangers of this action. Officers should check the exterior of buildings while on patrol—specifically fire escapes, doors, locks, windows, ladders, boxes, and equipment left out (Perry 1994, 38). They should have the knowledge to perform an inside security check as well.

One rural police department that used its Neighborhood Watch program to introduce community policing at the block level was Caldwell, Idaho, a city with a population of 22,000. The Neighborhood Watch was expected to identify three problems and come up with possible solutions. Neighborhood Watch captains were kept informed of criminal activities and civic events.

A study sponsored by the National Institute of Justice (Garofalo and McLeod 1988) concluded that Neighborhood Watch has the potential to reduce crime but that police need to find ways to ensure continuing citizen participation in existing programs and to make Neighborhood Watch more attractive for low-income, high-crime neighborhoods. The study made the following recommendations:

- The people who organize, lead, and participate in Neighborhood Watch should be encouraged to develop innovative practices and link Neighborhood Watch to other local concerns. Less emphasis should be placed on formal standards for Neighborhood Watch programs, primarily certification and recertification criteria. Alternative ways of operating Neighborhood Watch should be precluded by jurisdiction-wide predetermined criteria.

- Neighborhood Watch organizers and managers should make greater efforts to consider the characteristics and needs of specific neighborhoods and to tailor Neighborhood Watch efforts to them. Neighborhood Watch functions should be lodged within existing neighborhood or block associations when possible. Where a local association does not exist, organizers should be open to Neighborhood Watch as a possible point for one.
- Police departments, as the primary sponsors of Neighborhood Watch, need to strengthen crime prevention units by enhancing their organizational status, giving them sufficient resources, selecting personnel likely to work well with community groups, and providing appropriate rewards and a career path for such personnel.
- When citizen interest wanes, Neighborhood Watch sponsors should innovate rather then merely exhort. Flexibility and innovation in the basic Neighborhood Watch model can be enhanced by building on such existing tools as meetings and newsletters to create networks for exchanging promising ideas and information among jurisdictions. (4)

Citizen Patrols

In addition to Neighborhood Watch, communities can have citizen patrols that patrol a specific neighborhood. Citizen patrols can assist police by providing additional eyes and ears for the geographical area they are patrolling, since the police cannot be expected to patrol every city street or to know when every crime or neighborhood problem is occurring. Citizen patrols can be either on foot or in an automobile. Unlike the police, patrol members do not intervene physically or become involved in suspicious situations. Instead, they primarily offer surveillance and notify the police of any unusual or suspicious incidents. Usually, a citizen patrol consists of residents who volunteer to patrol their neighborhood for a specific time period. These patrols generally operate in the evenings or on weekends when the neighborhood volunteers are not working at their full-time jobs.

Business Watch

The National Sheriffs' Association (NSA) launched a Business Watch program in 1987. Various studies have uncovered that nearly a third of business failures are the result of losses from crime. Usually, small business are deficient in maintaining security systems or are unable to afford security personnel. Business owners, like residents, need to know about crime prevention strategies. Only when they become familiar with prevention tactics can they play a key role in the protection of their businesses. The National Sheriffs' Association championed a vicinity approach whereby merchants assisted each other, specifically in city blocks where many small shops and shopping malls were located.

Merchants involved in Business Watch used target-hardening tactics (installing physical barriers to prevent crime, such as locks, alarms, and fences) and were counseled through security checks by crime prevention specialists on how to make their shops less vulnerable to victimization. Police crime prevention specialists met with business owners and managers regularly to inform merchants about potential crime issues pertaining to their businesses, such as burglary prevention, bad check passing, robbery, and credit card fraud, and about various approaches they could put in place to avoid becoming crime victims. Close cooperation and communication between the crime prevention specialist and the businesspeople were extremely important for Business Watch to become a success (Cantrell 1988, 3).

Beep-a-Beat Program

The Beep-a-Beat Program (BBP) was developed by the Chicago Police Department to enhance communication between business merchants and members of the Chicago Police Department who were assigned to foot posts. Through BBP, preauthorized business merchants provide pagers for foot patrol officers so that citizens can contact foot patrol officers and relay messages. Foot patrol officers who receive a pager message requiring a response notify the police dispatcher that they are responding to a BBP message. They then give the business caller an appropriate response, record their resolution of the caller's message on their daily incident report form, and give the police dispatcher the appropriate code when they return in service. BBP keeps the lines of communication open between the merchants and the foot patrol officer. In a number of situations, BBP has helped in apprehending criminal offenders. For example, following a store robbery, a merchant contacted the foot patrol officer, who then had the merchant get into a squad car to canvass the geographical area. The victim spotted the offender, who was immediately arrested and charged with robbery. Cooperation between merchants and patrol foot officers contributed to a 25 percent reduction in robberies in 1995. The BBP program also led to foot patrol officers breaking up a ring of smash-and-grab robberies. Crime prevention programs like BBP can only enhance the image of the police in the eyes of citizens and should lead to greater cooperation and a reduction of crime. Programs like BBP make the citizen a partner in crime prevention (Officer J.A. Alcanter, Chicago Police Department, personal communication, 1998).

Police Station Automatic Teller Machines

In 1995 the Anne Arundel County Police Department in Maryland became the first police department to operate automatic teller machines (ATMs) in all of its county police stations. The idea developed after a series of ATM robberies. Because ATMs are open 24 hours a day and the locations of some ATMs leave customers prone to being robbed, the placement of ATMs in county police stations should provide greater security to citizens using ATMs, particularly in the wee hours of the morning and in some sections of the city. Los Angeles and Philadelphia have ATMs located in police stations, but only in a few areas. One key advantage of placing ATMs in police stations is that it provides opportunities for positive citizen-police contact so that citizens do not interact with police only in unpleasant situations such as receiving a traffic citation. This program should work to the advantage of both the community and the police.

Crime Stoppers

In 1976, an Albuquerque, New Mexico, police officer developed the Crime Stoppers program as a joint effort by which law enforcement agencies, the media, businesspeople, and community residents could increase citizen involvement in solving crimes. Crime Stoppers is a crime control mechanism that pays for information leading to the indictment of felony offenders.

Crime Stoppers provides anonymity and monetary rewards as inducements to obtain information from the general population regarding specific crimes. Those who associate with offenders and become privy to needed information are given the chance to exchange that information for cash while remaining protected from retaliation.

A program coordinator generally provided by a police agency runs the daily operations of the Crime Stoppers program. The coordinator's responsibilities include interacting with an established board of directors to determine the amount of all rewards; maintaining and operating a round-the-clock telephone hotline; preparing crime synopses for distribution as news releases; selecting and videotaping "Crime of the Week" simulations for showing on local television; and interacting with an established finance committee for fund-raising efforts.

Monetary awards for information relating to specific offenses generally range from $100 to $2,000, with the maximum amount normally paid only for capital crimes. As calls are received on the hotline in response to Crime Stoppers publicity, information is immediately forwarded to the appropriate police agency, with follow-up written reports distributed once each week. To collect a reward, tipsters must contact the Crime Stoppers program coordinator and submit a claim, citing their private alphanumeric code.

As community service efforts, Crime Stoppers programs are generally dependent upon the business community for financial support and, to a great extent, upon the local news media for publicity. As mechanisms to assist in crime control, these seem to be very worthwhile. A study by the National Institute of Justice (Rosenbaum et al. 1986) termed Crime Stoppers "a highly productive program from the taxpayer's point of view." The study found that on a national average, "Each crime solved recovered more than $6,000 in narcotics and stolen goods, and each felony arrest cost only $73 in reward money."

Innovative ideas such as Crime Stoppers can stimulate participation in crime control by all segments of the community and have a tremendous impact on reduction of crime rates and apprehension of offenders.

Operation Identification

Most of us want to mark our belongings for identification. As a crime prevention tactic, the identification of items was first used in Monterey Park, California, in 1963.

To deter a rash of hubcap thefts, the police chief recommended that residents engrave their license numbers on their hubcaps for identification purposes and eventually recommended that they engrave their household items as well.

When implementing an operation-identification program, the crime prevention specialist should require that the numbers selected be permanently engraved on the property marked. The numbering system used should be a standardized system that is acceptable to the National Crime Information Center (NCIC). The crime prevention specialist has to establish a record system that will provide for an inventory and description of the property that also includes serial numbers if they are available. In addition, the crime prevention specialist has the obligation to make the community aware that property that cannot be engraved should be photographed. All the rooms of a house should be photographed to show the items that are in them (Chuda 1990, 4–8).

Operation CLEAN

Since the mid-1980s, the crack cocaine drug epidemic has had a devastating impact on our nation's cities. Neighborhoods have been taken over by drug dealers. To check the deterioration of city neighborhoods, the Dallas Police Department initiated a program called Operation Community and Law Enforcement Against Narcotics (CLEAN) to reclaim neighborhoods from drug dealers and users. With the curbing of the neighborhood's drug problem, it was hoped that the crime rate attributed to the drug environment would decrease substantially. This operation required the support not only of the entire Dallas Police Department but also of the city's administration and a variety of other city agencies (Hatler 1990, 22–23). The CLEAN team was composed of several municipal departments with specific responsibilities:

- The police department is responsible for the removal of drug dealers, crime prevention training, intensive 24-hour personnel deployment, and coordination of Operation CLEAN activities.
- The fire department checks properties for fire code violations and orders the closing of unoccupied buildings with safety violations.
- The streets and sanitation department is responsible for the general clean-up of target areas: clearing alleys, trimming trees, and removing discarded furniture used by drug dealers.
- The housing and neighborhood services department is responsible for strictly enforcing applicable city codes, referring unsalvageable properties to the Urban Rehabilitation Standards Board for demolition, and working with outside groups to obtain vacant lots for housing units.
- The city attorney's office provides vigorous prosecution of code violations and aggressively seizes abandoned properties. (Hatler 1990, 23)

Operation CLEAN demonstrated that police departments can work with other city departments to enhance the quality of life of residents living in drug-infested neighborhoods. Strategies like Operation CLEAN are excellent examples of programs that can be used to decrease crime in a specific geographical area. The cooperation, respect, and support of law-abiding citizens can be gained when police agencies implement strategies that assist in improving the quality of life of a neighborhood's residents.

Security Surveys

Security surveys are conducted to discover existing physical and procedural deficiencies and to recommend security devices, procedures, and methods that can either eliminate or reduce potential criminal opportunities. A good security survey allows the crime prevention specialist to inform apartment and housing residents and businesses about the tactics and strategies they can use to keep from becoming crime victims. Even if no major security problems are found, the crime prevention specialist may find ways to increase the security of the premises. Two types of security surveys can be conducted, business surveys and residence surveys. The National Crime Prevention Institute (1986) has developed security survey guidelines for business establishments and for residences. The guidelines for businesses are as follows:

    I. Possible Maximum Loss through Criminal Attack (Assessment of Targets)
       A. Personal safety
       B. Cash
       C. Merchandise
       D. Damage factors
    II. Direct Protection of Targets
       A. Safe
          1. Rating
          2. Anchored
          3. Lighted
          4. Visible from street or mall
       B. Cash Registers
          1. Visible from street or mall
          2. Open at night
          3. Limited cash accumulation
          4. Locked when unattended
          5. Limited access
       C. Merchandise
          1. Controlled storage room
          2. Removed from windows
          3. Controlled displays
          4. Inventory accountability
          5. Identifiable tags or marking system

D. Deposits
  1. Prepared in protected area
  2. Made daily
  3. Armored service
  4. Made by two or more employees
  5. Varied times and routes
III. Employee Training
  A. Shoplifting
  B. Robbery
  C. Checks and credit cards
  D. Internal controls (opening and closing procedures, cash handling, purchasing and receiving, etc.)
IV. Building Surfaces
  A. Front, left side, right side, rear
    1. Construction
    2. Doors
    3. Windows
    4. Vents
    5. Lighting
  B. Roof
    1. Construction
    2. Skylights
    3. Vents and ducts
    4. Lighting
  C. Floor
    1. Construction
    2. Cellar
    3. On grade
    4. Post and lift
V. Outside and Perimeter
  A. Trees and shrubs
  B. Loading docks
  C. Trash storage
  D. Roof access
  E. Fences and gates
  F. Police access
VI. Surveillance
  A. Lighting
  B. Visibility
  C. Cameras
VII. Intrusion Detection
  A. Sensors
  B. Control
  D. Annunciation
  E. Response (106–107)

The guidelines for residents are as follows:

  I. Possible and Probable Maximum Losses through Criminal Attack
      (Assessment of Targets)
      A. Personal safety
      B. Cash
      C. High-value personal possessions
      D. Damage factors
 II. Direct Protection of Targets
      A. Cash storage
      B. Storage of jewelry, cameras, guns, silver, etc.
III. Family Member Training
      A. Telephone/door answering procedures
      B. Locking responsibilities
      C. Intruder-in-the-house procedures
 IV. Building Surfaces
      A. Front, left side, right side, etc.
         1. Construction
         2. Doors
         3. Windows
         4. Vents
         5. Lighting
      B. Roof
         1. Construction
         2. Doors
         3. Skylights
         4. Vents and ducts
         5. Lighting
      C. Floor
         1. Construction
         2. Cellar
         3. On grade
         4. Post and lift
  V. Outside Perimeter
      A. Trees, shrubs, hedges
      B. Roof-garage-cellar access
      C. Fences and gates
      D. Visibility
 VI. Surveillance
      A. Lighting
      B. Visibility
VII. Intrusion Detection
      A. Sensors
      B. Controls
      C. Annunciation
      D. Response (107–108)

Programs for the Elderly

In some communities, crime prevention specialists may have to develop and manage crime prevention strategies aimed specifically to the elderly. National studies have found that the elderly have a smaller chance of becoming victims of a crime than any other age group. However, when the elderly are victimized, they are more likely to be the victims of property crimes than any other age group.

Older citizens may become victims of crimes because of either their poor physical health or their isolation from family, friends, and acquaintances. Many elderly live in high-crime areas where they are statistically more likely to become victims of a crime. Despite some of these problems, the elderly can still be involved in crime prevention strategies. The crime prevention specialist can work with senior citizens to instruct them on how to deter crimes such as muggings and purse snatching. The crime prevention specialist has the responsibility to instruct the older citizens concerning the precautions they can observe when they take public transportation or walk in public, their personal safety at home, safety on the telephone, and how to protect money and valuables. The crime prevention specialist can give valuable advice on confidence games and consumer fraud, because senior citizens are frequently the victims of unscrupulous individuals and their scams.

Police departments that serve a substantial senior citizen population should initiate crime prevention strategies for the elderly. When senior citizens become aware of crime prevention techniques and strategies, they can make an excellent contribution to crime prevention.

Cellular Watch Programs

In 1994, the Metro-Dade Police Department in Florida initiated the Neighborhood Cellular Watch program in 11 county neighborhoods, according to a plan that won the acceptance of the police department and of the neighborhoods. The neighborhoods provided volunteers who were issued cellular telephones to report instantaneously any crime observed or any other behavior in their neighborhood that led to deterioration of the "quality of life" in their neighborhood. Neighborhoods were selected on the basis of their cultural diversity, degree of crime problems, involvement in other community programs, and, most important, the willingness to participate in the program. Cellular One of Florida underwrote equipment and air time for the cellular telephones. At the end of nine months, an evaluation by Florida International University's College of Urban and Public Affairs found that in the combined 11 neighborhoods,

- Burglaries decreased 33 percent, from 341 to 229.
- Robberies decreased 24 percent, from 41 to 31.
- Thefts decreased 9 percent, from 77 to 70.
- Auto thefts increased 9 percent, from 108 to 118. (Ellison 1995, 14)

Examining the 11 neighborhoods individually, the evaluation found the following:

- Burglaries increased in two neighborhoods, decreased in seven, and remained unchanged in two.
- Robberies increased in two neighborhoods, decreased in five, and remained unchanged in four.
- Auto thefts increased in one neighborhood, decreased in seven, and remained unchanged in one.
- Thefts increased in three neighborhoods, decreased in six, and remained unchanged in two. (Ellison 1995, 14)

The neighborhood that was most successful in reducing crime was in a high-crime area, an apartment complex. Crime decreased in three of the four categories, increasing only in auto theft. Cellular Watch may have contributed to a reduction of 69 burglaries, or 35 percent. Robbery decreased by 14 percent, and theft decreased by 17 percent. Auto theft increased by 57 percent (Ellison 1995, 21).

Generally, Cellular Watch volunteers believed that crime had decreased in their neighborhood. According to volunteers, the in-progress calls were responded to faster by the police. Also, the volunteers reported that because of their involvement in the Cellular Watch program, they were more willing to telephone the police about suspicious persons and other situations. The cellular telephones were considered to be beneficial because residents could have instant contact with the police. The cellular telephone had "taken Crime Watch out of the house and into the streets" (Ellison 1995, 16).

The use of cellular telephones for neighborhood improvement can only be as effective as the participants in the program. However, Cellular Watch programs can be another method of developing a police-neighborhood partnership in crime prevention.

Crime Prevention Electronic Bulletin Board

A computerized bulletin, like other bulletin boards, gives users the opportunity to post messages, but instead of using pins placed on a corkboard, it uses a computer, modem, and host system. The purpose of a crime prevention bulletin board is to provide information and keep users up to date on events occurring in the neighborhood and community. In addition, mail files can be transferred, and questions can be answered. One of the key uses of a crime prevention bulletin board is to provide updated crime information and to cover crime prevention concepts relevant to a specific crime. Because many young people are familiar with computers, the computer is an excellent way for the crime prevention specialist to reach young people. Crime prevention tips relevant to young people can be provided in cyberspace.

Defensible Space

The term *defensible space* can be traced back to a conference held in 1964 at Washington University in St. Louis, Missouri, where two sociologists and two architects, along with police officers from the St. Louis Police Academy, discussed physical features that produced security for public housing (Newman 1973a, 1). It was defined in terms of a range of security mechanisms, including "real and symbolic barriers, strongly defined areas of influence, and improved opportunities for surveillance" (Newman 1973b, 3). According to Newman (1973b), who coined the term,

> Defensible space is a model for residential environments which inhibits crime by creating the physical expression of a social fabric that defends itself. All the different elements which combine to make a defensible space have a common goal—an environment in which latent territoriality and a sense of community in the inhabitants can be translated into responsibility for ensuring a safe, productive, and well maintained living space. (3)

For residents of public housing, defensible space is a milieu of security that includes areas other than residents' individual apartments, such as lobbies, hallways, playgrounds, and adjacent streets. Four elements of physical design can contribute to defensible space:

1. The territorial definition of space in developments reflecting the areas of influence of the inhabitants. This works by subdividing the residential environment into zones toward which adjacent residents easily adopt proprietary attitudes.
2. The positioning of apartment windows to allow residents to naturally survey the exterior public areas of their living environment.
3. The adoption of building forms and idioms which avoid the stigma of peculiarity that allows others to perceive the vulnerability and isolation of the inhabitants.
4. The enhancement of safety by locating residents in functionally sympathetic urban areas immediately adjacent to activities that do not provide continued threat. (Newman 1973b, 8–9)

These four elements can all influence the extent of disorder and crime in the area.

To a great extent, public housing in the late 1980s and 1990s, with grants from the Department of Housing and Urban Affairs, incorporated the four elements of defensible space. In a joint effort with their local police departments, residents of public housing reclaimed their territory—their living environment—from drug dealers, drug users, and other criminal offenders. Their ability to survey the area from their apartments helped them to identify the legitimate residents and distinguish them from law violators. Playgrounds were reclaimed so that children could again play on them. This led residents to have more positive feelings about their place of residence. Graffiti was removed from walls, and grass was planted where none had existed. Finally, in some projects, public housing sites were dispersed across safer, middle-class neighborhoods.

In the 1990s, the U.S. Department of Housing and Urban Affairs employed Newman to implement defensible space strategies for three projects. Newman (1996) reviewed the projects as follows:

1. *The Five Oaks community in Dayton, Ohio.* This consisted of the reorganization of an urban grid of residential streets to create mini-neighborhoods in downtown Dayton, Ohio. The successful reorganization of the existing urban grid of streets . . . is now sweeping the country. The phenomenon flows from our well publicized study of the private streets of St. Louis about 20 years ago. The creation of cul-de-sacs at the end of streets is a useful mechanism not only for reducing crime and traffic, but for stimulating reinvestment and the occupancy of previous vacant units. . . .

2. *The Clason Point Project, South Bronx, New York City.* The [physical] modification of a rowhouse public housing project . . . to reassign the previously public grounds to individual residents to use, control, and maintain. This reduced both crime and maintenance costs. . . . I used iron fencing and curbs to reassign all the grounds to individual residents. This removed gang turf and gave the drug dealers nowhere to operate. The resurfacing of buildings and the provision of new paths, lighting and play equipment improved the look of the project and encouraged residents to assume new responsibilities. This reduced maintenance costs and increased occupancy levels.

3. *Dispersion public housing in Yonkers, New York.* This project involved dispersing high rise public housing residents into scatterer-site townhouses in middle-class neighborhoods. (4–6)

Such environmental techniques for preventing crime have since been used for residential areas, commercial businesses, schools, and parking garages.

Crime Prevention through Environment Design

Like Oscar Newman, J. Ray Jefferies, the theorist who developed the concept of crime prevention through environmental design (CPTED), believed that crime prevention involved the physical design of buildings along with citizen involvement and the effective use of police agencies. According to Jefferies (1977), the physical environment of each building and each room in a building should be examined. Jefferies correctly claimed that usually only a small area of the city is responsible for the majority of crime, so that analysis of crime must scrutinize house-by-house or block-by-block variations in crime rates. He advocated that for the purpose of crime prevention, crime data be used to determine in what areas of a community crime generally occurs.

The CPTED model suggests that the physical environment be orchestrated to prevent potential offenders from committing crimes as well as to improve the quality of life. It involves creating a physical environment in which people can be free from the fear of crime and in which individuals prone to criminal activity can be checked in this inclination by physical barriers. The CPTED model draws on not only architecture but also sociology, psychology, and law enforcement.

Timothy D. Crowe (1991), an authority on CPTED, outlined three overlapping strategies in CPTED: "natural access control, natural surveillance, and territorial reinforcement" (30). The purpose of *access control* is to reduce the opportunity to commit a crime. Methods used to control access to an area can include gates, guards, locks, and shrubbery. The aim of this strategy is to create an illusion that the risk of attempting a criminal act is greater than the opportunity. The purpose of *natural surveillance* is to maximize opportunities to observe outsiders or intruders into the neighborhood. When strangers are under surveillance, the risk of being caught while committing an offense increases. Strategies of surveillance include police and security guard patrol, lighting, and strategic placement of windows. The third strategy, *territorial reinforcement,* creates a sense of ownership or influence through better design of the physical environment. Potential offenders can recognize the residents' ownership or influence over a geographical area. Natural access control and natural surveillance also contribute to a feeling of territoriality (30–32).

CPTED, to be successful, must be practical and be understood. Residents, store owners, school officials, and parking garage owners must visualize the benefits of CPTED before they will be able to implement it. Crowe (1991) advocated a "three D" approach to space assessment to be used as a guide for the nonprofessional. This approach is based on the three functions or dimensions of human space:

1. All human space has some designated purpose.
2. All human space has social, cultural, legal, or physical definitions that prescribe the desired and acceptable behaviors.
3. All human space is designed to support and control the desired behaviors. (33)

According to Crowe, space may be evaluated by asking the following questions, which use the three D's as a guide:

1. Designation
   - What is the designated purpose of this space?
   - What was it originally intended to be used for?
   - How well does the space support its current use? Its intended use?
   - Is there conflict?
2. Definition
   - How is the space defined?
   - Is it clear who owns it?
   - Where are its borders?
   - Are there social or cultural definitions that affect how that space is used?
   - Are the legal or administrative rules clearly set out and reinforced in policy?
   - Are there signs?
   - Is there conflict or confusion between the designated purpose and definition?
3. Design
   - How well does the physical design support the intended function?
   - How well does the physical design support the definition of the desired or accepted behavior?
   - Does the physical design conflict with or impede the productive use of the space or the proper functioning of the intended activity?

- Is there confusion or conflict in the manner in which the physical design is intended to control behavior? (33–34)

Situational Crime Prevention

Situational crime prevention, like the earlier approaches developed by Newman and Jefferies, is about reducing opportunities for crime. It is "(1) directed at highly specific forms of crime [in ways] (2) that involve the management, design, or manipulation of the immediate environment in as systematic and permanent way as possible (3) so as to reduce the rewards as perceived by a wide range of offenders" (Clark 1992, 4). Techniques used in situational crime prevention can include a wide variety of crime prevention measures, such as burglary alarms, fenced yards, graffiti cleaning, street lighting, baggage screening, closed-circuit TV systems, identification cards, and credit card photographs.

Situational crime prevention holds that specific crime situations have unusual features that can be appraised and examined for solutions. The focus is on finding solutions to specific problems (Brantingham and Brantingham 1990, 25). Crime analysis is the first step in this process. The following questions are asked: "What are the detailed characteristics of the problem? Who might be committing the offenses or causing the difficulty and why? What elements in the socio-physical background environment could be contributing, in an immediate way, to the observed crime pattern?" (Brantingham and Brantingham 1990, 26). Situational crime prevention uses 12 techniques:

1. *Target hardening*. The most obvious way of reducing criminal opportunity is to obstruct the vandal or the thief by physical barriers, such as locks, safes, screens, or reinforced materials.
2. *Access control*. Access control, or procedures to keep unauthorized persons from entering a facility, is now widely practiced by large employers in offices and factories, particularly in cities.
3. *Deflecting offenders*. This technique involves channeling inappropriate behavior in more acceptable directions: for example, by placing public urinals in areas that have a persistent problem of people urinating in public parks/streets.
4. *Controlling facilitators*. The low rates of homicide in Britain, where handguns are much less readily available than in the United States, provide one reason for believing that effective gun control can reduce levels of violent crime.
5. *Entry/exit screening*. Entry screening differs from access control in that the purpose is less to exclude people than to increase the likelihood of detecting persons who are not in conformity with entry requirements. Exit screens serve primarily to deter theft by detecting objects that should not be removed from the protected area, such as items not paid for at a shop.
6. *Formal surveillance*. Personnel such as police, security guards, and store detectives, whose main function is to furnish a deterrent threat to potential offenders, are the principal providers of formal surveillance. Their surveillance

role may be enhanced by electronic hardware, such as by burglar alarms, radar speed traps, and closed-circuit television (CCTV).

7. *Surveillance by employees.* In addition to their primary role, some employees, particularly those dealing with the public, also perform a surveillance role. They include shop assistants, hotel doormen, park keepers, parking lot attendants, and train conductors.

8. *Natural surveillance.* Households may trim bushes at the front of their homes, and banks may light the exterior of the premises at night, in attempts to capitalize upon the "natural" surveillance provided by people going about their everyday business.

9. *Target removal.* This can include the removal of change makers, which are frequent targets of thefts; the imposition of cash limits at convenience stores; and the requirement of exact change on public transportation.

10. *Identifying property.* Cattle branding is a crude but effective way to identify property. Modern organizations pursue essentially the same logic when they mark their property. Property marking can be extended to household valuables through "operational identification."

11. *Removing inducement.* In certain parts of New York City, it is unwise to wear gold chains in the streets or to leave unattended cars, such as a Chevrolet Camaro, which is highly attractive to joyriders. Some inducements are less obvious. For example, extensive experimental research has suggested that the mere presence of a weapon, such as a gun, can induce aggressive responses in some people.

12. *Rule setting.* To protect themselves from crime, all organizations find it necessary to regulate the conduct of their employees. (Clark 1992, 12–20)

One retail company that employs many of the 12 situational crime prevention techniques is L.L. Bean, a nationally known retail outlet. Shoplifters who have obtained merchandise by theft frequently attempt to obtain a cash refund without ever leaving the store. L.L. Bean has a special coding system to detect this scam. An extensive CCTV monitoring system is in place in the outlet store that allows cameras to zoom in anywhere in the store and parking lot. This system can produce instant photographs from an image on a video screen. It allows store detectives to have photos of a shoplifter within minutes so that a store detective can make an apprehension. Also, to keep internal thefts at a low rate, L.L. Bean offers its employees anywhere from 50 to 90 percent off its merchandise. The idea behind such a liberal reduction policy is the belief that it would be foolish for employees to steal merchandise when they can get it for almost nothing. It appears that this policy has worked: L.L. Bean does not have an internal theft problem (Hoffmann 1994, 94–96).

Another organization that has employed situational crime prevention is the Downers Grove, Illinois, Police Department. The department determined that the best strategy to prevent residential burglaries was to offer homeowners a home security survey conducted by a crime prevention specialist. The police department was able to have included in the water bill to residents a notification that a free home security survey was available from a crime prevention specialist. Then the police department initiated a telemarketing approach to reach residents who had not volunteered for the survey. The survey, which took approximately one hour, allowed the crime prevention specialist to evaluate home security risks, such as exterior lighting, locks, doors, windows, and landscaping, and to give the homeowner advice that would make his or her property and possessions less vulnerable to burglars. The home security survey was inexpensive and required no special equipment (Rechenmacker 1991, 8–9).

One problematic aspect of situational crime prevention is that crime may merely be displaced to an adjacent geographical area. When opportunities to commit crime are blocked in one area, criminals may simply shift their activity to another area. However, some studies of situational crime prevention have found no resulting displacement effect to nearby areas (see the section "Seattle's Community Crime Prevention Program").

SPECIFIC COMMUNITY CRIME PREVENTION PROGRAMS

Seattle's Community Crime Prevention Program

In 1972, the City of Seattle's Law and Justice Planning Office initiated a crime prevention program with a project to identify which crimes the program would do best to target. Three variables were taken into consideration:

1. *Frequency of occurrence.* Data were taken from Seattle Police Department's reports of crime incidents.
2. *Severity and level of public fear and tolerance.* This element was added to allow the rating plan to weigh an aggravated assault more heavily than a shoplifting incident. Data were collected through interviews with city residents.
3. *Potential for crime reduction.* Priority crimes must also be amenable to prevention strategies. For example, homicide is not a priority crime. Although it scores high on other measures, research indicated that the bulk of homicides occurred between acquaintances and that the crime is not particularly amenable to reduction strategies. (Lindsay and McGillis 1986, 47)

Priority was given to burglary, rape, and robbery as serious crimes that have the greatest opportunity of being prevented.

This section will discuss only the burglary prevention program. The program selected the area with the highest burglary rate and overall crime rate. Civic organizations and church organizations were invited to assist in the project. The staff developed a community profile that included demographic data and crime information about the residents. The first contact with citizens was by mail and the second by canvassing the neighborhood. The prevention program included block watch organizing, property identification, and security inspections. Forty percent of the area's residents ended up participating.

The block watch was the most important aspect of the burglary prevention program. A block watch consisted of a group of approximately 10 to 15 neighbors who would keep a watch on their neighbors' homes to prevent burglaries. Each block watch was provided with an engraving instrument to mark property. Members of the block watch were given window decals to warn intruders that property on the premises was marked. Finally, each block watch home was inspected for burglary vulnerability.

Seattle's community crime prevention program benefited from a single focus—residential burglary. An evaluation of the burglary prevention program reveals that those involved had a lower burglary rate than those not involved in the program. Although not conclusive, the evaluation of adjacent areas indicated that displacement, or the transfer of crime to adjacent neighborhoods, did not occur.

Community Response to Drug Abuse

In the 1990s, the Bureau of Justice Assistance, Office of Justice Programs, U.S. Department of Justice, launched a program called Community Response to Drug Abuse (CRDA), which was administered by the National Crime Prevention Council. Ten communities were funded for the program. All 10 sites funded were expected to meet the following program guidelines:

- *The active involvement of law enforcement* as a community partner in planning and implementing the local initiative
- *A multisector approach* including all key stakeholders—individuals representing those with a vested interest in the community's well-being, such as residents, schools, social services, businesses, law enforcement, and other local agencies
- *A locally designed workplan* with short-term and long-term goals and objectives to meet needs identified by the targeted community. For example, in the Bronx, one short-term objective was to close drug houses; the corresponding long-term goal was to convert those properties into low-income housing. (National Crime Prevention Council 1992, 5)

The objective of the guidelines was to obtain general strong community support and to produce effective action. CRDA hoped to place prevention efforts in the hands of the local residents who would be willing to create an effective community strategy that would reduce drug use in their neighborhood. The National Crime Prevention Council and the Bureau of Justice Assistance established the following overall program goals for CRDA initiative:

- Empower community residents to feel more comfortable and less fearful in their communities, so they would become more willing to participate in community life
- Provide community residents with knowledge of resources that can be of assistance to their community
- Test a variety of drug reduction activities which empower communities to take action and implement prevention programs
- Introduce effective drug reduction activities which empower communities to take action and implement prevention programs

- Develop a local process to help build ongoing working relationships among residents, law enforcement, and other key services, e.g., local government, religious institutions, social services, housing authorities, schools
- Establish indices of success which relate to each community's specific workplan, e.g., active involvement of community residents and local agencies in program planning and implementation, community policing initiated, drug houses closed, recreation programs begun (6)

Generally, it took community groups a year to develop a work plan and to develop a cooperative relationship with government entities and law enforcement. Establishing a common agenda and developing active partnerships usually required two and a half years. The working together of key community members led to them sharing a stake in the success of the program. CRDA found that a working relationship between community groups and law enforcement and a local sense of ownership were critical to the success of the community-based drug prevention program.

Ideas about community-based strategies against drug abuse can be adapted to suit specific conditions of each neighborhood. Neighborhood drug prevention programs can reduce the fear of residents by showing that the program is supported by legitimate institutions, assuring residents that there is safety in numbers, beginning with strategies that are successful, and publishing successes. Another important aspect of community response to drug abuse programs is the development of good relations with local news media. Frequent positive publicity about CRDA activities and accomplishment can encourage residents not involved in the program to become involved. CRDA can work and reduce drug abuse if neighborhood residents become involved.

There are many community crime prevention programs, but to discuss all of them here would be an impossible task. Crime prevention specialists and community policing officers should review the literature on various community crime prevention programs to get ideas for their own programs.

MARKETING CRIME PREVENTION

The crime prevention specialist and/or community policing officer has to know and understand the community—its socioeconomical makeup, its crime problems, and its needs. Do robberies, burglaries, and other serious crimes occur there? Does the community have a domestic violence problem, a child abuse problem, or a drug problem? Once this information is obtained, crime prevention programs can be developed that meet the community's needs.

As any knowledgeable police officer knows, the police cannot succeed in solving or preventing crime without the community's assistance. Crime prevention programs have to involve the community and deal with community-identified problems to be effective. When crime prevention programs are successful, they usually have the full support and participation of the community. The crime prevention specialist must sell prevention programs to the police department and to the citizens of the community. The police prevention specialist must work to overcome any resistance to crime prevention programs by educating the community on how the program works. The crime prevention specialist is an extremely important link between the police and the community.

Since 1980, the Bureau of Justice Assistance within the U.S. Department of Justice has supported the National Citizen's Crime Prevention media campaign, which works to make the public aware of crime prevention strategies and programs that allow citizens to protect themselves and their property. The goals of the media campaign are to achieve the following:

1. Change unwarranted feelings and attitudes about crime, drug use, and the criminal justice system.
2. Generate an individual and community sense of responsibility for crime and drug prevention.
3. Initiate individual and community action toward preventing crime and illicit drug use.
4. Mobilize additional resources for crime and drug prevention efforts.
5. Enhance existing crime and drug prevention programs and projects conducted by national, state, and local agencies and organizations.
6. Develop organizational capacities to implement crime and crime prevention programs. (O'Keefe et al. 1996, 21)

The success or failure of mass media campaigns rests with the willingness of mass media programmers to disseminate crime prevention announcements and of local police officials to support crime prevention efforts. Findings indicate that both police officials and media programmers support crime prevention media campaigns (O'Keefe et al. 1996, 58).

At the local level, crime newsletters can provide specific crime information targeted for certain neighborhoods. A newsletter can supply detailed news about crime problems and offer measures that people can take to avoid becoming victims. The crime newsletter can include the classes of crime committed, where the offenses were committed, and descriptions of offenders if available. Another use of media has been the Crime Stoppers program, which was already discussed in this chapter. In this program, the media inform the general public about a specific crime, and persons who have knowledge of the crime are encouraged to phone in information on the individual or individuals who may have committed the offense. The caller can remain anonymous and can receive a reward if an offender is arrested.

In the 1990s, television shows re-creating a crime have been successful in apprehending fugitives. We are familiar with many of these shows, including *America's Most Wanted* and *Unsolved Mysteries*. At the local level, the sheriff of Sedgewick County/Wichita, Kansas, announces on cable television his "Ten Most Wanted" list. A picture of the wanted person along with a physical description is provided to the viewing audience. The Wichita Police Department on cable television announces the names of individuals having warrants for their arrest. The local cable network of

Sedgewick County/Wichita also places on television the names and ages of those offenders convicted of drunk driving.

As we can see, the news media, though they cannot be expected to provide all the information needed by citizens to keep from becoming a crime victim, can play an influential role in crime prevention strategies.

SUMMARY

One of the key objectives for the police when Sir Robert Peel implemented the Metropolitan London Police in 1829 was crime prevention. But police researchers have found that many of the strategies incorporated into policing are not necessarily effective in preventing crime. Reactive strategies, such as the use of 911 by citizens to call the police after a crime has been committed, do not prevent crime and must be supplemented by proactive strategies. All police officers should be knowledgeable about crime prevention techniques. Especially in those departments adopting the community policing philosophy, an understanding of crime prevention is essential.

There are many views about how the responsibility for crime prevention should be distributed. The approach most consonant with community policing is that the police, neighborhoods, and individuals share this responsibility. Obviously, the police have an obligation to concentrate on crime prevention, but they cannot do it alone; they need the cooperation of the neighborhood and its citizens. Crime prevention cannot be successful unless it involves the participation of all formal and informal structures and the willingness of all neighborhood residents to learn about how to avoid becoming victims of crime.

No other component of the criminal justice system has the vantage point that the police possess in working with the community. The police are the first to know if a crime has been committed. They have to investigate crimes, solve crimes, arrest criminal offenders, and recover stolen property. By the nature of their positions, they also know how offenders commit crime—knowledge that can be used to prevent crime from occurring.

Crime prevention and community policing are closely related. A number of police agencies have integrated their crime prevention efforts into community policing. Crime prevention specialists can give the community information on techniques and strategies to reduce and prevent crime. Both physical and social aspects of the neighborhood must be addressed by crime prevention.

Successful crime prevention strategies and programs involve police-community cooperation. The police crime prevention specialist or community policing officer works with the community to develop and implement crime prevention programs. The crime prevention specialist has to be involved in security surveys, crime analysis, neighborhood watches, operational identification, and a multitude of other crime prevention strategies.

Situational crime prevention refers to reducing opportunities for crime. It can employ a wide variety of crime prevention measures, including burglar alarms, fenced yards, graffiti cleaning, street lighting, baggage screening, closed-circuit TV systems, identification cards, and credit card photographs. Situational crime prevention focuses on specific crime situations in order to come up with specific solutions for them.

The police can neither solve nor prevent crime without the assistance of the community. Even when crime has been reduced, crime prevention participation by citizens needs to continue if crime is not to take an upward turn.

KEY TERMS

Beep-a-Beat Program (BBP)	computerized mapping	Neighborhood Watch
Business Watch	crime analysis	Operation CLEAN
Cellular Watch	Crime Stoppers	operation identification
citizen patrols	defensible space	police station automatic
computer-aided dispatch (CAD)	direct control of crime	teller machines
	electronic bulletin board	security surveys
	indirect control of crime	

REVIEW QUESTIONS

1. Explain the meaning of crime prevention.
2. Explain the police role in crime prevention.
3. Explain how crime prevention and community policing are interrelated.
4. Describe some crime prevention strategies.
5. Explain what is meant by situational crime prevention.
6. Explain what is meant by community crime prevention.
7. Describe how the media have been used in crime prevention.

REFERENCES

Bayley, D.H. 1994. *Police for the future.* New York: Oxford University Press.

Bouza, A.V. 1990. *The police mystique.* New York: Plenum.

Brantingham, P.L., and P.J. Brantingham. 1990. Situational crime prevention in practice. *Canadian Journal of Criminology* 32, no, 1: 25.

Bursik, R.J., Jr., and H.G. Grasmick. 1993. *Neighborhood and crime.* New York: Lexington.

Cantrell, B. 1988. A commitment to crime prevention. *FBI Law Enforcement Bulletin* 57, no. 10: 3.

Chang, S., et al. 1979. *Crime analysis support system.* Gaithersburg, MD: International Association of Chiefs of Police.

Chuda, T.J. 1990. *Basic crime prevention curriculum.* Columbus, OH: International Society of Crime Prevention Practitioners.

Clark, R.V. 1992. *Situational crime prevention: Successful case studies.* New York: Harrow & Heston.

Crime prevention and community policing: A vital partnership, part I. 1996a. *Washington Crime News Service* 2, no. 1.

Crime prevention and community policing: A vital partnership, part II. 1996b. *Washington Crime News Service* 2, no. 1.

Crime prevention and community policing: A vital partnership, part III. 1996c. *Washington Crime News Service* 2, no. 4.

Crowe, T.D. 1991. *Crime prevention through environmental design: Application of architectural design and space management concepts.* Boston: Butterworth-Heinemann.

Ellison, S.L. 1995. The Dade County (Miami, Florida) Neighborhood Cellular Watch Project: Evaluation of an innovative approach. Unpublished report, Florida International University, College of Urban and Public Affairs.

Garofalo, J., and M. McLeod. 1988. *Improving the use and effectiveness of Neighborhood Watch programs.* Washington, DC: National Institute of Justice.

Hatler, R.W. 1990. Operation CLEAN: Reclaiming city neighborhoods. *FBI Law Enforcement Bulletin* 59, no. 9: 22–23.

Hoffmann, J.W. 1994. Crime prevention officers can improve retail security. *Law and Order* 46, no. 6: 94–96.

Jefferies, C.R. 1977. *Crime prevention through environmental design.* Beverly Hills, CA: Sage.

Kelling, G., et al. 1974. *The Kansas City preventive patrol experiment: A summary report.* Washington, DC: Police Foundation.

Lab, S.P. 1997. *Crime prevention: Approaches, practices, and evaluations.* 3rd ed. Cincinnati, OH: Anderson.

Lindsay, B., and D. McGillis. 1986. Citywide community crime prevention: An assessment of the Seattle program. In *Community crime prevention: Does it work?* ed. D. Rosenbaum. Beverly Hills, CA: Sage.

National Advisory Commission on Criminal Justice Standards and Goals. 1973. *Police.* Washington, DC: Government Printing Office.

National Crime Prevention Council. 1989. The success of community crime prevention. *Canadian Journal of Criminology* 31, no. 4: 488.

National Crime Prevention Council. 1992. *Creating a climate of hope: Ten neighborhoods tackle the drug crisis.* Washington, DC: National Crime Prevention Council.

National Crime Prevention Institute. 1986. *Understanding crime prevention.* Boston: Butterworth.

Newman, O. 1973a. *Architectural design for crime prevention.* Washington, DC: U.S. Department of Justice.

Newman, O. 1973b. *Defensible space: Crime prevention through urban design.* New York: Collier.

Newman, O. 1996. *Creating defensible space.* Washington, DC: U.S. Department of Housing and Urban Affairs.

O'Keefe, G., et al. 1996. *Taking a bite out of crime: The impact of the National Citizens' Crime Prevention Media Campaign.* Thousand Oaks, CA: Sage.

Perry, T. 1994. *Basic patrol procedures.* Salem, WI: Sheffield.

Rechenmacker, D.I. 1991. Telemarketing crime prevention. *FBI Law Enforcement Bulletin* 60, no. 10: 8–9.

Rich, T.F. 1988. The use of computerized mapping in crime control and prevention programs. *National Institute of Justice Research in Action.* Washington, DC: Government Printing Office.

Rosenbaum, D.P., et al. 1986. *Crime stoppers: A national evaluation.* Washington, DC: National Institute of Justice.

Sherman, L.W., et al. 1997. *Preventing crime: What works, what doesn't, what's promising.* ncrj.org/works/chapter8.htm.

Walker, S. 1989. *Sense and nonsense about crime.* 2nd ed. Pacific Grove, CA: Brooks/Cole.

Whisenand, P.M. 1977. *Crime prevention.* Boston: Holbrook.

Chapter 5

Concepts, Strategies, Experiments, and Research Findings That Have Influenced Community Policing

CHAPTER OBJECTIVES

1. Be familiar with the police-community relations concept.
2. Be familiar with the team policing concept.
3. Be familiar with the Kansas City Preventive Patrol Experiment.
4. Be familiar with the directed patrol concept.
5. Be familiar with the response time concept.
6. Be familiar with the ministation concept.
7. Be familiar with the value of research to the evolution of community policing.

INTRODUCTION

This chapter discusses various concepts, strategies, experiments, and research findings that have influenced community policing. Community policing should not be considered a completely new philosophy. Many of its concepts and strategies are based on concepts and strategies that have been used by police departments in the past. For instance, community policing has adopted ideas about police-community relations that have been propounded since the 1960s. The team policing concept developed in the 1980s could also be considered a forerunner of the philosophy of community policing. The strategy of directed patrol, or patrolling guided by analysis of crime patterns and directed toward the solution of specific neighborhood problems, was developed in the 1970s and is still used in community policing today. Ministations, or police substations that cover a small geographical area, have been in use for decades as a means of putting police officers in closer contact with the people they serve. Experiments in bringing back foot patrols as an alternative to motorized patrolling were first conducted in the 1970s and early 1980s. They have helped decrease the "fear of crime" and to develop rapport between patrol officers and their neighborhoods and have been recognized as an important aspect of community policing. The "broken windows theory," first advanced in 1982, has been extremely important in exploring how disorder creates a fear of crime and may even lead to a neighborhood's decay and an increase in crime. The theory of broken windows has played an important part in developing strategies for community policing. Finally,

116

research on policing over the last several decades has corrected many misconceptions about policing and has provided a great deal of support for the community policing concept. Further research will be required to determine which aspects of community policing work.

POLICE-COMMUNITY RELATIONS

Police-community relations are an important focus of policing today, especially under the community policing model. The International City Managers' Association (1967) defined it as

> a police-department initiated program designed to offer an opportunity for police and other public agencies and individuals in the community to dis-cover their common problems, ambitions, and responsibilities and to work together toward the solution of community problems and the formulation of positive community programs. . . . It is not merely a problem-solving de-vice. It is a problem-avoidance methodology which, when correctly recog-nized, can create healthy community attitudes. (3)

Raymond Momboisse, in his book *Community Relations and Riot Prevention* (1967), defined police-community relations as meaning

> exactly what the term implies—the relationship between members of the police force and the community as a whole. This includes human, race, pub-lic and press relations. The relationship can be bad, indifferent, or good, depending upon the attitude, action and demeanor of every member of the force both individually and collectively. (97)

The concept of police-community relations developed during the 1960s, a period of turmoil in America's cities. There were demonstrations against the war in Vietnam, civil rights demonstrations, and numerous riots, and the police often came into direct confrontation with the people they were to serve. Because of the chaos in American cities and alienation that existed between the police and the community, President Lyndon Johnson appointed a Commission on Law Enforcment and Administration of Justice in 1965. In 1967, the commission published a report stating that police indif-ference, police mistreatment of citizens, and citizen hostility toward the police were disruptive influences on the community (President's Commission 1967a). According to the report, police officers who work in a community that is hostile toward them will have difficulty providing police protection to that community, because citizens who are hostile toward the police will not report crimes to the police or provide them with the information necessary to solve crimes. When a community has negative feelings toward the police, tension rises, and aggressive actions against the police begin to occur, which in turn can trigger irrational behavior on the part of police officers (100).

The commission therefore stressed the importance of establishing good police-community relations. According to the commission,

- A community relations program is not a *public relations* program to "sell the police image" to the people. It is not a set of expedients whose purpose is to tranquilize for a time an angry neighborhood by, for example, suddenly promoting a few Negro officers in the wake of racial disturbance.
- Community relations are not the exclusive business of specialized units, but the business of an entire department from the chief down. Community relations are not exclusively a matter of special programs, but a matter that touches on all aspects of police work.
- The need of good community relations and of effective law enforcement will not necessarily be identical at all times.
- Improving community relations involves not only instituting programs and changing procedures and practices, but re-examining fundamental attitudes. The police will have to learn to listen patiently and understandingly to people who are openly critical of them or hostile to them, since those people are precisely the ones with whom relations need to be improved.
- The police must adapt themselves to the rapid changes in patterns of behavior that are taking place in America. This is a challenge when traditional ideas and institutions are being challenged with increasing insistence. (100)

The need for good relations between the police and the communities they serve is as important today as it was in the 1960s when the President's Commission on Law Enforcement and Administration of Justice published its report. In our multicultural society, African Americans, Hispanics, Asians, and gay men and lesbians want to be respected and will not tolerate abuse, either verbal or physical, by the police. All groups demand fair treatment and want the opportunity to present their viewpoints to the police. Excellent police-community relations are required to maintain the stability of the community, to solve the community's problems, and to allow the police to be successful in doing their jobs.

Conversely, as the commission's report pointed out poor police-community relations weaken the police ability to solve crime and make arrests because community residents who have hostile feelings toward the police will not report crimes or come forward as witnesses when they observe a crime take place. It also influences police officers' attitudes and behavior in the field. Officers who feel unappreciated by the citizens they serve tend to be more abrasive and even more abusive.

The following police-community operations should be goals for the police officers to implement in order to be effective:

1. Closer relations with underprivileged and minority neighborhoods, where the need is greatest for police understanding and involvement
2. More effective and more open communication between the police and the community
3. Increased citizen involvement in crime prevention and solving of social problems as a means of reducing crime
4. Improved understanding between the police and the community, with both gaining greater recognition of each other's problems

5. Creation of awareness among police of police-community relations problems, and encouragement of officers to help solve them
6. Direction of all department efforts toward improving relations with the total community, whether these involve crime prevention, public relations, or neighborhood problem solving (Neiderhoffer and Smith 1974, 5–6)

This last point deserves special emphasis. Police-community relations should be a consideration in all aspects of police work, including departmental policy, supervision, personnel procedures, records and communications, acceptance of complaints against departmental members, and planning and research. Ideally, it is an attitude, a way of perceiving police responsibilities that permeates the entire police organization (Niederhoffer and Smith 1974, 3).

Community policing fosters good community relations in that it emphasizes listening to people rather than telling them what to do. It stresses that the police need to take seriously the concerns and the input of citizens who live and work in the community. And the police are listening, because they now recognize that

1. citizens may legitimately have ideas about what they want and need from the police that may be different from what police believe they need
2. citizens have information about the problems and people in their areas that police need in order to operate effectively
3. police and citizens each hold stereotypes about the other that, unless broken down by nonthreatening contacts, prevent either group from making effective use of the other (Wycoff 1988, 105–106)

The police can also improve their relations with the community by establishing citizen police academies and community advisory councils. Citizen police academies are schools operated by police departments that offer courses to citizens on police subjects once a week for 15 weeks. Citizens walk away with a better understanding of police problems. Graduates of citizen police academies also constitute a source of police volunteers who can perform a number of services that leave officers free to perform police tasks.

Community advisory councils meet with police operations personnel to work on decreasing the crime rate, eliminating disorder, and reducing the fear of crime. A partnership between a community advisory council and the police can be linked to improving the quality of life for community citizens. The community advisory council can identify and prioritize problems within the community that a majority of residents want rectified. Once the problems are identified, strategies can be developed to solve the problems.

Finally, police departments need to be more open if they hope to gain the confidence of the people they serve. They must recognize that not all police information need be kept confidential, even though criminal investigations and other confidential information should obviously not be discussed with citizens or even with officers who have no need to know. The police must agree that they should be accountable not only for misconduct but also for any policies, procedures, or activities that are questionable. As communities and police develop a partnership, we may expect fewer complaints against the police and a greater success rate in solving crime.

TEAM POLICING

Early Systems of Team Policing in Great Britain

Team policing began in Aberdeen, Scotland, following World War II. The Aberdeen experiment was an effort to counter low morale and boredom among police officers patrolling their beats. This project provided for 5- to 10-member teams of foot patrol and car officers assigned to different areas of the city. The number of officers assigned to an area was determined by the volume of service calls and the crime rate. The team policing method eliminated loneliness and boredom for the foot beat officer.

A further development in team policing was unit beat policing (UBP), introduced in Lancashire, England, during the summer of 1966 and later spreading to other areas of England. It was introduced as a means of

- improving police-community understanding
- increasing clearance rates by encouraging information and intelligence flows within the department
- creating a more challenging and attractive beat for the beat officer
- utilizing manpower more efficiently by combining resources
- minimizing response time (Gregory and Turner 1968, 42)

In creating the UBP the English hoped to design a patrol system that would reduce response time while maintaining a level of police-community relations. The English had a high regard for community cooperation and communication so they devised a system with a quick response time. Until the implementation of UBP, the primary method of patrol was the foot patrol. The use of patrol car beats increased substantially under UBP, and foot beat constables were all provided with two-way radios to communicate with dispatchers. Beat car constables were encouraged to leave their vehicles periodically to make a closer examination of conditions on their beat. The UBP strategy, although adding a beat car to an area to respond to emergency calls, still assigned a beat constable to the area to maintain community relations, gather intelligence information, and provide general police work. The patrol beat constable was given the option to work flexible hours and to wear either a uniform or plainclothes. He was also responsible for performing minor investigations for major investigations, homicide and those covering a large geographical area, handled by the central investigative division of the police department (Gay et al. 1977b, 45–46).

Under this scheme, a UBP team was assigned to a geographical area and was responsible for the beat on a 24-hour-a-day basis. The team was decentralized, and the personnel assigned to the team were assigned specialized tasks. Each beat had a sergeant as a supervisor who would plan the team's activities. The foot constable had considerable discretion to perform a variety of functions. Three constables were assigned to the car beat to work 24-hour coverage. In some cities, the beat had an investigator. In contrast to the American policing system, the English emphasize foot patrol over motorized patrol. The English believe that crimes are prevented and can be detected and cleared when a constable knows his community and the people of the community. The UBP system recognizes that problems exist with specialization. Investigations are assigned to the beat to improve clearance rates and to enlarge the constables' role. The English recognized that police-community isolation usually exists when patrol officers are all motorized and are expected to react to 911 calls. UBP recognized that a mobile patrol force is needed along with strong police-community relations. Therefore, both motorized and foot patrol are a part of the UBP (Gay et al. 1977b, 45–47).

1967 President's Commission on Law Enforcement

The President's Commission on Law Enforcement (1967b) made recommendations to improve not only police-community relations but also police personnel assignments and the quality of police personnel. Team policing was suggested as a means to overcome problems created in urban police agencies by centralization and job specialization, to improve the cooperation between the patrol and investigative forces, and to attract more competent police recruits. The commission's task force report (President's Commission 1967b) urged that medium and large police departments create three classes of police personnel—police agents, police officers, and community service officers. The community service officer rank, which would be the lowest rank, was proposed

1. to improve police service in high crime rate areas
2. to enable police to hire persons who can provide a greater understanding of minority group problems
3. to relieve police agents and officers of lesser duties
4. to increase the opportunity for minority group members to serve in law enforcement
5. to tap a new reservoir of manpower by helping talented young men who have not been able as yet to complete their education to qualify for police work (123)

The next rank would be *police officer.* Police officers would perform the traditional duties of enforcing the law and investigating crimes that could be solved by immediate investigative follow-up. In addition, they would perform routine patrol duties, perform emergency services, and investigate traffic accidents. The highest rank would be police agent. Police agents would have the responsibility to investigate major crimes and to handle the more complex issues of policing. The police agent, police officer, and community service officer would function as a team (122–123).

The National Advisory Commission on Criminal Justice Standards and Goals

The 1973 Commission on Criminal Justice Standards and Goals (National Advisory Commission 1973) recommended that both large and small police departments adopt a *team policing* strategy, which they defined as "essentially . . . assigning police responsibility for a certain area to a team of police officers. The more responsibility this team has, the greater the degree of team policing" (154). For example, team policing units that have investigative duties are more complete than those lacking these duties. Ideally, teams develop strategies specifically geared to the neighborhood they serve. The goal of team policing was to overcome the isolation of the police from the community under the traditional model of policing in order to facilitate crime control (154–156). Team policing would combine the line operations of patrol, traffic, and investigations for a specific geographical area into one unit under one supervisor. Teams would include generalists and specialists who would be permanently assigned to the area and would have responsibility for all police services for the area.

Elements of Team Policing

A study of team policing in seven cities by the Police Foundation found that police departments that implemented team policing were interested in improving three basic operational elements:

1. *Geographical stability of patrol (i.e., permanent assignment of teams of police to small neighborhoods).* The geographical stability of patrol is the most basic element. The only city which did not assign its team permanently to a neighborhood was Richmond, California. There, teams were assigned as units on staggered shifts. Each team remained on duty for eight hours, and a new team came on duty every four hours. . . . Richmond [was included] in this study because the city is small enough to function as a neighborhood and because the patrol officers function as team members in much the same way as those geographically based teams, despite the assignment by time.

2. *Maximum interaction among team members, including close internal communication among all officers assigned to an area during a 24-hour period, seven days a week.* The element of encouraging interaction among team members was evident in all the team policing cities, but with considerable variation. Implicit in the concept of maximum interaction is exchange of information. One of the simplest means of accomplishing this exchange is through the scheduling of team conferences at regular intervals. Analogies may be found in the case conferences conducted by social workers or doctors, in which each professional describes several difficult cases of the previous week and opens them to discussion with his colleagues, soliciting criticism and advice. The police teams which followed a similar route with their conferences found that, in many instances, the cases were continuing problems covering more than one shift and required cooperation among several police officers. Those teams which did not have formal con-

ferences had to rely on informal ways of communicating—a practice which was more successful when the team was stationed and thereby isolated in a separate building than when sharing a station house with the regular patrol units. The other critical factor in communication was the team leader. When he encouraged sharing of information and was able to instill a sense of teammanship, the members communicated more frequently and informally.

3. *Maximum communication among team members and the community.* The third element, maximum communication among team members and members of the community, seemed to be aided by regular meetings between teams and the community. These meetings were a means of emphasizing the cooperation aspects of the peacekeeping function, facilitating the flow of information, and assisting in the identification of community problems. Such conferences have been a vehicle for eliciting community involvement in the police function. Another technique, participation of community members in police work, has been accomplished through auxiliary patrols, supply of information leading to arrests, and community voice in police policy-making. Such participation was designed to bring the police and community together in a spirit of cooperation. Finally, maximum communication among teams and the community has also been enhanced by an efficient system of referral of non-police problems (e.g., emotional problems, garbage collection, drug addiction) to appropriate service agencies. Teams that have developed their own neighborhood lists of social service units and names of social workers have made referrals far more quickly than through centralized traditional channels. (Sherman et al. 1973, 3–5)

The Police Foundation study found that those departments that were successful in implementing the three operational elements had strong organization support in the following areas: unity of supervision, lower-level flexibility in policy making, unified delivery of services, and combined investigative and patrol functions (Sherman et al. 1973, 5–6). The *unity of supervision* principle (patrol officer reports to one supervisor) led to effective team performance. When a police officer has more than one supervisor, he or she is often caught in the middle, with the two supervisors giving conflicting orders. *Lower-level flexibility* in policy making (the ability of low-ranking officers to make their own decisions) was of extreme importance for teams' productivity. Several police departments pushed decision making down to the teams because teams had the most accurate information about their neighborhood; this increased the authority of the lower levels of the hierarchy and responsibility and accountability of the team officers.

Some police departments put in place a *unified delivery of services,* in which the team had complete control over the delivery of all police services in the team neighborhood. The unified delivery of services included the authority to decide when specialized police units could be assigned to the neighborhood. The idea was to recognize the knowledge of the team officers about their neighborhood and to give police generalists more authority as decision makers for their neighborhood.

The fourth ingredient, *combined investigative and patrol functions* (investigators and patrol officers working as a unit), increased information sharing among team members, thereby increasing officers' familiarity with the neighborhood and making investigations more effective.

Today, the term *team policing* can encompass any of a variety of elements, depending on the community where it has been put into operation. Table 5–1 lists various team policing elements and the activities usually undertaken to operationalize them.

Table 5–1 Program Aspects of Team Policing

Organization and Team Building	
Elements	*Activities*
Team organization	Permanent assignment of officers to teams of 14 to 56 officers
	Permanent team assignment to shift or 24-hour responsibility for neighborhood
	Manpower allocation based on crime analysis and patrol workload
Enlarged job role and decentralization	Generalist/specialist officers
	Participation in team planning and decision making
Altered supervisory role and decentralization	Supervisor as planner/manager/leader
	Unified command structure
	Development of policy guidelines
	Participant and decentralized decision making
	Team meetings to plan operations
	Team information coordination

Community Relations	
Elements	*Activities*
Stable geographic assignment	Officers work in a defined neighborhood for an extensive period
Service orientation and increased citizen contact	Referral and "special" services
	Storefront headquarters
	Officer participation in community activities
	Walk-and-talk programs
	Foot and scooter patrol
	Nonaggressive patrol tactics
	Informal "blazer" uniforms
	Specially marked cars
Increased citizen participation in law enforcement	Citizen volunteer programs
	Crime prevention programs
	Citizen advisory councils
	Community meetings

Source: Reprinted from W.J. Gay et al., *Issues in Policing: A Review of the Literature*, p. 4, 1977, National Institute of Law Enforcement Assistance Administration, United States Department of Justice.

These elements fall into two categories: (1) organizational and team building character, and (2) neighborhood community relations (Gay et al. 1977b, 3).

Types of Team Policing Programs*

Team policing programs can be classified according to the functional responsibilities assigned to teams. There are four categories:

1. basic patrol teams
2. patrol-investigative teams
3. patrol-community service teams
4. full-service teams (Albright and Siegel 1979, 8–17)

The simplest form of team policing employs *basic patrol teams*. This program decentralizes the patrol function by reorganizing patrols into teams responsible for preventive patrol, radio, and traffic duties. Teams are instituted to improve manpower allocation, reduce response time, and increase clearance of service calls. This program hearkens back to the traditional police precinct organization, for example, in New York City.

Programs using *patrol-investigative teams* connect the basic features of the basic patrol team with the assignment of follow-up investigative responsibilities to the team. The Rochester, New York, Police Department is an example of the patrol-investigative approach. The entire police department—patrol officers, supervisors, and administrators—contributes to the investigate process. The City of Rochester, New York, developed this approach in response to a rising crime rate, a poor clearance rate for crime cases, and the apathy and ineffectiveness of the police department's investigative division. The city recommended

- Making patrol officers and detectives responsible for manageable-sized areas of the city (districts) so that they could become more familiar with the neighborhood aspects of the crime problem
- Giving responsibility for preliminary and follow-up investigations of crimes in a district to the unit commander for that district
- Enabling patrol officers to work more closely with detectives to improve preliminary investigations by placing both groups under the same unit commander and having them work out of the same quarters and serve the same designated area of the city
- Encouraging patrol officers to expand their role as crime solvers
- Using unit commanders to improve the patrol and investigative functions under their command (*MCT Supervisor's Manual* 1970, 31–34)

---

*Source: Data from E.J. Albright and L. Siegel, *Team Policing: Recommended Approaches,* pp. 8–17, 1979, Government Printing Office.

The police department was therefore divided into seven neighborhood districts, each with a captain as unit commander. Each district had eight investigators and 32 patrol officers assigned to it, in addition to supervisory personnel. The detective division was decentralized, with investigators being transferred to the districts. The centralized detective unit still maintained responsibility for the Physical Crimes Unit, which handled serious crimes such as homicide and rape; the Persons Unit, which handled juveniles; the Property Crimes Unit, which processed licenses and the service of warrants; and the Check and Fraud Squad.

A former Chief of the Rochester Police Department stated, "we have been decentralizing detectives in the belief that patrol officers and detectives operating as a unit can be more efficient in the solution of crimes" (*Crime Control Digest* 1975, 9). An audit report prepared by the Urban Institute supported this belief: it confirmed that patrol officers and detectives working in teams were more successful in solving crimes than detectives functioning in traditional ways. Rochester has been the only police department to implement a patrol-investigative team that was successful in improving clearance rates.

Patrol-community service teams are basic patrol teams that also have the responsibility of handling community relations. The Albuquerque Police Department, which used this approach, found that police attitudes toward the community improved but that the team was unable to give a higher level of service to the community. The San Diego Police Department, however, found that patrol-community service teams improved police attitudes toward the community as well as increasing community service.

Full-service teams are the most complex form of team policing. They involve the decentralization of patrol, investigative, and community relations, and sometimes traffic duties as well. The transfer of personnel from centralized bureaus to the full-service team usually involves investigators, typically three or four to each team, and to a lesser extent community relations and traffic personnel. Full-service multispecialist teams in Albany, Arbor, and Los Angeles have been generally successful. These teams have produced improvements in workload management, investigative effectiveness, and police attitudes toward the community. The Los Angeles full-service multispecialist team has been credited with lower crime rates and improving police-community relations.

Comparison of Neighborhood Team Policing with Traditional Policing

As Table 5–2 shows, neighborhood team policing strategies are different from those of traditional policing. In team policing, a team of 20 to 40 police officers establish objectives to be accomplished, with all members of the team being allowed to provide suggestions. The team commander is responsible for all aspects of the team around the clock. The team provides all police services to its neighborhood, and officers are assigned to teams for an extended period of time. Special police units (such as tactical units) inform themselves of team goals and whenever possible consult local team policing commanders in advance. Team policing considers community relations as an essential function that consists of providing good police service, making friendly on-street contacts, and attending various neighborhood meetings. Team policing provides for decentralization of planning (e.g., of crime prevention programs).

Table 5–2 Comparison of Neighborhood Team Policing with Traditional Policing

Traditional Policing	*Team Policing*
Smallest patrol unit (precinct or division) has 100 to 250 officers	Team has 20 to 40 officers
Quasi-military supervision	Professional supervision, with consultation, setting of objectives, an in-service training program, encouraging suggestions, permitting the exercise of responsibility within necessary limits
Shift responsibility (eight-hour tours with only unit commanders—captains or inspectors—responsible for around-the-clock operations)	Team commander responsible for all aspects of police service on an around-the-clock basis
Assignment of the first available car to call for police service, with priority for emergency calls	Team provides all police service for its neighborhood. Nonteam members take calls in the neighborhood only in emergencies
Officers rotated to new divisions or assignments	Officers given extended assignments to a neighborhood
Special police units (tactical, detective, etc.) operate in local neighborhoods without informing local police officials	Special police units inform themselves of team goals and, whenever possible, consult in advance with the local team commander
Community relations as "image building" (special units for community relations plus speaking engagements for officials)	Community relations is an essential patrol function, planned by the team commander and the team and consisting of good police service, friendly on-street contacts, and attendance at meetings of various community groups
Reactive policing (responding to calls) or aggressive policing (stop-and-frisk and street interrogation)	Proactive policing (crime analysis, use of plainclothes or special tactics, investigations, preventive programs, referral programs, service activities)
Centralized planning (innovation through orders from the chief or other important officials)	Decentralized planning (innovation by team commanders, subject to review by their supervisors)

Source: Reprinted with permission from P. Bloch and D. Specht, *Prescription Package, Neighborhood, and Team Policing,* © 1973, Urban Institute.

Traditional policing, unlike team policing, mandates quasi-military supervision, with a watch commander responsible for an eight-hour tour and unit commanders (e.g., captains) responsible for around-the-clock operations. The first available car is assigned to calls, with priority given to emergencies, so that police unfamiliar with a neighborhood may end up dealing with that neighborhood. Officers are periodically rotated to new divisions or assignments. Special police units (such as tactical units) can come into a beat or district unannounced and stir up the citizens in the neighborhood. Traditional policing is reactive—responding to calls—or aggressive—using stop-and-frisk tactics. Police operations are centralized, originating from the chief and being passed down to the beat patrol officer.

Many of the strategies of team policing have been incorporated into community policing because it requires the police to be responsive to the communities and neighborhoods where they are assigned and to develop rapport and cooperate with the community. Like community policing in general, team policing requires organizational change: it cannot be successfully implemented unless the police agency is completely reorganized and the authoritarian centralized management style is abandoned.

Evaluations

Team policing has been an elaborate experiment in American law enforcement. During the 1970s, approximately 60 police departments established team policing programs. Many of these programs were evaluated, and most of these evaluations are available for review. This chapter could not possibly cover all the evaluations of team policing programs. Therefore, it will review evaluations of two major programs—the Community Sector Team Policing experiment of Cincinnati, Ohio, and the Richmond, California, Team Policing Program—selected because they are often referred to in books and articles about team policing.

The Cincinnati Team Policing Report

Community Sector Team Policing (COMSEC) was initiated in Cincinnati in March 1971. This program did not receive substantial funds for planning and training. In 1973, a second COMSEC program was implemented that was the focus of an Urban Institute study, sponsored by the Police Foundation, from March 1973 to September 1975 (Schwartz and Clarren 1977).

During the COMSEC experiment, Cincinnati had a city population of approximately 500,000 and an area of 78 square miles. Team policing was carried out in Police District 1, a 3.7-square-mile area with a population of 35,000 people. The central business district lay within District 1 and attracted 250,000 shoppers, tourists, and nonresidents on weekdays. Also, District 1 had 25 percent of the city's crime and was composed of diverse neighborhoods. Under COMSEC, District 1 was divided into six sectors, with a team responsible for each sector. The sectors included:

- two predominantly black, high-crime, low-income residential areas
- a low-income, mixed residential and business area
- a predominantly white, middle-class residential area
- a racially mixed (black and Appalachian white), low-income, high-crime, largely residential area
- the central business district (Schwartz and Clarren 1977, 4)

The COMSEC program permanently assigned officers to specific neighborhoods. Emphases of the program included communication among team members, unity of supervision, decentralized operations, autonomous team decision making, and the unified delivery of all police service except homicide investigations. Team members operated as *generalists,* performing both investigative and patrol functions. The major findings of the evaluation were as follows:

1. Effects of team policing on crime during the first 18 months
   - COMSEC was more successful than policing in other parts of the city in reducing burglary, as well as in controlling other kinds of crime.
   - The proportion of small businesses struck by burglary and robbery decreased significantly in District 1 but not elsewhere.
   - Small businesses in District 1 reported to the police a larger percentage of the crimes that occurred than they had reported before COMSEC.
   - With the deterioration of the program and the fall in morale among officers, the victimization rates for businesses in District 1 returned to pre-COMSEC levels. Burglary also appeared to be increasing in District 1 at 30 months, the end of the evaluation.
2. Police-community relations
   - Fewer citizens in District 1 felt "very unsafe" walking in their neighborhoods at night.
   - District 1 citizens believed that officers were more likely to arrive when called.
   - Citizens and businessmen in District 1 noticed more frequent use of foot patrol, and more of them recognized the officers who worked in their neighborhoods.
   - Citizens' support for the idea of team policing, which was high before COMSEC, increased under the program.
3. Officers' jobs and their attitudes toward their jobs
   - Officers reported positive changes in the breadth of their jobs (task scope) and in their influence over decisions, although most of the reported gain in job breadth was lost by 18 months.
   - Satisfaction with the amount of freedom available and with supervisors rose after 6 months, then fell again by the end of 18 months. Satisfaction with work showed a similar pattern. (Schwartz and Clarren 1977, 6–7)

The evaluation of COMSEC revealed many positive effects of the program. Nevertheless, the program did not last. The biggest problem appeared to be the attitude of police administrators. They did not like the autonomy and decision-making authority given to the patrol officers. Police administrators felt they were losing control and did not want to decentralize police operations. They were comfortable with the centralized bureaucracy in which they had control over all police operations. They did not trust their officers and did not believe that the officers could learn from their mistakes. Eventually, team policing died in Cincinnati, as it would in other cities. Those police administrators who support or claim to support community policing can learn a great deal from the history of team policing.

Evaluation of the Richmond, California, Team Policing Program

In June 1968, the Richmond, California, Police Department switched from the traditional "watch" system to team policing. Unlike Cincinnati and other major urban centers, Richmond is a small city with 168 police officers. The objectives of the Richmond team policing strategy were twofold: to distribute manpower in accordance with frequency of calls for service and to improve officer morale, decrease specialization, and establish closer supervision. Under the new program, teams usually consisted of a sergeant and five patrol officers. There were five teams to cover a 24-hour period and two relief teams to cover for days off and vacations.

The evaluation of the program found several strengths but also several weaknesses:

1. Positive Results
   * Increased manpower during peak periods of activity
   * Closer and more effective supervision, with a sergeant supervising 8 to 12 officers and using his influence to affect their working habits
   * Improved communication between sergeants and police officers
   * Specialists (e.g., evidence technicians) placed with the teams
   * Increased responsibility for each patrol officer in handling a case from its inception to its prosecution
   * Tendency of patrol teams to assume the style and personalities of their supervisors, which can be beneficial when the supervisor is aggressive in controlling crime
   * Greatly enhanced training potential—a team could be trained together by the sergeant so that every member of the team had the same understanding of how and when to apply the techniques learned
2. Negative Results
   * Relief teams having fewer officers than the teams they relieved, although the understaffing was not that severe (one or two fewer officers on relief teams than on the teams that they relieved)
   * Favoritism on the part of team sergeants toward their own officers: sergeants were not always quick to take disciplinary actions against team officers and often tried to arrange for promotions or good assignment transfers for team members
   * Lack of clarity of the job function of patrol lieutenant; lieutenants overlapped team schedules so that a team would work with more than one lieutenant, and line supervision was unclear (Phelps and Harmon 1972, 2–5)

The Richmond team policing program had more positive than negative effects. The problems that developed could have been worked out with some planning. They were not that serious—the favoritism of the sergeants of team members could be checked, and the role of the lieutenant could clearly be defined. Thus, it appears that the team policing strategy for Richmond was a success.

Overview of Team Policing Program Evaluations

The purpose of evaluating a program is to collect information about its short- and long-range effects. Students who have taken methodology know that program evaluations can be of different types and that the type chosen depends on what information researchers wish to obtain. Types of methods include case studies, quasi-experiments, and random sampling. Several problems can occur with program evaluations, such as poor design, incorrect implementation, inappropriateness of the research to the program being evaluated, invalid data, or misinterpretation of data. The accuracy of the information obtained from a program evaluation can determine whether a specific program will continue or be disbanded.

In 1977, the Law Enforcement Assistance Administration (Gay et al. 1977b) sponsored a review of evaluations of team policing. It stated that

> evaluation studies of team policing have been few in number and varying in quality. Evaluation reports have ranged from anecdotal records of the impressions of participants written up by the police chief and case study descriptions, to detailed reports of large scale, multi-year evaluations conducted by outside evaluators making use of expensive and systematic data collection methodologies and experimental designs. (22)

Gay et al. found that

> because of the fluctuating quality of the evaluation studies, the results reported are of questionable validity. What results do exist range from reports of positive and rapid changes of the type anticipated, to lack of results and results contrary to expectations. Most results reported, however, have been of a positive nature. (23)

According to the review, several of the evaluations failed to measure police performance, and others failed to measure goal attainment. Gay et al. listed four problem areas pertaining to team policing:

1. *Inadequacy of measures of goal attainment.* Team policing has a number of goals, primarily the reduction of crime and the improvement of police-community relations. The crux of the evaluation problem is getting valid and reliable criteria of goal attainment.
2. *Confounding: The problem of intervening variables.* Another significant measurement problem for team policing is that of the confounding of dependent and independent variables, resulting in an inability to distinguish the program or strategy to which the evaluation results should be attributed. This has been caused by a number of factors in the implementation of team policing programs, including the introduction of team policing as only one of several concurrently initiated innovations; the uniqueness of team policing programs; the introduction of team policing programs as demonstration projects in only one section of most cities; and the novelty of the programs.

3. *Costliness of experimental evaluations.* The type of systematic, experimental evaluation required to get valid information about the effects of team policing is very expensive. This is not only because of the tendency to use victimization studies to get a more accurate view of the "real" crime rate but also because of the necessity of using control groups.

4. *Political constraints and lack of evaluation impact.* Team policing evaluations have been subject to numerous political constraints. Such constraints are inherent in all evaluations to a greater or lesser degree. The operation of these constraints can, however, be a critical factor influencing the validity and use of evaluation results. Since a main purpose of evaluation is to provide information for decision making, the degree to which the results of an evaluation affect the decision to retain or modify a program is a prime factor in assessing the usefulness of an evaluation effort. (Gay et al. 1977b, 23–24)

Many of the problems associated with the evaluation of team policing are also associated with the evaluation of community policing. Community policing has been called a success without an evaluation. Measures of goal attainment are difficult to determine. For many police departments, community policing is a specialized program for high-crime areas, and the community policing concept has not been implemented citywide. Politics can influence validity and use of evaluation results.

Supporters of community-oriented policing can learn a great deal from team policing. If they examine its history, they may be able to avoid some of the pitfalls that police administrators who directed team policing made.

PREVENTIVE PATROL

Traditional Preventive Patrol

Generally, police departments assign 60 to 70 percent of their police officers to patrol. The patrol unit has been called the "backbone" of the police department. Uniformed officers are assigned either on foot or in vehicles to patrol a specific geographical area, often referred to as a beat. Although other units such as investigations may be accorded more prestige, the patrol division has the most influence on public perceptions of the police. During the 1970s, police departments began to review the goals and functions of patrol operations. Traditionally, patrol operations function to provide a multitude of services, both crime and noncrime related.

The police functions can be divided into four categories: calls for service, preventive patrol, officer-initiated activities, and administrative tasks. Calls for service are usually referred to as reactive policing. A citizen calls the police for service—for example, about a dog barking in the wee hours of the morning—and the police respond. In most cases, the patrol officer is expected to correct the situation and to satisfy the citizen's complaint. When a patrol officer is not reacting to a citizen's call, he or she should engage in preventive patrol—that is, walking or driving around the area and keeping a lookout for potential problems. The notion behind preventive patrol is the belief that patrol prevents and deters crime. While on patrol or perhaps even while eating lunch, a patrol officer could stumble upon or observe a crime occurring. The patrol officer has the responsibility to initiate corrective action on his or her beat regardless of whether it is a criminal or noncriminal matter. Finally, all patrol officers have administrative tasks that they must perform, including writing reports, running department errands, appearing in court, transporting prisoners, or having the police cruiser cleaned. Traditional preventive patrol has five goals: (1) deterrence of crime, (2) apprehension of criminals, (3) satisfaction of the public demands for services related to crime, (4) development of a sense of security and confidence in the law enforcement agency, and (5) recovery of stolen property (Szynkowski 1981, 169). Patrol officers have more opportunity to recover stolen automobiles on their beats than many other types of stolen property.

All patrol officers have been taught in the police academy that they are out on patrol to deter crime. However, they do not know if they are a deterrent. They realize that most individuals will not commit a crime in their presence. Most seasoned patrol officers realize that an individual will commit a crime when police leave the areas if the desire is there. Arrests may require not only hard work but a great deal of luck.

Patrol officers are asked or given the responsibility to provide services that other governmental agencies are unwilling or unable to perform. Also, they may be left with performing a variety of services simply because they are available 24 hours a day.

Kansas City Preventive Patrol Experiment

The Kansas City, Missouri, Police Department, under a grant from the Police Foundation, initiated in 1972 an experiment to determine the effectiveness of routine preventive patrol. The experiment involved 15 preventive patrol beats within Kansas City, randomly divided into three groups of five beats. Group 1 beats were *reactive:* officers were instructed to respond only to calls for 911 service. Group 2 beats were *controls:* standard preventive patrol was maintained at its usual level of one car per beat. Group 3 beats were *proactive:* regular preventive patrol was intensified by two or three times the usual level by assignment of additional patrol cars and by the frequent presence of cars from reactive patrols (patrols looking for crime, such as drug buys) beats (Kelling et al. 1974, 3). The experiments were developed to test the following hypotheses:

1. Crime, as reflected by victimization surveys and reported crime data, would not vary by type of patrol.
2. Citizen perception of police service would not vary by type of patrol.
3. Citizen fear and behavior as a result of fear would not vary by type of patrol.
4. Police response time and citizen satisfaction with response time would vary by experimental area.
5. Traffic accidents would increase on the reactive beats. (3)

The Kansas City Preventive Patrol Experiment found that type of patrol—reactive, control, or proactive—appeared not to have any effect on crime rates, delivery of services, or citizens' feelings of security. Specifically:

- As revealed by victimization surveys, the experimental conditions had no significant effect on residence and nonresidence burglaries, auto thefts, larcenies involving auto accessories, robberies, or vandalism—crimes traditionally considered to be deterrable through preventive patrol.
- In terms of rates of reporting crime to the police, few differences and no consistent patterns of differences occurred across experimental conditions.
- In terms of departmental reported crime, only one set of differences across experimental conditions was found, and this was judged as likely to have been a random occurrence.
- Few significant differences and no consistent pattern of differences occurred across experimental conditions in terms of citizen attitudes toward police services.
- Citizen fear of crime, overall, was not affected by experimental conditions.
- There were few differences and no consistent patterns of differences across experimental conditions in the number and types of anticrime protective measures used by citizens.
- In general, the attitudes of businesspeople toward crime and police services were not affected by experimental conditions.
- Experimental conditions did not appear to affect significantly citizen satisfaction with the police as a result of their encounters with police officers.
- Experimental conditions had no significant effect on either police response time or citizen satisfaction with police response time.
- Although few measures were used to assess the impact of experimental conditions on traffic accidents and injuries, no significant differences were apparent.
- About 60 percent of a police officer's time is typically noncommitted (available for calls); police officers spent approximately as much of this time on non–police-related activities as they did on police-related activities.
- In general, police officers were given neither a uniform definition of preventive patrol nor any objective methods for gauging its effectiveness.
- While officers tended to be ambivalent in their estimates of preventive patrol's effectiveness in deterring crime, many attach great importance to preventive patrol as a police function. (3–4)

Several of the preceding findings directly challenge the assumptions of traditional preventive patrol. But this study should not be interpreted as meaning that patrol officers should be removed from their patrol beats or that police presence may not be of value in reducing crime. Nor should the study be used to justify reducing the levels of policing. Also, the fact that a majority of a police officer's time is typically spent on noncriminal matters does not mean that the officer's role in preventing crime is not important.

The findings of the Kansas City Preventive Patrol Experiment were not welcomed by all police administrators and police officers; indeed, they are still being debated today. But the study is important because it tested a long-held belief that preventive patrol deterred crime. Students of community policing should be aware of this research because it may affect decision making pertaining to community policing. Also, the Kansas City Patrol Experiment has been given credit for the initiation of *directed patrol* programs.

Directed Patrol

Directed patrol developed in the 1970s out of a sense that random preventive patrol was ineffective in addressing the growing crime problem and was an inefficient use of limited police budgets. The idea behind directed patrol was that patrol officers could function more effectively if they received formal guidance and detailed crime analysis. Directed patrol has three major components:

1. identification, through rigorous crime analysis, of the places and times when crimes are occurring and are most likely to occur in the future
2. preparation of written directions describing in detail the way problem areas are to be patrolled
3. activation of these patrol directions at specific times determined by crime analysis (Gay et al. 1977a, 13)

Crime analysis may perhaps be the most critical component of directed patrol. The collection and analysis of crime information to detect crime patterns can assist in establishing patrol procedures to combat specific crimes. The crime analysis unit identifies order maintenance, traffic, and crime problems, and the information produced is used to guide police operations in solving criminal or order maintenance problems. Directed patrol units can target specific crime or traffic areas and can become more effective and efficient in solving problems on their beats. In Kansas City, for example, the police department established a Crime and Intelligence Unit that gave patrol officers information on the most active criminal and problem offenders (Directed Patrol Project 1979, 41–43).

Directed patrol uses uncommitted patrol time for implementing its activities and strategies. Citizens' calls for service are prioritized so that patrol officers have more time to carry out directed patrol strategies. Citizens are encouraged to telephone in reports or walk in to the precinct station and make a report instead of calling 911. Calls that are not serious can be handled by a delayed response. In Charlotte, North Carolina, the police department created an "expeditor unit" to conserve the time of patrol officers who had the task of taking reports over the telephone. Directed patrol also involves scheduling and deploying patrol officers by computer geographically, in ways that are matched to workload conditions.

The National Institute of Law Enforcement and Criminal Justice offered the following recommendations for directed patrol:

1. Patrol administrators should integrate the crime analysis and evaluation processes as means to develop, modify, and improve directed patrol operations.
2. Every department should monitor the extent to which the patrol force is deployed according to workload demands. Periodic deployment adjustments

should be made when the city workload and deployment patterns are out of phase.

3. Patrol administrators should periodically review the way in which calls are serviced and adjust service response patterns to provide "blocks of patrol time" for directed activities.

4. Patrol officers should be required to keep a log of all directed patrol assignments. These data should be reviewed and compiled by sergeants and watch commanders and used as a patrol planning tool.

5. Every department should analyze reported crime data to measure the prevention and deterrence effectiveness of various patrol tactics. Where possible, data from directed patrol areas should be compared with periods before the program began and with areas patrolled randomly.

6. Every department should analyze data concerning the circumstances that led to an arrest. Wherever possible, the number of arrests that can be attributed to a directed patrol tactic or other patrol actions should be evaluated. This can be used to rate the effectiveness of specific patrol in planning directed patrol assignments. (Gay et al. 1977a, 167–168)

In the 1970s, the National Institute of Law Enforcement and Criminal Justice (currently the National Institute of Justice) provided funding to three cities—Albuquerque, New Mexico; Charlotte, North Carolina; and Sacramento, California—to determine the extent to which the patrol function could be effectively directed, in what was called the Managing Patrol Operations (MPO) experiment (Howlett et al. 1971, 34–43). The Charlotte, North Carolina, Police Department implemented such elements of directed patrol as computer-developed work schedules, computer-assisted resource allocation, call prioritization, an expeditor unit, crime analysis, and specific, formal guidance of patrol officers according to crime analysis information (36).

The experiment in Charlotte improved police operations and increased productivity. It also led to a closer scrutiny and analysis of patrol activity by police administrators and line officers and consequently a better understanding of the police organization and the community (43).

Like team policing, directed patrol has fallen by the wayside. Both directed patrol and team policing were 1970s interventions that died in the late 1970s and early 1980s. But now many of the strategies of directed patrol and team policing are being reintroduced into community policing. Administrators of police departments should examine why police administrators and politicians allowed these valuable programs to become defunct.

RESPONSE TIME

One of the key objectives of the professional model of policing was a rapid response time. Police administrators believed that a rapid response time would increase arrests, produce more witnesses, and produce greater citizen satisfaction with the police, and they proclaimed this belief to the general public as a fact. Police chiefs have considered rapid response so important that they have tried to enable their patrol officers to respond to a call within three minutes of a dispatcher's receiving the call. In a study of 1,100 emergency calls dealt with by 24 police departments, patrol officers arrived within 3 minutes 50 percent of the time, within 5 minutes 75 percent of the time, and within 11 minutes 95 percent of the time (Attunes and Scott 1981, 165–179). Police have accepted the premise of O.W. Wilson, a leader of the professional model of policing, that "the likelihood of arrest is far greater if a patrol officer arrives within three minutes of the time a serious crime takes place" (Wilson and McLaren 1977, 345).

Wilson's premise certainly holds for crimes in progress, when it might be possible for the patrol officer to make an arrest. I myself have been in situations where a rapid response time has led to an arrest. But will a rapid response be of any use when a homeowner who has been on vacation for a week comes home to find his or her home burglarized? What value does a rapid response time have in this situation?

One study on rapid response time in five cities found that fewer than 3 percent of rapid police responses resulted in an on-scene response-related arrest. This study also found that 75 percent of crimes committed are discovered after the fact and that in many cases citizens wait an average of 10 to 15 minutes before notifying the police (Spelman and Brown 1980, 48). Citizens delay reporting crime to the police for the following reasons:

1. *Apathy.* Citizens exhibiting this pattern typically indicated that they did not think the incident was personally important or that they did not want to get involved in the incident or take the responsibility of calling the police.
2. *Being unsure about police assistance.* Most frequently, citizens cited the feeling that the police could not help because there was no evidence. A second justification for this delay was that the police might think the incident was unimportant or would not want to help.
3. *Contacting security.* This action was commonly taken because it was company policy to contact security or security guard prior to reporting the crime to the police, although almost as many citizens reported taking this action first in the absence of any company policy.
4. *Investigating the incident scene.* This delay commonly resulted from citizens trying to enumerate missing articles, search for missing property, etc., prior to telephoning the police.
5. *Telephoning another person or receiving a call.* Citizens generally indicated that they called a second party (or another person called them) in order to obtain advice, assistance, or additional information concerning the incident.
6. *Waiting or observing the situation.* Often the reason for waiting or observing the situation was related to a search for additional information about the seriousness of the incident and the need for police assistance.
7. *Injury.* This problem occurred when physical injuries to the reporting party or the necessity of giving first aid or transporting another person to the hospital precluded immediate reporting.

8. *Not being informed or being misinformed about the incident.* In almost all cases with this problem, the reporting parties indicated that the delay was due to the fact that they had not been immediately informed of the crime by the person who had discovered or who was involved in it. (Van Kirk 1977, 32–33)

The same study found that it took anywhere from 3 to 6 minutes to report an aggravated assault or robbery after the victim or witness was free from the involvement in the crime; 4 to 7 minutes for victims or witnesses involved in a burglary, larceny, or auto theft after they were free to notify the police; and 10 to 15 minutes for the average rape victim or witness. A crime discovered was reported to the police between 10 and 15 minutes (Spelman and Brown 1980, 59).

Response time–related arrests were most likely to be made for property crimes. For property crimes in progress, chances were excellent for an arrest if the police response was rapid. Property crimes constituted between 10 and 15 percent of the crimes in the five cities examined. Anywhere from 45 to 70 percent of the robberies and assaults could not be cleared by a response-related arrest. Less than 5 percent of Part I offenses (serious crime of homicide, robbery, larceny, aggravated assault, arson) resulted in an on-scene arrest, and police response time had no effect on on-scene arrest in 70 to 85 percent of these crimes because they were discovered after they had occurred. Police response time also had no effect on 50 to 80 percent of the rest of crimes, because victims or witnesses were too slow in reporting the incident to the police or failed to take the responsibility of calling the police.

Thus, the study came to four basic conclusions about response time. First, although some patrol strategies do affect police response time, a large proportion of Part I crimes are not affected by rapid police response. Second, whether rapid response produces arrests depends on the time that citizens take to report the crime. Third, factors that cause reporting delays by citizens are due to citizens' attitudes and the actions they take rather than to any uncontrollable problem that they confront. Fourth, if a citizen does not take an exorbitant amount of time to notify police, a prompt police response can have a significant impact on certain categories of crimes, but for the most part it has a limited impact on crime outcomes in general (Spelman and Brown 1980, 40).

According to the findings of the study, the value of rapid response to service calls is questionable. Most Part I crimes (62.3 percent) are discovered after the crime has taken place. The chance that a police officer will be able to make an arrest at the scene of the crime or even locate witnesses is not assisted by a rapid response time. Citizens who discover crime incidents realize that a rapid police response will have had no effect on the outcome and are not disappointed with the slower response time. The study also addressed the significance of the time that citizens take to report a crime in progress, which precedes the dispatchers' assigning an officer and the officer's travel time. Ideally, the crime should be reported in less than five minutes, and reporting of a crime in less than two minutes will increase the likelihood of an arrest by 10 percent. Because of the time it takes for citizens to report a crime in progress, on-scene apprehension is possible in only about 50 percent of crimes (Spelman and Brown 1980, 40–42).

To summarize, rapid response time can lead to an arrest on a crime in progress if the police are notified within five minutes. The biggest source of delay is victims' delay in notifying the police. Reporting a crime after the fact usually will not be helped by a rapid response time. A rapid response time should not be expected for citizens for all service calls. Rapid response will not be of value to a week-old crime. Citizens should not expect a rapid response to dog calls, noise complaints, or other nonemergency situations.

MINISTATIONS

One of the goals of community policing is to get the police into closer contact with the public. To meet this goal, police departments in big cities like New York, Chicago, and Philadelphia have always divided the city into districts, each of which has its own substation. More recently, many midsized and small cities have established district stations as well. Some cities have gone further and established ministations, or substations that cover a smaller geographical area than a district, such as a mall or other business area or a public housing complex.

Detroit

In the mid-1970s, the Detroit Police Department established ministations under a grant from the Law Enforcement Assistance Administration. The city installed 52 ministations, with each police precinct having a minimum of three. The ministations were considered to be neighborhood police stations and were staffed 24 hours a day. An officer was permanently assigned to a ministation and was not required to respond to 911 calls or to patrol an assigned beat. Neighborhood ministations encouraged walk-in calls for police service and the filing of crime reports and were charged with crime prevention duties (Skolnick and Bayley 1986, 53–57). An evaluation of the ministations came up with several positive findings:

- A detailed analysis of reported crime by category revealed positive program effects in terms of both increased reporting (for assaults, larceny, rape, and robbery) and decreased incidents (auto theft and, marginally, business burglary) for those areas in which stations were opened. In addition, it was found that ministation areas had a higher percentage of crimes cleared by arrest.
- The greatest effects were felt for both reporting and incidence of crime nearest the station and smaller effects further away. One-half mile seemed to be the maximum radius in which measurable impact on any of the crime categories can be ascertained, and this range varied with the different types of crime.
- The most effective ministations were staffed by officers who became actively involved in the surrounding community, knew more community residents, involved themselves in youth and community programs, and participated less in traditional patrol activities.
- Race had an effect as well in that more active and effective stations seemed to be staffed to a greater degree by black officers. (Green et al. 1976, x–xi)

Public Housing Ministations

Savannah, Georgia, was one city among many that initiated ministations to restore order in public housing. In the late 1980s with the introduction of crack cocaine, Savannah experienced a dramatic increase in crimes committed by offenders using drugs—both crimes against persons and crimes involving drugs. The Savannah Police Department identified 10 areas, by citizen complaints and arrests, as being heavily drug infested; five were public housing areas.

In these areas, drug use and sales were conducted openly. The crime rate was high, and residents were fearful for their safety. Public service utility crews could not openly walk in these neighborhoods without being harassed or approached to buy drugs. Drive-by shootings averaged one to two per week, and police usually confiscated weapons in encounters with suspicious persons.

In 1990, the City of Savannah received a Drug Control and Systems Grant from the federal and state governments, and the Housing Authority of Savannah received a grant from the Department of Housing and Urban Affairs. The city and the Housing Authority, working together, contributed funds to a program entitled "Project Shield," whose purpose was to implement police ministations to control crime and eliminate drugs in public housing and to enable tenants to become active in neighborhood revitalization. The initial public housing area selected for Project Shield was Garden Homes because it had one of the worst drug problems in the city. Six officers and one sergeant were assigned to Garden Homes full time. Officers were divided into two overlapping shifts of three officers each, from 4:00 p.m. to midnight and from 9:00 p.m to 5:00 a.m. The sergeant overlapped the two shifts, working from 8:00 p.m. to 2:00 a.m.

Police officers assigned to the ministation at Garden Homes were instructed in problem-oriented policing, departmental resources, and community resources. The officers and sergeant assigned to the ministation were hand picked by the lieutenant responsible for Garden Homes in order to select officers who were able and willing to work in a public housing complex. The officers assigned to the ministation saturated the complex with walking patrols to eliminate drug dealers and nontenant drug customers. Residents involved in drug activity were informed by the Housing Authority's resident management that they were subject to the eviction process already in place.

Like other residents of public housing, Garden Homes residents were more vulnerable to being victims of crime than other segments of the population. They also were disturbed by the disorderly behavior of rowdy young people, addicts, and drug merchants and by the deterioration of their physical environment, as manifested by graffiti, broken windows, and poorly-maintained apartments. With the implementation of the ministation in the Garden Homes public housing complex, the "fear of crime" substantially decreased, and the residents were able to take back their neighborhood. The children were able to play in the playground, which had become the domain of drug merchants. Drive-by-shootings were all but eliminated, and residents' victimization by crime was greatly reduced. Outsiders who did not reside in Garden Homes and who caused most of the problems within

Garden Homes were chased away by the police. Those individuals who wished to buy drugs within Garden Homes were unable to do so. Graffiti was removed, grass was planted, and vacant apartments were remodeled.

Project Shield provides a model for other police departments and communities. It shows cooperation of the police department with another city agency, the use of problem-oriented policing, the development of a partnership with community residents, and the value of the ministation in curbing crime and disorder in a small geographical area (Palmiotto 1998).

FOOT PATROL

Patrol has often been called the backbone of the police department. Generally, other services provided by the police department evolve from the work of the patrol division. The patrol units are the primary line units that have as their mission the prevention and control of criminal conduct. If the aphorism "thin blue line" is to be considered accurate, then it has to be applied to the patrol force. The patrol force, more than other police operations units, provides the first line of defense against the street criminal and predatory crimes. The delivery of police service to the community by the patrol force has become expected by the community. Most police-citizen contacts occur at the patrol level.

The oldest procedural method, one that has received much attention and has continued to be proven valuable, is the foot patrol. Before the advent of the automobile, foot patrol was the primary means of patrolling the community. When the automobile became popular with the police department as a patrol technique, foot patrol lost favor. Foot patrols are useful for special events such as parades, dignitary visits, and public relations. They are also valuable for patrolling shopping malls, beaches, apartment complexes, schools, and areas where motorized vehicles cannot gain access.

Officers on foot can observe more than officers in vehicles. By being on the street, an officer's sense of smell and sense of hearing are improved. Foot patrol is "an effective method of improving face-to-face communication between community residents and the police. The increased communications often culminates in the exchange of information needed to prevent and solve crime" (Trojanowicz 1984, 49). In addition, foot patrol officers, making personal contact with citizens on their beat, can function as community organizers, as dispute mediators, and as valuable links between social services agencies and the community (Trojanowicz 1984, 47).

The Flint Foot Patrol Experiment

During the late 1970s, the Flint, Michigan, Police Department primarily operated preventive motorized patrols. In 1979, with a grant obtained from the Charles Mott Foundation, a community-based foot patrol experiment was implemented. Its goals were

1. to decrease the amount of actual or perceived criminal activity
2. to increase the citizen's perception of personal safety

3. to deliver to Flint residents a type of law enforcement service consistent with the community needs and the ideas of modern police practice

4. to create a community awareness of crime problems and methods of increasing law enforcement's ability to deal with actual or potential criminal activity effectively

5. to develop citizen volunteer action in support of, and under the direction of, the police department, aimed at various target crimes

6. to eliminate citizen apathy about reporting crime to police

7. to increase protection for women, children, and the aged (Payne and Trojanowicz 1985, 4)

The Flint Patrol Program began with 22 foot patrol officers assigned to 14 geographical areas that included approximately 20 percent of the city's residents (Trojanowicz and Barnes 1985, 4). Police patrolled residential areas as well as business areas. They not only provided full law enforcement services but organized neighborhood watches and functioned as a link to government and other social service agencies, making referrals and intervening when appropriate. They were involved in problem solving, attempting to find solutions to community problems on their beats, such as safety and security, victimization of the elderly, and juvenile problems. Because they had a permanent beat that they worked day in and day out, they were able to develop a rapport with the people on their beat that led to a good working relationship with the community (5).

The Flint Foot Patrol Experiment found that crime decreased in the experimental area whereas crime was increasing in the other areas of the city. Residents perceived that foot patrol reduced crime in their areas. The vast majority of citizens felt safe with foot patrol officers, who were well known and extremely visible. They mostly knew the duties of the foot beat officers and were basically satisfied with what the officers had accomplished. The experiment decreased calls for service, because residents would talk to the foot officer about the specific problem that concerned them. Foot patrol officers considered enforcing laws and keeping public acceptance of the police department as top priorities. The foot patrol officers encouraged citizens to report crime, and the neighborhood valued their involvement in counseling and assisting people on their beats (Trojanowicz and Bucqueroux 1990, 214–220).

The findings of the Flint Foot Patrol Experiment provide useful information on how to produce a close police-citizen partnership that can lead to a possible decrease in crime rates. This is the foundation of community policing.

The Newark Foot Patrol Experiment

The stimulus for the use of foot patrol in New Jersey communities was the passage of the Safe and Clean Neighborhood Act by the New Jersey legislature in 1973, which provided funding to establish foot patrols in Newark. At that time, the reactive approach to policing was predominant. It was believed that the primary purpose of patrol units was to respond to citizens calls, and the 911 syndrome controlled patrol operations.

Foot patrol beats were held in low esteem and were primarily seen as a public relations gimmick to keep the merchants happy. Foot patrol beats were used by some police departments as punishment posts. In one northeastern city, squad car officers who made too many mistakes on their reports were reassigned to walking posts.

An evaluation of Newark's Foot Patrol Experiment was conducted from February 1978 through January 1979. The activities of foot patrol officers were obtained from a daily activity log maintained by the officers. Officers reported that they made few arrests and filed very few reports but did issue summonses, primarily for traffic violations. One of their main activities was not reported on their daily activity sheet: the amount of time talking to citizens and visiting business establishments.

Although the foot patrol experiment did not reduce crime, it created a feeling of safety for residents and merchants in the areas patrolled. However, there is no reason why the crime rate could not be reduced if foot patrol was completely integrated into patrol operations.

The Newark Foot Patrol Experiment made the following recommendations on how to improve the use of the foot patrol strategy in police operations:

1. Raise the status of foot patrol officers to equal that of other units. The rationale behind this recommendation is that if foot officers are to make their maximum contributions to a complete patrol strategy, their work must be seen as being at least as important as motor patrol. If it is, there are indications that many officers would be drawn to foot patrol because of their inherent characteristics of the work and the potential for regular assignment. . . .
2. Increase the use of foot officers to respond to calls for service. Research into police response to calls for service indicates that rapid response rarely is warranted. Citizens properly handled by telephone are comfortable with predictable delays. The use of foot officers to respond to all but those rare calls when speedy response is justified will increase the familiarity of the officers with the citizens and vice versa. In addition to having important consequences for citizens' attitudes, the use of foot officers to respond to calls for service in their beats can increase their stock of information about citizens, crimes, and victims. This has a crime reduction potential.
3. Provide specific training for foot patrol reflecting its functions. Although not codified, we believe that knowledge and skill exists about foot patrol that could be systematically taught to officers both pre- and in-service. . . .
4. Attempt to find ways of using the information foot officers can get about criminal activities and individual criminal events as a result of their closeness to a neighborhood. . . .
5. Emphasize closer integration of officers into neighborhood activities. This is not to be confused with recommendations that officers move into communities or become public service officers. It suggests instead that officers become neighborhood consultants regarding crime and public order issues. . . .
6. Increase the flexibility of hours so that officers are in beats at times of highest street activity, and when residents most want to use the streets. (*Newark Foot Patrol Experiment* 1981, 126–127)

Like the Flint Foot Patrol Experiment, the Newark Foot Patrol Experiment provides valuable information on how foot patrol beats can establish citizen-police cooperation in solving crime and disorder. The foot patrol experiments should be seen as another stage in the development of community policing. Many of the suggestions of the Newark Foot Patrol Experiment have been incorporated into community policing. The foot patrol experiments of the 1970s gave walking beats legitimacy. Foot patrol has become recognized as an important aspect of community policing.

BROKEN WINDOWS

The *broken windows* theory was first set forth in a 1982 *Atlantic Monthly* article by James Q. Wilson and George Kelling. Based on the results of the New Jersey Safe and Clean Neighborhood Program and the findings of the Newark Foot Patrol Experiment, it argues that minor public disorder offenses such as vandalism, littering, rowdy behavior, public drunkenness, and prostitution can start a downward spiral of neighborhood deterioration and fear of crime that leads to more deterioration and more serious crime.

> Social psychologists and police officers tend to agree that if a window in a building is broken and is left unrepaired, all the rest of the windows will be broken. This is as true in nice neighborhoods as in run-down ones. Window-breaking does not necessary occur on a large scale because some areas are inhabited by determined window-breakers whereas others are populated by window-lovers; rather, one unrepaired broken window is a signal that no one cares, and so breaking more windows costs nothing. (Wilson and Kelling 1982, 31)

When rowdy behavior is not checked, community controls break down: residents no longer take care of their homes, watch out for and correct one another's children, and scowl at unwanted intruders. When they abandon property, windows are broken and weeds and debris are allowed to grow, creating an environment of "not caring." When drunks, panhandlers, prostitutes, drug addicts, unruly teenagers, loiterers, and the mentally ill throng the streets, frightened residents no longer go out much, and their control over the public realm is lost. When gangs spray-paint street signs and the streets department takes three months to remove the graffiti, the impression is given that the gang has more power than the city, the police, or the residents. When teenagers are allowed to block city streets and be rowdy, the disorderly behavior will escalate to even more serious criminal behavior (Wilson and Kelling 1989, 47–48).

Conversely, police enforcement of public order can help to revitalize a neighborhood and ultimately reduce crime. As officers patrol their beats and enforce informal rules that are familiar to all the beat residents—for example, telling teenagers to keep the noise down or vagrants not to bother people at bus stops—they restore public order and provide a sense of security to residents, encouraging residents' higher presence and involvement in their neighborhoods.

Wilson and Kelling (1989) argued that

> we must return to our long-abandoned view that the police ought to protect communities as well as individuals. Our crime statistics and victimization surveys measure individual losses, but they do not measure communal losses. Just as physicians now recognize the importance of fostering health rather than simply treating illness, so the police—and the rest of us—ought to recognize the importance of maintaining, intact, communities without broken windows. (38)

Neighborhood crime for the most part has been found to be committed by offenders who reside near their victims. This makes it primarily a local problem that must be locally solved. Wilson and Kelling (1989) provided several examples of how the police and neighborhoods together can handle disorderly behavior constructively:

- When local merchants in a New York City neighborhood complained to the police about homeless persons who created a mess on the streets and whose presence frightened away customers, the officer who responded did not roust the vagrants but instead suggested that the merchants hire them to clean the streets in front of their stores every morning. The merchants agreed, and now the streets are clean all day, and the customers find the stores more attractive.
- When people in a Los Angeles neighborhood complained to the police about graffiti on walls and gang symbols on stop signs, officers assigned to the Community Mobilization Project in the Wilshire station did more than just try to catch the gang youths who were wielding the spray cans; they also organized citizens' groups and Boy Scouts to paint over the graffiti as fast as they were put up. (48)

Clearly, the concept of broken windows fits well into problem-solving approaches and into the overall philosophy of community policing. The theory of broken windows has been incorporated into the community policing philosophy of many police departments. In their 1996 book, *Fixing Broken Windows*, George Kelling and Catherine Coles compared the broken windows approach with the approach of traditional criminal justice process in dealing with crime.

VALUE OF RESEARCH

The importance of research to improving police strategies and tactics cannot be overemphasized. According to Brown and Curtis (1967),

> Many practitioners within criminal justice have met with repeated failure over the years because they relied upon only their common sense. Thus, millions of dollars have been spent on police patrol efforts that do not reduce crime, judicial practices that are widely perceived as unfair, rehabilitation programs that do not rehabilitate offenders and countless other failures. (3)

Police operations have often been based on untrue or outdated assumptions. The myth that the police are primarily crime fighters who deter crime has been deeply ingrained into our American culture. This myth has been reinforced for several decades not only by the police agencies promoting of their interests but also by the news and entertainment media. Not only do the general public believe it, but the police have bought into it themselves. Young people are often drawn to police work because of the glamour of the crime-fighter image, and they are reinforced in this attitude by their recruit trainers. In general, the training curriculum emphasizes the law enforcement aspect of policing. In fact, many police administrators, middle managers, supervisors, and police officers consider community policing to be social work or "touchy-feely" activity and not "real" police work. A good number of police officers have told me that they did not enter policing to "hug" citizens but to make arrests, perform investigations, and deter crime. Even those police administrators who claim to support community policing may stress or at least claim they are crime fighters.

The myth of the police as crime fighters has shaped traditional patrol operations. For example, the automobile was initially used to increase the geographical patrol area and to enable officers to move swiftly from one area to another. Preventive patrol evolved from the notion that officers could be everywhere on their beat or at least create that impression. Such patrolling, along with fast response to calls, was sold as a useful means of controlling crime. But, in reality, this approach had little effect on controlling criminal activity, and an examination of crimes solved by the police indicates that the police are not very successful. Research has found that most police departments spend most of their time in service-related activities, such as dealing with noisy youngsters, settling domestic disputes, and handling derelicts.

A vast number of police studies have corrected misconceptions about policing. They have also provided support for the community policing approach in that they highlight the importance and effectiveness of strategies that improve police-community relations and develop a police-community partnership in solving and preventing crime and in maintaining public order and arresting community decline.

SUMMARY

The concept of police-community relations developed during the 1960s, a period of turmoil in American cities. This decade saw the police in direct confrontation with the community they were to serve. In 1967, the President's Commission on Law Enforcement and Administration of Justice argued that when a community has negative feelings toward the police, tension rises and aggressive actions against the police begin to occur, which can in turn trigger irrational behavior on the part of the police. Further, citizens will not report crimes to the police or give the police the information they need to solve crimes.

Thus, it was recommended that police-community relations be considered in all police operations, including departmental policy, supervision, personnel procedures, records and communications, the acceptance of complaints against departmental members, and planning and research. Ideally, police-community relations emphasizes listening to people rather than telling them what to do and taking seriously the concerns and input of community members.

The goal of team policing was to overcome the isolation that police had experienced under the traditional model of policing. Basically, team policing assigns responsibility for a certain geographical area to a permanent team of police officers. The more responsibility this team has, the greater the degree of team policing. In some cities, teams include generalists and specialists and provide all police services to an area, including patrol, traffic, and investigations, under one supervisor.

Team policing, like the community policing of the 1990s, requires the police to be responsive to the neighborhoods where they are assigned and to develop rapport and cooperate with the community. Many of the strategies of team policing have been incorporated into community policing. In addition, both concepts require organizational change and a shift away from the traditonal authoritarian management style.

Another set of innovations concerns preventive patrol, or the regular patrolling of an area, either in cars or on foot, with the purpose of preventing crime and spotting and responding to problems. Generally, police departments assign 60 to 70 percent of their police officers to patrol; thus, the patrol unit has been called the "backbone" of the police department. When patrol officers are not responding to a call or performing an administrative task, they are expected to engage in preventive patrol.

The Kansas City Preventive Patrol Experiment was a study conducted in 1972 to determine if preventive patrol was effective. The study found that preventive patrol did not have any effect on crime rates, delivery of services, or citizens' feelings of security. These findings, and increasingly tight budgets, led police departments to institute changes in preventive patrolling to make it more efficient and effective. The new system was called directed patrol.

The idea behind directed patrol was to provide officers with guidance that was lacking in random preventive patrol. Patrol officers would be scheduled and deployed by computer in ways that matched workload conditions in an area. Crime analysis units would identify specific problems for police to target and would give patrol officers information—for example, on active offenders—that could guide patrol activities. Like team policing, directed patrol was abandoned in the late 1970s and early 1980s, but many of its strategies are being reintroduced into community policing.

A rapid response time to calls for service has long been considered a key objective of policing. Police chiefs have considered rapid reponse time so important that they have tried to ensure, or at least claim, that their patrol officers would respond to a call within three minutes of a dispatcher's receiving the call. They have assumed that rapid response will increase arrests, produce more witnesses, and produce greater citizen satisfaction with the police. But a major study conducted during the 1970s found no support for any of these assumptions. It found that (1) most Part I (serious) crimes are discovered after the crime has taken place, so that rapid response does not enable the officer to make an arrest at the scene of the crime or even, in most cases, to locate witnesses; (2) citizens typically delay calling the police for so long that a rapid police response becomes irrelevant; (3) reporting delays are due to citizens' attitudes and actions rather than to any uncontrollable problem they confront; and (4) if the citizen does not take an exorbitant amount of time to notify police, a prompt police response can have a significant impact on some property crimes but otherwise has a limited impact on crime outcome.

Another innovation of the 1970s was the ministation, or a police substation that covers a smaller area than a district, such as a mall or other business area or a public housing complex. Ministations are used to bring the police into closer contact with the public. They can decrease crime and increase reporting of crime and citizen cooperation with police in the areas where they are opened.

Still another innovation of the 1970s and early 1980s was the replacement of some car patrols with foot patrols that provided full law enforcement services for a neighborhood, referred individuals to social services, and organized crime prevention activities cooperating with the community. Foot patrol officers who had a permanent beat were able to develop good rapport with citizens and increase citizens' feelings of security. Many of the suggestions of the foot patrol experiments have been incorporated into community policing. Whereas foot patrols under traditonal policing were low-prestige assignments, the foot patrol experiments of the 1970s gave walking beats legitimacy.

A theory that has greatly contributed to community policing is the broken windows theory of Wilson and Kelling (1982). According to this theory, public order offenses such as vandalism and rowdy behavior, if not checked by the community, can create a downward spiral of neighborhood deterioration and fear of crime that leads to more deterioration and more serious crime. This is because graffiti, broken windows, abandoned properties, and public drug use, drunkenness, and prostitution give the impression that "nobody cares" about the community and that residents, police, and the city have lost control over it. Conversely, increased police presence and enforcement of informal rules of conduct as well as laws can make a neighborhood seem less chaotic and more safe, thereby encouraging heightened presence and involvement of residents in their community and lowering crime rates. The broken windows theory has been shown to work and has been incorporated into community policing.

This chapter has emphasized the value of research on policing in dispelling numerous misconceptions about policing that have guided police work and rendered it ineffective. Police research has led to the acceptance by progressive police administrators and students of policing of a need for police to develop a positive and cooperative relationship with the community in order to reduce crime. Most currently, this viewpoint has been expressed in the philosophy of community policing, which has incorporated many of the policing innovations of previous decades that this chapter has discussed.

KEY TERMS

basic patrol teams	COMSEC	patrol–community service teams
broken windows theory	directed patrol	patrol-investigative teams
citizen police academies	English unit beat system	random preventive patrol
community advisory councils	full-service teams	response time
	ministations	team policing

REVIEW QUESTIONS

1. Outline the principles of good police-community relations.
2. How have principles of police-community relations been incorporated into community policing?
3. What is team policing and how does it work?
4. Discuss the four types of team policing.
5. Describe random preventive patrol.
6. Describe directed patrol.
7. Why does rapid response time have little effect on solving or preventing most crimes?
8. Explain the broken windows theory.

REFERENCES

Albright, E.J., and L. Siegel. 1979. *Team policing: Recommended approaches.* Washington, DC: Government Printing Office.

Attunes, G., and E.J. Scott. 1981. Calling the cops: Police telephone operators and citizens calling the police. *Journal of Criminal Justice* 9:165–179.

Brown, S.E., and J.H. Curtis. 1967. *Fundamentals in criminal justice research.* Cincinnati, OH: Anderson.

Crime Control Digest. 1975. Volume 9, 5:9.

Directed Patrol Project: Kansas City, Missouri. 1979. Washington, DC: Law Enforcement Assistance Administration.

Gay, W.G., et al. 1977a. *Improving patrol productivity: Vol. 1. Routine patrol.* Washington, DC: National Institute of Law Enforcement and Criminal Justice.

Gay, W.J., et al. 1977b. *Issues in team policing: A review of the literature.* Washington, DC: Government Printing Office.

Green, R.L., et al. 1976. *Detroit Police Department Mini-Station, final report: Process and planning evaluation.* East Lansing: Michigan State University, College of Urban Affairs.

Gregory, E., and P. Turner. 1968. Unit beat policing in England. *Police Chief* 35, no. 7:42.

Howlett, J.B., et al. 1971. Managing patrol operations: During and after in Charlotte, North Carolina. *Police Chief* 48, no. 12:34–43.

International City Managers' Association. 1967. *Police community relations programs.* Washington, DC: Management Information Service Report no. 286.

Kelling, G.L., and C.M. Coles. 1996. *Fixing broken windows.* New York: Free Press.

Kelling, G.L., et al. 1974. *The Kansas City Preventive Patrol Experiment.* Washington, DC: Police Foundation.

MCT supervisor's manual. 1970. Rochester, NY: City of Rochester.

Momboisse, R.M. 1967. *Community relations and riot prevention.* Springfield, IL: Charles C Thomas.

National Advisory Commission on Criminal Justice Standards and Goals. 1973. *Task force: The police.* Washington, DC: Government Printing Office.

Niederhoffer, A., and A.B. Smith. 1974. *New directions in police community relations.* San Francisco: Holt, Rinehart, & Winston.

Palmiotto, M.J. 1998. A study of residents' perceptions of fear of crime in public housing. *Justice Professional* 10, no. 4.

Payne, D.M., and R.C. Trojanowicz. 1985. *Performance profiles of foot patrol officers versus motor officers.* East Lansing: Michigan State University.

Phelps, L., and L. Harmon. 1972. Team policing—four years later. *FBI Law Enforcement Bulletin*, December, 2–5.

President's Commission on Law Enforcement and Administration of Justice. 1967a. *The challenge of crime in a free society.* Washington, DC: Government Printing Office.

President's Commission on Law Enforcement and Administration of Justice. 1967b. *Task force: The police.* Washington, DC: Government Printing Office.

Schwartz, A.I., and S.N. Clarren. 1977. *The Cincinnati team policing experiment: A summary report.* Washington, DC: Police Foundation.

Sherman, L.W., et al. 1973. *Team policing: Seven case studies.* Washington, DC: Police Foundation.

Skolnick, J., and D.H. Bayley. 1986. *The new blue line: Police innovations in six American cities.* New York: Free Press.

Spelman, W., and D.K. Brown. 1980. *Calling the police: Citizen reporting of serious crime.* Washington, DC: Government Printing Office.

Szynkowski, L.J. 1981. Preventive patrol: Traditional versus specialized. *Journal of Police Science and Administration* 9, no. 2:169.

The Newark foot patrol experiment. 1981. Washington, DC: Police Foundation.

Trojanowicz, R.C. 1984. Foot patrol: Some problem areas. *Police Chief* 49.

Trojanowicz, R.C., and D.W. Barnes. 1985. *Perception of safety: A comparison of foot patrol versus motor patrol officers.* East Lansing: Michigan State University.

Trojanowicz, R.C. and B. Bucqueroux. 1990. *Community policing: A contemporary perspective.* Cincinnati, OH: Anderson.

Van Kirk, M. 1977. *Response time analysis: Executive summary.* Washington, DC: Law Enforcement Assistance Administration.

Wilson, O.W., and R.C. McLaren. 1977. *Police administration.* 4th ed. New York: McGraw-Hill.

Wilson, J.Q., and G.L. Kelling. 1982. Broken windows. *Atlantic Monthly,* March.

Wilson, J.Q., and G.L. Kelling. 1989. Making neighborhoods safe. *Atlantic Monthly*, February, 31–48.

Wycoff, M.A. 1988. The benefits of community policing: Evidence and conjecture. In *Community policing: Rhetoric or reality,* ed. J.R. Greene and S.D. Mastrosfski. New York: Praeger.

Chapter 6

Communities, Neighborhoods, and Multiculturalism

CHAPTER OBJECTIVES

1. To have an understanding of the term *community*.
2. To have an understanding of what produces and reinforces a sense of community and the role of community policing in strengthening community.
3. To have an understanding of the term *neighborhood*.
4. To have an understanding of how police can contribute to community empowerment.
5. To have an understanding of the term *multiculturalism*.
6. To have an understanding of how *multiculturalism* relates to community policing.

INTRODUCTION

The focus of this chapter is on communities, neighborhoods, and multiculturalism. The community policing officer works in a community, a neighborhood, and often a multicultural environment. According to the community policing philosophy, the police need to develop a partnership with the community. How can a partnership between the police and the community be developed if police have no understanding of the concepts of community, neighborhood, and multiculturalism?

For community policing to be successful, police officers must be open to new ideas and experiences and be willing to take risks. Unlike the traditional model of policing, in which the police rarely react to a crime or social problem, the community policing model requires police to be proactive, searching for crime and disorder problems to solve. Community policing officers serve their communities and neighborhoods not only by solving crime and social disorder problems but by preventing them before they occur. This requires understanding what communities and neighborhoods are and how they work.

In the last several decades, multiculturalism has received a great deal of attention. The police in general, and particularly community policing officers, must have an understanding and knowledge of multiculturalism in order to be respectful and appreciative of various cultures within the community and neighborhood where they work. For community policing to be successful, police departments and officers need to grasp what comprises a community and neighborhood. Once an understanding of these concepts has been established, then, and only then, can a successful partnership between the police and community take place. Many of the communities and neighborhoods in our country are multicultural, and, by attempting to understand multiculturalism, the police are showing the members of their communities and neighborhoods that they are reaching out to establish a partnership.

THE COMMUNITY

The term *community* has a variety of meanings. Generally, a community is conceived of as a group of people occupying the same geographical area by which they can identify themselves, and in which a degree of solidarity exists (McNall and McNall 1992, 179). Thus it is about both territorial settlement and social relationship. The word is often used to identify local, small geographical units with communal relationships, such as villages, towns, or neighborhoods (Gusfield 1975, 32–33), but it can also be used to describe larger units, such as whole cities. According to the sociologist Robert MacIver (1936), it means "an area of common life, village or town, or district, or country, or even wider area" (22). MacIver further stipulated that, "to deserve the name community, the area must be somehow distinguished from further areas, [and] the common life may have some characteristics of its own such that the frontiers have some meaning" (23). The community has been considered a territorial group that embraces all aspects of social life—a local area over which people usually are using the same language, conforming to the same mores, feeling more or less the same sentiments, and acting upon the same attitudes. According to Robert Nisbet (1966),

> The word, as we find it in much nineteenth- and twentieth century thought, encompasses all forms of relationship which are characterized by a high degree of personal intimacy, emotional depth, moral commitment, social cohesion, and continuity in time. Community is founded on man conceived in his wholeness rather than in one or another of the roles, taken separately, that he may hold in a social order. (47)

The term *community* can be used to describe people's associations around skills, professions, or interests independent of locality: for example, university professors belong to a "community of scholars." But in community policing, the term usually has a geographical component, because the areas that police patrol are geographically based.

Communities are commonly classified as rural or urban. Rural and urban communities are usually distinguished by population size, but this distinction has been criticized as arbitrary and devoid of significance. Other distinguishing criteria used have been density of population and legal status of the locality as rural or urban.

A rural community consists of a group of families living on contiguous land who generally think of themselves as living in the same locality, to which they give a name, and who interact with each other through visiting, borrowing and lending tools, exchanging services, or participating in social activities. Frequently, a community has a center of common activities that includes such institutions as a school or church. Community members know one another, and their relationships appear to be intimate.

The urban community usually consists of a large group of people living in a small geographical area who have their own local government and carry on various economic enterprises. Urban communities are often only loosely "communities" in the social sense. They are more likely than rural communities to be characterized by anonymity: an individual who travels two blocks from his or her home may not be known by anyone. Urban communities are also generally more heterogeneous. Urban people are likely to associate with each other chiefly on the basis of interest rather than locality, and there is less general association among people than occurs in rural communities.

Until recent years, it was generally accepted that urban communities had a higher rate of crime, violence, narcotics, divorce, desertion, prostitution, and other forms of deviance. But today rural communities are seeing an increase in violence, crime, narcotics use, and other forms of deviant behavior that are normally associated with the urban community.

Sense of Community

For community policing to be a success, there must exist a "sense of community." Officers involved in community policing have to understand, and, more important, strive to create and reinforce, a sense of community.

Elements of a Sense of Community

A sense of community can be defined as "a feeling that members have of belonging, a feeling that members matter to one another and to the group, and a shared faith that members' needs will be met through their commitment to be together" (McMillan and Chavis 1986, 9). It consists of four elements:

1. *Membership*—the feeling of belonging
2. *Influence*—a sense of mattering, of making a difference to a group and of the group mattering to its members
3. *Integration and fulfillment of needs*—the feeling that members' needs will be met by the resources received through membership in the group
4. *Shared emotional connection*—the commitment and belief that members have shared and will share history, common places, time together, and similar experience (McMillan and Chavis 1986, 9)

Membership. McMillan and Chavis (1986) have described five aspects of community membership—a sense of belonging, boundaries, emotional safety, personal investment, and a common symbol system. Community membership gives one a feeling of belonging. The sense of belonging and identification includes feelings, expectations, and beliefs that are consonant with those of other community members. The boundaries of a community allow some people to belong whereas others do not. The community gives its members a sense of structure and security. Members feel personally invested in the community. Finally, a common symbol system is important in creating and maintaining the sense of community (9–11).

Influence. In a community, the individual both influences and is influenced by the group members. McMillan and Chavis (1986) outlined the following postulates concerning influence:

1. Members are more attracted to a community in which they feel that they are influential.
2. There is a significant positive relationship between cohesiveness and a community's influence on its members to conform. Thus, both conformity and community influence on members indicate the strength of the bond.
3. The pressure for conformity and uniformity comes from the needs of the individual and the community for consensual validation. Thus, conformity serves as a force for closeness as well as an indicator of cohesiveness.
4. Influence of a member on the community and influence of the community on a member operate concurrently, and one might expect to see the force of both operating simultaneously in a tightly knit community. (12)

Integration and Fulfillment of Needs. Integration and fulfillment of needs is reinforcement of the shared values of the community. For a sense of community to exist, reinforcement must be used to bind members of the community together. According to McMillan and Chavis (1986),

1. Reinforcement and need fulfillment is a primary function of a strong community.
2. Some of the rewards that are effective reinforcers of communities are status, membership, success of the community, and competencies or capabilities of other members.
3. There are many other undocumented needs that communities fill, but individual values are the source of these needs. The extent to which individual values are shared among community members will determine the ability of a community to organize and prioritize its need-fulfillment activities.
4. A strong community is able to fit people together so that people meet others' needs while they meet their own. (13)

Emotional Connection. Emotional connection is the connection or identification of members of the community with the history of the community and with each other. McMillan and Chavis (1986) made the following assertions about emotional connection in communities:

1. *Contact hypothesis:* The more people interact, the more likely they are to become close.
2. *Quality of interaction:* The more positive the experience and the relationships, the greater the bond. Success facilitates cohesion.

3. *Closure to events:* If the interaction is ambiguous and the community's tasks are left unsolved, group cohesiveness will be inhibited.
4. *Shared valent event hypothesis:* The more important the shared event is to those involved, the greater the community bond. For example, there appears to be a tremendous bonding among people who experience a crisis together.
5. *Effect of honor and humiliation on community members:* Reward or humiliation in the presence of community members has a significant impact on attractiveness (or adverseness) of the community to the person.
6. *Spiritual bond:* This is present to some degree in all communities. (13–14)

Solid communities provide members with avenues to interact, events to share, and ways to recognize members. They also give members opportunities to invest in the community and to bond with other members (14).

Sense of Community and Community Policing

The concept of a sense of community has importance to the community policing strategy. Community policing will be much more successful in a place that has a sense of community than in one that does not. For community policing to be a success, the police and the community must work together to maintain or improve the quality of life of the community members.

Generally, communities organize around needs, and members of communities join communities that meet their needs. Many successful communities have associations to represent their interests (McMillan and Chavis 1986, 15–16). Such interests might include dealing with crime in the community. Community policing officers should be aware of the community associations and make contact with them to find out what they consider to be safety and crime issues. The community policing officer who becomes aware of a community's concern about safety and crime issues may be able to implement crime prevention strategies or initiate problem-solving strategies before a crisis occurs.

In a study of the relationship between a sense of community and individual well-being, it was found that people who had a strong sense of community scored high on social well-being and happiness and that a sense of community contributed to people's happiness. Also, people with a strong sense of community had a better grasp of how to use the resources of their community than people who had no sense of community (Davidson and Cotter 1991, 251). It is important that community policing officers recognize the contribution of a sense of community to the success and happiness of members of a community and that they encourage developments that enable members of the community to help each other.

Empowerment

Empowerment is "the process by which people organize, attain a collective objective, and learn about their own personal power" (Heskin 1991, 63). An empowered community is structured in a way that provides members with the opportunity to participate in community activities and to be responsible for the community, and its members are willing to use that structure (Heskin 1991, 64). Community empowerment begins with community organizations and groups that take an activist role with a common purpose. Such organizations have deep and strong roots in America. They draw on

- the American traditions of self-help and the building of voluntary associations
- the civil rights and self-determination movements among blacks, Hispanics, and other minority and ethnic groups
- the churches' quest for greater social justice and a renewed sense of community
- the neighborhood movement and people's determination to resist outside institutions that injure or neglect their neighborhoods and to build their own institutions
- the desire, throughout America, for new vehicles to give people greater control over the political process and economic decision making
- the desire for new, human-scale organizations and "mediating structures" in a society in which too much has been left to massive, unaccountable, alienating, and often ineffective institutions (Mott 1992, 12)

Community policing officers should be aware that community organizations empower the community and that this movement has a long tradition. Community organizations can make the community policing officer's job easier, and in turn the community policing officer can be helpful to community organizations. Consequently, the intelligent community policing officer will respect the organizations within the community, learn about their mission and goals, and work to develop a partnership with them. Police agencies that have adopted the community policing philosophy should recognize that community groups have solved some very tough problems and have been successful in taking on street crime and criminal justice issues. Community organizations have demonstrated that they can play major roles with great effectiveness, including

> 1) organizing and mobilizing people to press for improvements; 2) creating self-help programs to provide community services and develop housing and economic development projects; and 3) pursuing changes in public policy through voter registration, research, education, litigation, and legislative and political action. (Mott 1992, 12)

Organizations typically go through stages of community building before they can become fully empowered. Jason (1997) described four stages:

> First, there is pseudo-community, in which people avoid disagreements and just pretend to be a community. During the next stage (chaos), healers attempt to heal and convert others; this is a time of considerable fighting and struggle. During the third stage (emptiness), people begin to remove barriers to communication, such as expectations, prejudice, and ideologies. During the final stage—aptly called community—a peaceful, soft sense of quietude descends. People feel safe to share their vulnerability, their sadness and their joy. (73–74)

One way that community organizations can become more empowered is by periodically reappraising their current activities to determine if they are still relevant and effective. Mott (1992) offered several basic rules to be followed in this process:

1. *Learn, synthesize, and innovate.* Groups should look at [the reappraisal as an opportunity for] learning and synthesis, knitting together the strongest elements of the approaches that different organizations have followed.
2. *Let the leaders lead.* Groups should deeply involve their leadership and membership, as well as their staffs, in the appraisal. This should be seen not only as a matter of principle for community-based organizations, but also as a matter of hard pragmatism. People from the community are far more likely to take a tough, dispassionate look at their needs and the advantages of alternative courses of action than are staff people who have records, budgets, and jobs to defend.
3. *Go back to basics.* The reexamination should start by going back to the long-range, fundamental goals that led to creation of the organization. That vision undoubtedly is far larger than the sum of the organization's current activities. Is the organization on track, or should it make substantial changes to gain the power and capacity to achieve those goals over time?
4. *Go beyond localism.* The group should also raise its sights beyond the neighborhood. To gain the resources, allies, and institutional changes that the community will need if there is to be either development or social progress, the group must decide how it can have an impact that reaches beyond its borders and affects government and private institutional policies. (14)

Generally, members of the community are the core of community organizations, but nonmembers may also play a role, including organizational helpers such as consultants, professionals, and volunteers; government officials, such as public housing officials; and, most recently, community policing officers.

The role of community members in a given program as compared to that of outside professionals or officials will vary depending on the organization and its influence. Wandersman (1981) has described the following patterns:

- *Creation of parameters and objects:* [Community members have] the decision-making power and generate plans without preconceived parameters from professionals or others.
- *Self-planning:* [Community members] generate alternative plans within available parameters and [have] the responsibility and decision-making power for decisions. (The professional can play a consultant role.)
- *Choice:* [Community members] choose between alternative plans generated by professionals or government officials.
- *Feedback:* [Community members are] asked for [their] ideas and opinions about a plan. This information is evaluated by the professionals or government officials who [have] the responsibility and decision-making power.
- *No participation:* The decisions are made by the professional or government officials for the [community]. (45)

Generally, community policing officers should encourage a high degree of participation and responsibility for programs among community members. Techniques of participation such as surveys, public hearings, or game simulation all allow members of the community to have input on issues that they consider to be important in their community. Participation in community organizations can affect the individual and the community in a variety of ways:

1. Individual Level
   - *Evaluation of the community organization:* degree of satisfaction or liking of the organization and evaluation of specific qualities of the organization
   - *Evaluation of the community:* degree of satisfaction or liking of the community and evaluation of specific qualities of the community
   - *Attitudes toward participation:* favorable or unfavorable views about participation in community organizations generally
   - *Attitudes toward the authority figures:* attitudes about the planners', leaders', and politicians' concern for the individual
   - *Feelings about oneself and one's role in the community:* feelings of creativity, alienation, responsibility, anonymity, helpfulness, or being "a good citizen"
   - *Behavioral effects:* interactions with the social and physical environment of the community, including use of community services and facilities, investment in home improvements, mobility, and contact among neighbors
2. Organizational Level
   - *Organizational growth:* increased membership, attendance at meetings and actions, resources (e.g., money, in-kind contributions), and new chapters or affiliates
   - *Leadership development:* advancement of group leaders to more powerful positions within the community and city, development of secondary leadership, and emergence of leadership skills
   - *Legitimacy:* amount of publicity for the organization, representativeness of the organization, degree of credibility, and organizational image among nonmembers and elites
   - *Adaptiveness:* flexibility in response to changing the environment, development of collective problem-solving techniques, and creation of new organizational capabilities, new issues, committees, structures, etc.
   - *Congruence between the goals and the activities of the organization:* extent to which the organization is achieving its goals as well as any additional goals individual members may have (Pearlman 1979, 43–63)
3. Community Level
   - *Environmental effects*—changes, for example, in noise and density levels
   - *Ecological effects*—changes such as number of settings or facilities available
   - *Social effects*—changes in the social characteristics of the community, such as composition of the population, social networks, norms, and functioning (e.g., crime rates, quality of service delivery) (Wandersman 1981, 48–49)

Community policing officers should be aware of the various effects of participation between the various levels. The community policing officers should know the nature and extent of the interrelationships between the various levels. It is important that community policing officers become integrated into the community. The more the community policing officers know and understand the nuances of community organizations, the better they will be able to understand community organizations and to achieve the goals of the community policing philosophy.

The role of the police in community empowerment has perhaps been best summarized by Cheryl A. Steele, program coordinator for the community-oriented policing program in Spokane, Washington, and founder of Spokane's first volunteer substation. She tells the police that "it is all right to make a commitment to do community policing as an agency—but don't think you are going to do it *to* your community. You must get community members to drive the initiative for themselves. In doing so, the stewardship of public safety is then totally owned by the entire community" (personal communication, June 17, 1998). Rather than trying to fix all the social ills of a community, police should concentrate on what they know the most about—law enforcement and crime prevention. The police need to call community meetings, release the information, and then participate in the changes for which they have the resources. Steele calls this the "EMOM" model—Educate, Mobilize, Organize, and Move out of the way!

THE NEIGHBORHOOD

The study of urban centers by sociologists has tended to focus on neighborhoods, or small geographical areas that are embedded in a larger area, such as a town or city, and that include residential dwellings. By one definition,

> The most sensible way to locate the neighborhood is to ask people where it is, for people spend much time fixing its boundaries. Gangs mark its turf. Old people watch for its new faces. Children figure out safe routes between home and school. People walk their dogs through their neighborhood, but rarely beyond it. Above all, the neighborhood has a name: Hyde Park or Lake View in Chicago; Roxbury, Jamaica Plain, or Beacon Hill in Boston. (Williams 1985, 33)

An individual might describe his or her neighborhood by referring to such landmarks and geographic features as rivers, lakes, expressways, parks, schools, and malls. The terms *neighborhood* and *community* are often used synonymously, but the term *community* can be used to describe larger areas as well, such as an entire town or city that contains many neighborhoods.

Bursik and Grasmick (1993) have stated that a neighborhood has some degree of "collective life that emerges from the social networks that have arisen among the residents and the sets of institutional arrangements that overlap these networks. That is, the neighborhood is inhabited by people who perceive themselves to have a common interest in that area and to whom a common life is available" (6). They also state that "the neighborhood has some tradition of identity and continuity over time" (6). But neighborhood residents do not necessarily share a culture, associate cooperatively, or form many kinds of ties with each other.

Wellman and Leighton (1988) have discussed several problems with the assumption that a neighborhood is, automatically, some kind of "container for communal ties." They argued that "even though space-time costs encourage some relationships to be local," it does not necessarily follow that the neighborhood has a well-developed social organization. Further, "even the presence of many local relationships does not necessarily create discrete neighborhoods," because "there may well be overlapping sets of local ties, the range of these ties being affected by the needs and physical mobility of the participants. And "there are important ties outside of the neighborhood even in the most 'institutionally complete community' "—especially work relationships (59).

In other words, a sense of community does not necessarily coincide with the divisions marks on maps. As Wellman and Leighton (1988) argued, "administrative officials have imposed their own definitions of neighborhood boundaries upon the maps in attempts to create bureaucratic units. Spatial areas, labeled and treated as coherent neighborhoods, have come to be regarded as natural phenomena" (58). Further, "territory has come to be seen as the inherently most important organizing factor in urban social relations rather than just one potentially important factor" (58).

A neighborhood can develop interpersonal ties, networks of sociability and support, sentiments of solidarity, and shared activities and, if residents focus their energies in this direction, it can become a political entity if residents perceive it as a base for building power. Community policing officers are often assigned to specific neighborhoods. They need to be aware of the extent to which the neighborhood they serve is a cohesive social unit.

A neighborhood that offers many opportunities for sociability, from block parties to front-porch visiting, for the exercise of interpersonal influence through opinion sharing, for mutual aid among neighbors in time of need or emergencies, for the discussion of common concerns, for the development of local pride, and for the achievement of status among neighbors already has many resources and strengths that a community policing officer can draw upon (Williams 1985, 37).

Neighborhood Crime

The U.S. Department of Justice, through its Office of Justice Programs, has surveyed neighborhood residents from 1985 to 1995 (surveying every other year beginning in 1985) on their perceptions of neighborhood crime. In 1995, 7 percent of neighborhood households identified crime as a neighborhood problem, but the figure was higher for central city residents and particularly for black central city residents. For public housing residents specifically, figures were still higher: 25 percent of black public housing residents and 13 percent of white public housing residents. The perception of crime vulnerability usually reflected the actual likelihood of criminal victimization. From 1992 to 1995, a larger percentage of black urban households (9 percent) than white urban households (7 percent) experienced nonlethal violent crimes, and in 1995, 8 percent of public housing households were victimized by violent criminal acts compared to 6 percent of non–public housing households (De Frances and Smith 1998, 1–2).

Neighborhood organizations are frequently established to combat the crime problem. The neighborhoods that are most likely to formally establish neighborhood anti-crime organizations tend to have the following characteristics:

1. They are of moderate size in the urban environment, several square blocks to perhaps a square mile.
2. Their boundaries are fairly clear.
3. Their inhabitants are somewhat homogeneous in background, lifestyle, and income.
4. Residents perceive the neighborhood as a good place to live, at least more desirable than bordering neighborhoods. Their neighborhood in fact may be quite deteriorated, however.
5. Residents feel threats to their neighborhood's stability.
6. Residents are somewhat optimistic about the future of their neighborhood; this optimism is based on a sense that they control the neighborhood social order to an extent.
7. There are preexisting organized subgroupings, although they must interact very little with each other. (Williams 1985, 43)

There are several theories about the causes of crime in neighborhoods. One is that crime could be caused by the unwillingness or inability of neighborhood residents to exert social control by informal means over the public behavior of residents or non-residents in their neighborhood. Another is that economic deprivation could possibly be a cause of crime by operating as a justification for alienation and reducing an individual's future opportunities. On the basis of both theories, action taken to alter the neighborhood environment can reduce crime (Williams 1985, 176). And if the neighborhood environment plays a role in crime, then neighborhood organizations, aided by community policing officers, have an important role to play. Neighborhood organizations can deter crime in several ways:

> First, they organize residents to watch out for each other (for children, the elderly, one another's homes, and so forth) on an informal basis. A variant of this is the Whistlestop Program, in which residents buy whistles for a dollar or so at a block meeting. One blows the whistle if one is attacked; the goal is to startle and repel the offender while at the same time summoning help.
>
> Second, some neighborhood organizations create more formal civilian patrols. These groups work closely with police, carry walkie-talkies, and do not attempt to apprehend suspects. This last point differentiates them from vigilante groups. They may walk or ride in cars; they may be tenants in a

housing project or taxi drivers. They function as people taking some formal responsibility for attending to public behavior. When behavior improves, these patrols often cease.

Third, the ordinary operations of the neighborhood organization in creating block clubs, disseminating information about the neighborhood to its residents, and involving them in its neighborhood-wide activities, all help to weave social fabric. . . . The sense of neighborhood identity—that one not only resides in an area, but that one belongs to and therefore cares about it— is a less tangible but extremely important outcome of everything the neighborhood organization does.

Fourth, neighborhood organizations squelch crime in cooperation with the police. This is perhaps what residents might think of first but for a variety of reasons is the most difficult to accomplish. Police assistance is essential, of course, for those situations involving public behavior or outright criminal activity that cannot be handled by residents alone. (Williams 1985, 177–178)

Traditionally, neighborhood cooperation with the police has been hard to achieve in the low-income neighborhoods that are likely to have the most crime because police and residents have had a relationship much like that between white colonizers and black African natives: police have been seen as "motorized bwanas cruising, hunting, and foraging among the natives in the urban jungle" (Williams 1985, 178). Under the traditional model of policing, police have distanced themselves from the neighborhood residents, locking themselves in police cruisers and developing a siege mentality of "us versus them." But the community policing model, unlike the traditional policing model, which segregated the police from the neighborhood, gets the police involved in the neighborhood. Foot patrols are one means of accomplishing this. Also, police act as consultants in helping residents to solve crime and disorder problems. They recognize that residents have a stake in their own safety and can do a great deal to secure it through community action.

Fear of Crime

Not only crime itself but fear of crime is a major issue facing neighborhoods. Fear of crime, as stated earlier, often reflects actual rates of victimization by crimes against persons and crimes against property. But there also exists a third category of offenses that creates a fear of crime among neighborhood residents: the disorderly offenses, which include loud noises, loitering, defacing public property with graffiti, drunkenness, prostitution, drug trafficking, gambling, and rowdiness (Wilson and Kelling 1982, 30). Thus, the fear of crime may in reality be much greater than the actual statistics on crime. Fear of crime can be related to both neighborhood characteristics and individual characteristics such as socioeconomic status, gender, and age. At the neighborhood level, it is often related to the perception that the neighborhood is on the decline (Austin 1991, 81). As discussed in Chapter 5 in the section on the broken windows theory, "Fear of crime has been hypothesized as one factor, perhaps in a chain of other factors, that can lead to abandonment and deterioration of neighborhoods" (Brown and Wycoff 1987, 71).

The fear of crime can lead to a sense of anomie that has four components:

1. A sense of neighbors as strangers
2. A sense of physical, social, and psychological distance from the police, who, especially in a rapidly changing environment, may have an even greater responsibility for being the visible symbol of social control
3. A feeling of powerlessness caused by the sheer size of the city, with the subsequent physical distance from City Hall and the involvement of local government with a vast array of problems, many of which do not bear directly on the neighborhood where a given individual lives
4. A lack of accurate information about the nature and extent of crime in neighborhoods (Brown and Wycoff 1987, 74)

Community policing officers can break down this anomie by reducing their physical, social, and psychological distance from neighborhood residents, and they can reduce feelings of distance from neighbors and of neighborhood powerless by encouraging neighborhoods to organize to protect themselves and develop more familiarity and solidarity among themselves. They can also provide accurate information on neighborhood crime.

Community policing officers can dispel many of the inaccuracies of residents' perceptions of crime in their neighborhoods. But they can only learn about residents' perceptions by interacting with residents and listening to them talk about crime in their neighborhood. Neighborhood police officers should be able to obtain crime statistics from the crime analysis unit of their department for their neighborhood and for specific areas where a reported crime has occurred. They should freely share these statistics with neighborhood residents and organizations. If the neighborhood members or organizations consider the crime statistics to be inaccurate, the officer can inform the residents that the statistics reflect crimes reported by citizens to the police but, that if crimes occur and are not reported to the police, it cannot be expected that the police will have a record of the crime. Therefore, all residents need to report all crimes to the police, no matter how minor, so that the police can have an accurate record of crime for the neighborhood. The police can only concentrate on solving, preventing, and controlling crimes when residents report the crime.

Communities that want to successfully control crime and disorder must develop programs that strengthen neighborhoods by developing strong levels of social control either informally or formally through a network of associations among neighborhood residents and between residents and local governmental and private agencies. The police can assist in creating this network and in channeling outside resources to the neighborhood.

MULTICULTURALISM*

Why Such a Discussion

Multiculturalism is an approach to cultural diversity that emphasizes understanding, respecting, and adapting to other cultures and cultural differences. In today's world, cultures are coming into contact more often and more rapidly. The issues surrounding such contact have major implications for community policing. In addition, changing population distributions in the United States, with increases in percentages of minority populations and a concurrent decrease in the percentage of the majority population, have led to questions about the interaction of these populations in the workplace and in the larger society.

Some futurists have already predicted that by the year 2050, the current minority/majority distribution will be radically changed, with the minority population representing the majority of United States citizens (Henry 1990, 4). These predictions are based on two population phenomena. First, the traditional migration sources for the United States have shifted from the European continent to the Asian Pacific rim, to Central and South America, and to Africa and the Middle East. Changes in immigration law and the percentages of immigrants from these areas have changed the distribution of whites and people of color in the immigration streams. Second, whites and nonwhites have different birth rates. These changes have had a major impact on the composition of the workforce, with a decrease in the number of white males predicted to enter the workforce and an increase in the number of females and nonwhites over the next several years.

These immigrants bring cultural beliefs and practices that may or may not be compatible with the mainstream American culture. What happens when people engage in various forms of cross-cultural contact, and how does this affect community policing?

What Is Culture?

As discussed in Chapter 2, culture includes artifacts, concepts, and behaviors. Humans must display four features in order to be considered as having a culture. There must be an identifiable group, whether identified by geography, language, and/or religion; the group must have a shared set of beliefs and experiences; the group must attach feelings of worth and value to those beliefs and experiences; and there must be a shared interest in a common historical background (Brislin 1981).

Culture is not innate, it is learned. Humans born into one culture but adopted into another culture will be products of the adopted culture. This is especially evident when children are adopted cross-racially. To be learned, culture must therefore be transmissible. Transmission occurs by words and by nonverbal actions. The libraries of a culture exist in the minds of the people. The best way to learn about a culture is through daily, long-term socially intimate contact. Because most humans do not have

*Source: Copyright © 1999, Dr. Anna Chandler.

access to different cultures on this level, it is difficult to learn well all the facets of a different culture. Culture is dynamic, never static; it evolves. It changes through inventions, such as the wheel, the automobile, or the computer; infusion of elements, such as the adoption of words in a language, foods in the diet, or religious practices; and calamities such as war, plague, or natural disasters.

Each culture is selective. Of the myriad possible human behaviors, each culture selects those behaviors that will be acceptable and rewards its members who engage in those behaviors. Conversely, the culture punishes members who engage in behaviors unacceptable to the culture. Because all behaviors cannot be acceptable, different cultures will select different combinations of behaviors as acceptable. Each culture selects the behaviors that are important to that culture. Differences in the acceptability of human behaviors define the boundaries between cultures. Facets of culture are interrelated. Adjustments in any one part of the culture then require adjustments in other parts of the culture.

How Culture Affects Values, Perceptions, and Behaviors

To understand culture, one must recognize that the cultural behaviors exhibited by individuals are the product of two interrelated "histories": the history of the group or culture of which an individual is a member and the personal history of the individual, or the individual's exposure to concepts and his or her treatment by others. These are events over which the individual has little control. Napoleon Bonaparte defined history as "a set of lies agreed upon." History in the United States can be defined as an accident that occurred at an intersection with five corners: the five geopolitical groups of African Americans, Asian Americans, European Americans, Hispanic Americans, and Native Americans. The way in which the history is then reported and recorded is dependent on which corner an individual or group occupies. Thus, historical events such as the Civil War will have different labels, heroes, causes, and perceived outcomes depending on who reports and records them.

Culture also is the mechanism that gives the individual the processes and frameworks for organizing and negotiating the real world. Just as all behaviors cannot be acceptable in a culture, all physical phenomena cannot be processed by individuals in a culture. The culture determines which events in the world must be attended to and dealt with. Three major areas of culture determine how individuals view the world and how individuals treat people different from themselves. These are (1) the language that people are expected to learn, (2) the way in which people are reared and disciplined, and (3) education and the mass media.

Language

There is a relationship between language and interaction with others. Language gives people concepts, structures, and vocabulary that form the basis for reacting to others, especially to outgroups and to strangers. Limits are placed on interaction by language. People speaking different languages experience the world in very different ways. For example, the Eskimo have many words for snow, because their survival is dependent upon being able not only to accurately label the type of snow but to engage in the type of behavior that will ensure their survival in that type of snow. Cultures in less wintry climates have no need to have such precise definitions. So language gives individuals and the culture the labels of various physical phenomena in the world, the descriptions of them, and the expected behavior when encountering them. Unless the phenomena can be labeled, described, and reacted to, then for all intents and purposes they do not exist.

Culture includes values, technology, and religious practices as well as a particular and peculiar cognitive structure. Language, then, extends beyond being able to recognize and reproduce sounds and words. It is more than mere vocal skills; it is the worldview of the group. Langston Hughes said, "The limits of my language are the limits of my world" (Brislin 1981, 34).

Some aspects of language that involve outgroup members define the treatment of outgroups. In English, certain structural characteristics facilitate the development of concepts that have an effect on how outsiders are perceived and help define how they should be treated. First, words that are used to describe individuals can be used to describe groups. This allows ease in expressing generalizations from the observation of one individual to the group to which that individual belongs. It also makes it easy to form and to think in terms of stereotypes. Second, the use of collective nouns without qualifications contributes to stereotypes and generalizations and masks any variation of individuals within the group. Because any qualification would require explanation and complex grammatical construction, people find it easier to talk about groups in the same way that individuals are talked about. It is also easier to use categories that listeners are certain to understand.

Color words are also a source of reference for treatment of outgroups. Certain color words used to label skin color have negative connotations in English. *Black* is one such word. Children learn the connotation as part of the language and then transfer the connotation to people who have the skin color. Color symbolism in the Western tradition reveals a tendency to use *black* when connoting evil and *white* when connoting good. Concepts associated with *black* include woe, gloom, darkness, dread, death, terror, horror, curse, mourning, and mortification. It is no accident that mourners wear black to funerals, or that the day of the stock market crash was labeled Black Tuesday, or that the series of virulent diseases that wiped out much of the population of Europe was labeled the Black Plague.

Other color words have positive connotations. *White* in the Western tradition has been associated with triumph, light, innocence, joy, divine power, purity, regeneration, happiness, gaiety, peace, chastity, truth, modesty, femininity, and delicacy (Gergen 1997, 397). It is no accident that brides wear white, or that souls are washed whiter than snow, or that the heroes wear white hats. In the 1950s, Chrysler Corporation had an advertising slogan, "All us good guys wear white hats!" When questioned, their reply was that they were following the Western movie tradition in which the hero rode in on a white horse wearing a white hat. They acknowledged that some heroes wore black hats but stated that the white hat made the heroes instantly recognizable. This use of color to connote good and evil extends even into the Saturday morning cartoons, where the villains are always in black and the heroes are always in white. One manifestation of this connotation is that among darker skinned people, those whose skin is somewhat less dark are more desirable marriage partners (Brislin 1981, 32).

Affective connotation exists in all languages and has three dimensions: evaluation, potency, and activity. People from different cultures rely on these same dimensions when explaining the meaning of concepts. "Outgroup" is a concept, and members of the outgroups are reacted to as other concepts. When encountering a member of an outgroup, people pose three questions based on the three dimensions: (1) Does this person seem good or bad? (2) Does this person seem weak or strong compared to me? and (3) Do the actions of this person require passivity or activity on my part? Depending on the perceived appearance of the person, contact or no contact will occur (Osgood and Osgood 1975, 102).

Because the decision for contact cannot be based on long-term socially intimate interaction, it must be based on the concept that this person matches. That concept has a label, a description, and a behavior already embedded in the culturally prescribed worldview of the group. It should come as no surprise that women will clutch their purses when confronted by young black males. The culturally prescribed worldview is that they are deviant, dangerous, and to be avoided and feared. This may or may not be true of individuals, but the decision to find out has already been made by the cultural definition of them. Conversely, contact may be attempted with a person from a group, such as bankers or lawyers, who are labeled as safety by the culturally prescribed worldview but who actually may be deviant, dangerous, and to be avoided at all costs.

Phrases that use the names of the different ethnic groups as referents contribute to the tendency to attribute specific characteristics to ethnic groups. Such phrases as "Jew him down" reinforce stereotypes that discourage interaction with outgroup members. A study done by the National Conference of Christians and Jews (1994) supported the existence of this use of referents and pointed to its universality among the ethnic groups of the United States. All ethnic and religious groups were found to have stereotypical views of all other ethnic groups and/or religions and to base their decisions about the treatment of those groups and members of those groups on these stereotypes.

Child-Rearing Practices

There is a relationship between parents' child-rearing practices, especially discipline practices, and attitudes toward outgroups (Brislin 1981, 34). Adults acquire their initial attitudes about outgroups from their parents. Parents control access to outgroups. Given the housing and other population distribution patterns in the United States, children generally have little long-term socially intimate contact with outgroups. Children raised in cultures with traditional and conservative norms generally exhibit negative attitudes toward outgroups. These norms include such practices as shunning or declaring the child dead to the group if behavior deemed deviant, such as marriage to an outgroup member, occurs. In cultures where there is little encouragement for children's self-development and where children are to follow the occupational and living patterns of the parents without question, there may be negative attitudes toward outgroups. A pattern of dominance by adults such that the child is in a subservient position and punishment is ensured if the child does not conform with the parents' wishes has also been associated with negative attitudes toward outgroups. Also, when adults ignore children, keep them from becoming an important part of the family, and disregard their needs, negative attitudes toward outgroups may develop. Children have not developed the sophistication or had the experience to allow them to compare their treatment with the treatment of others in the larger world. Adults who experience mistreatment as children acquire the point of view that such treatment is normal and treat outgroup members accordingly (Brislin 1981, 35).

Much has been written and debated about the legacy of child abuse. It is speculated that children who are abused often go on to become abusers. Because most of such abuse occurs within the confines of family, the attitudes developed by children in such situations have major impact on how they will interact with other members of society, especially outgroup members.

Education and the Mass Media

Two features of schools have a major impact on children's attitudes toward outgroups. One is the organization and operation of the schools themselves. The age-stratified organization of schools reflects nothing of real life. Also schools with very punitive discipline policies and disparate application of those policies generally have been shown to produce students with negative attitudes toward outgroups (Stephan and Rosenfield 1978, 802). Some experts have gone so far as to suggest that the organization of schools and the treatment of children in schools are tantamount to the teaching of racism ("U.S. Schools Teach Racism" 1992, 1D, 3D).

However, the feature of schools that has the most far-reaching and long-lasting impact on children's attitudes is the school curriculum and the materials used to teach it. Beginning with the initial introduction to reading and mathematics, the materials, counting rhymes, and illustrations teach children that some ethnic groups are to be treated as objects. Also, the illustrations used in the material bear little resemblance to real life in America. They are based on nuclear families with the father employed and the mother as a full-time homemaker. Currently only about 30 percent of students in America's schools reflect this family organization.

Furthermore, life in the United States is portrayed as easy and comfortable. The people, who generally are white, blond, and of northern European origin, are kind and generous. The underlying suggestions are that there may be other kinds of people in the world, but they live in faraway places and in faraway times, and they certainly have no place in the American scene (Klineberg 1964). When any attempt is made to include other ethnic groups, it is usually unrealistic, such as one black or Asian family in the all white neighborhood. Also, in the teaching of the development of the country, little or no mention is made of groups other than Europeans in the building of the United States beyond the mention of slavery, the Civil War, and Martin Luther King, Jr.

Finally, the following five myths of America are taught and inculcated into students' psyches every day:

1. *America is great because of individualism or reliance on oneself.* We hold up icons such as Horatio Alger as the hallmark of all that is good and proper in this

country. Competition rather than cooperation is the guiding value. Contrary to this myth, cultures in this country know and understand that cooperation is the key to survival, and the idea of competition is antithetical to them. If these icons were truly examined by an unjaundiced eye, one would know and understand that they, too, built on cooperation.

2. *We must distrust foreign alliances and be concerned only with internal affairs.* From the isolationist views prior to World War I to the current debate of our role in conflicts in Europe, Africa, the Middle East, and the Far East, the sentiments expressed reflect this stance. Growing out of a sense of betrayal and a sense that foreign alliances would require participation in other peoples' wars that surfaced during the time of the Revolutionary War, this debate has guided the country's foreign policy and the conduct of that foreign policy to the present.

3. *America is a "melting pot," and all cultures need to divest themselves of "foreignness" and become "Americans."* In fact, it was once predicted that by the year 2000 all Americans would be a uniform color somewhat akin to weak tea. But this can scarcely happen if certain groups are not allowed into the mix. Further, because culture defines not only who we are but also how we view, organize, and behave in our world, the melting-pot ideal is unattainable.

4. *It is acceptable to impose one's views on others.* This myth is evident in the doctrine of "manifest destiny" that governed the settlement of the American West and in incidents that led to the Spanish-American war and the Monroe Doctrine. It also undergirded the entire institution of slavery, Indian removal, and the importation of Asian laborers into the West.

5. *The status quo is perfectly acceptable in race and in gender relations.* The question is often asked, "What do 'you people' want?" This rhetorical statement reflects the lack of knowledge and understanding of the questioners. (Brislin 1981, 18–30)

The mass media reinforce the same stereotypes and display unrealistic attitudes about outgroups that are found in school materials. The depiction of minorities in the media is most often a caricature. A caricature may be defined as a pictorial representation in which the subject's distinctive features or peculiarities are intentionally distorted or exaggerated to generate a comic or grotesque effect (*Webster's II* 1984, 231). This caricature of minorities dehumanizes, but it is inspired and made acceptable by earlier dehumanizing influences, namely an absence of information on who minorities are and where they have been (Brislin 1981, 30).

In analyses of the depiction of minorities in print media in one midwestern city, it was found that Native Americans and Asian Americans were virtually invisible in the local paper, whereas Hispanic Americans were somewhat more visible and African Americans were highly visible. However, the areas in which African Americans had a major presence were crime, sports, and entertainment; their visibility in business and government was low. This was in a city with sizeable populations of the four above-mentioned minorities. Ironically, the city itself had been named for a Native American nation and was located in a state named for a Native American nation. This is hardly surprising when in 1993 just 5 percent of reporters in print media were minority and 45 percent of newspapers in the United States had no minority reporters on staff (Chideya 1995, xv).

But the most influential area of the media is television. As of 1990, 95 percent of all homes in United States had at least one television set. Children spend more time watching television than they spend in any other activity except sleep (Brislin 1981, 37). It is estimated that by the age of 18, children in the United States will have spent more hours in television viewing than in classroom instruction. Television reinforces the same stereotypes and displays the same attitudes toward outgroups that are found in school materials and in the print media.

Not only the content but the structure of television material has a major effect on how children view the world and especially outgroups. Issues are simplified to be covered in a small amount of time, and problems can be solved within a time slot. Happy endings are common, and complex problems are resolved to everyone's satisfaction. Characters are overdrawn, larger than life, and clearly labeled. The world of television bears little resemblance to real life (Brislin 1981, 37).

Cross-Cultural Communication

Much of the breakdown in cross-cultural communication has to do with cultures' differing organization of time and space. Organization of time includes the concepts both of how time is organized and of how it should be used. Different cultures have distinctive ways of interacting with time. Even the verbs used to describe how time operates differ from culture to culture and reflect the definite emphasis and value each culture places on time (Armas 1975). Other areas of difference concern whether the culture is more monochronic or polychronic in the use of time—that is, whether only one activity is planned for a given time frame or whether many activities occur in the same time frame—and whether the planned activity consumes the amount of time allotted or consumes as much time as required to complete the task (Levine 1998, 20–25).

Organization of space includes both physical and psychological space. Issues to be considered are positioning of bodies in conversation, touching of people not closely related, taking turns in conversation, and the like. Psychological space includes eye contact, forms of address, deference to elders and persons in authority, and body language and gestures (Garvey 1992, 58–65).

In an issue of the *FBI Law Enforcement Bulletin,* Weaver (1992) argued that

> to better serve citizens from increasingly diverse backgrounds, law enforcement officers need to understand the cultural aspects of communication and behavior. Frustration will only mount if the criminal justice community ignores diversity or assumes that it can continue to function according to traditional expectations and norms. In short, officers need to know the dynamics of cross-cultural communications. (1)

Weaver listed four assumptions that keep the law enforcement community from being effective in a diverse society: (1) as the workforce becomes more diverse, differences become less important; (2) we're all in the same American "melting" pot; (3) it's just a matter of communication and common sense; and (4) conflict is conflict, regardless of culture. He explained the fallacies of these assumptions and listed steps that the law enforcement community can take to be a more culturally aware workforce:

1. Know your own culture and the cultural biases from which you operate.
2. Learn about the different cultures found within the agencies and the community, while avoiding the cookbook approach that lumps similar cultures together.
3. Understand the dynamics of cross-cultural communication, adjustment, and conflict; this will give officers a sense of control and allow them to develop coping strategies.
4. Develop cross-cultural communicative, analytical, and interpretative skills through ongoing experience with the cultural groups within the community.

The issue, according to Weaver (1992), centers not on eliminating diversity but rather on managing it and learning from it (2–7).

SUMMARY

A community is a cluster of people living within a continuous geographical area who share, to some extent, a way of life. It is a local territorial group that is also a social group. A community can be a small aggregation, such as a neighborhood, village, or town, or a large one, such as a city. A neighborhood is a small geographical area that is embedded in a larger area, such as a town or a city, that includes residential dwellings. As is the case for communities in general, the extent to which it is a cohesive social unit will vary.

For community policing to be successful, there must exist a "sense of community." Officers involved in community policing must have an understanding of what factors create a sense of community, how police can assist in creating a sense of community, and the extent to which the area they serve functions as a cohesive social unit.

Neighborhood organizations have an important role to play in deterring crime. They can organize residents to watch out for each other and form civilian patrols. They can work to clean up the neighborhood, thereby sending the message that residents care about their community and will not tolerate crime and disorder. Finally, by organizing neighborhood activities, they can increase neighborhood solidarity and residents' familiarity with and concern for each other.

The role of community policing officers in these efforts is to encourage neighborhood organizing, provide information and guidance related to crime and crime prevention, and assist in channeling outside resources to the neighborhood and networking with outside agencies. Multiculturalism is an approach to cultural diversity that emphasizes understanding, respecting, and adapting to other cultures and cultural differences. It is important for all police personnel, especially those involved in community policing, to become aware of and sensitive to diversity issues and to develop cross-cultural communication skills.

KEY TERMS

community empowerment   multiculturalism   sense of community
community organizations   neighborhood

REVIEW QUESTIONS

1. What is a community?
2. What is meant by *sense of community*?
3. What is meant by *empowerment*?
4. What does the term *neighborhood* mean?
5. Why are neighborhood organizations important?
6. How can police help to develop a sense of community in the areas they serve?
7. Why is multicultural awareness important in community policing?

REFERENCES

Armas, J. 1975. Antonia and the mayor: A cultural review of the film. *Journal of Ethnic Studies,* Fall.

Austin, M.D. 1991. Neighborhood attributes and crime prevention activities. *Criminal Justice Review* 16, no. 1:18.

Brislin, R.W. 1981. *Cross cultural encounters: Face to face interaction.* New York: Pergamon.

Brown, L.P., and M.A. Wycoff. 1987. Policing Houston: Reducing fear and improving service. *Crime and Delinquency* 33, no. 1:71.

Bursik, R.M., Jr., and H.G. Grasmick. 1993. *Neighborhoods and crime: The dimensions of effective community control.* New York: Lexington.

Chideya, F. 1995. *Don't believe the hype: Fighting cultural misinformation about African Americans.* New York: Penguin.

Davidson, W.B., and P.R. Cotter. 1991. The relationship between sense of community and subjective well-being: A first look. *Journal of Community Psychology* 19 (July):251.

De Frances, C.J., and S.K. Smith. 1998. *Perceptions of neighborhood crime, 1995.* Washington, DC: U.S. Department of Justice.

Garvey, R. 1992. Foreign exchange. *USAir Magazine* (June):58–65.

Gergen, K. 1997. The significance of skin color in international relations. *Daedalus* 96:397.

Gusfield, J. 1975. *Community: A critical response.* New York: Harper & Row.

Henry, W.A., III. 1990. The browning of America. *Time* 137, no. 4:4.

Heskin, A.D. 1991. *The struggle for community.* San Francisco: Westview.

Jason, L.A. 1997. *Community building: Values for a sustained future.* Westport, CT: Praeger.

Klineberg, O. 1964. *The human dimension in international relations.* New York: Holt, Rinehart & Winston.

Levine, R. 1998. Re-learning to tell time. *American Demographics* 20 (January):20–25.

MacIver, R.M. 1936. *Community: A sociological study.* London: Macmillan.

McMillan, D.W., and D.M. Chavis. 1986. Sense of community: A definition and theory. *Journal of Community Psychology* 14 (January):9–16.

McNall, S., and S.A. McNall. 1992. *Sociology.* Englewood Cliffs, NJ: Prentice Hall.

Mott, A.H. 1992. The decades ahead for community organizations. *Social Policy* 16, no. 4:12.

National Conference of Christians and Jews. 1994. *Taking America's pulse: The National Survey on Inter-Group Relations.*

Nisbet, R. 1966. *The sociological tradition.* New York: Basic Books.

Osgood, M.W., and M.M. Osgood. 1975. *Cross-cultural universals of affective meaning.* Urbana: University of Illinois Press.

Pearlman, J. 1979. Neighborhood research: A proposed methodology. *South Atlantic Urban Studies* 4:43–63.

Stephan W., and D. Rosenfield. 1978. Effects of desegregation on racial attitudes. *Journal of Personality and Social Psychology* 36:802.

U.S. schools teach racism, educator tells conference. 1992. *Wichita Eagle,* September 11, pp. 1D, 3D.

Wandersman, A. 1981. A framework of participation in community organizations. *Journal of Applied Behavioral Sciences* 17, no. 1:43–53.

Weaver, G. 1992. Law enforcement in a culturally diverse society. *FBI Law Enforcement Bulletin* 61, no. 9:1.

Webster's II: New Riverside University Dictionary. 1984. Boston: Houghton Mifflin.

Wellman, B., and B. Leighton. 1988. Networks, neighborhoods, and communities: Approaches to the study of the community question. In *New perspectives on the American community,* 5th ed., ed. R.L. Warren and L. Lyon. Chicago: Dorsey.

Williams, M.R. 1985. *Neighborhood organizations: Seeds of a new urban life.* Westport, CT: Greenwood.

Wilson, J.Q., and G.L. Kelling. 1982. Broken windows. *Atlantic Monthly* 249, no. 3:30.

Chapter 7

Problem-Oriented Policing

CHAPTER OBJECTIVES

1. Be familiar with the concepts of problem-oriented policing and the aspects of traditional policing that it was reacting against.
2. Be familiar with SARA.
3. Be familiar with the value of problem-oriented policing for community policing.
4. Be familiar with the various aspects of implementing problem-oriented policing.

INTRODUCTION

Problem-oriented policing is an approach to policing that encourages all members of a police agency to engage in proactive problem solving (Eck and Spelman 1988, 5). For all practical purposes, the philosophy of community policing has as its foundation the problem-oriented police design. The primary idea behind problem-oriented policing is police accountability to the community and a focus on addressing community concerns. Community-oriented policing draws on this idea but takes it even further by emphasizing full community participation in the problem-solving process.

BACKGROUND OF PROBLEM-ORIENTED POLICING

Problem-oriented policing first emerged in the late 1970s out of a dissatisfaction with policing under the professional (traditional) model. It can best be understood by examining critiques of traditional policing by the first problem-oriented theorists.

The professional model of policing, as described in detail in Chapter 1, was characterized by a centralized police structure, top-down transmission of orders, a mission of crime fighting, a code of secrecy, adherence to written policies and procedures, close supervision over patrol officers, and separation of the police from citizens. In many police departments, patrol officers would be assigned a different beat each day or night they worked so that they would not become too familiar with the residents and business owners of a particular beat and possibly be corrupted by local influences. This approach placed a heavy emphasis on new operating technologies, such as "(1) patrol forces equipped with cars and radios to create an impression of omnipresence and to respond to incidents of crime and (2) investigative units trained in sophisticated methods of criminal investigation, such as automated fingerprint identification and the use of criminal histories," as means of solving and preventing crime (Moore and Trojanowicz 1988, 5).

In 1979, Herman Goldstein laid the groundwork for the concept of problem-oriented policing. He claimed that attempts to improve policing had for decades been focusing primarily on administrative improvements, such as recruiting better personnel, modernizing equipment, and incorporating business practices into police operations. These improvements were necessary: police departments in the earlier decades of this century were often disorganized, poorly trained, and inefficient (238–239). But under the professional model, the means of efficiency and routinized procedures had become ends in themselves, and police agencies had lost sight of the *effectiveness* of organizational improvements in reducing crime and meeting community needs. Police were devoting most of their resources to responding to calls from citizens, reserving too small a percentage of their time and energy for acting on their own initiative to prevent or reduce community problems. Further, they were failing to draw on the resources of the community, which had "an enormous potential, largely untapped, for reducing the number and magnitude of problems that otherwise become the business of the police" (Goldstein 1990, 15). And within the agencies themselves, they were ignoring "another huge resource: their rank-and-file officers, whose time and talent [had] not been used effectively" (Goldstein 1990, 15).

In 1988, Eck and Spelman offered a similar critique of professional-model policing that focused on its reactive approach to problems. In that approach, the job of the police was simply to respond to service calls. A citizen would call the police dispatcher, and an officer on patrol would be sent to address the citizen's complaint. When officers were not busy answering calls, they were expected to do random preventive patrol, but they could not leave their beat without authorization from a supervisor, and they were not expected to take initiative to solve problems of crime or public disorder. Thus, patrol officers were slaves to 911. The professional model of policing is so ingrained in police culture that many police officers and citizens are still convinced that answering calls is the most important aspect of police work. Citizens also may expect the police to respond to every call for police service, and police departments respond to these expectations. But, as discussed in Chapter 5, there is no need for the police to respond to every call: many calls can be handled over the telephone, or the citizen can be encouraged to come to the police station to make a complaint.

Eck and Spelman coined the term *incident-driven policing* to describe the traditional reactive approach of police patrols. They asserted that incident-driven policing had four characteristics:

> First, it is reactive. Incidents that have already occurred control the workloads of patrol officers and detectives. Most of these incidents are reported by citizens, though some are first detected by the police. There are some exceptions; the work of vice officers and crime prevention officers, for example, is usually proactive. However, these exceptions make up a small portion of police work.

Second, patrol officers and detectives gather information primarily from victims, witnesses, and suspects. The goal of information-gathering is to resolve the incident, sometimes by leading to the identification and arrest of suspects.

Third, the threat of enforcing laws by invoking the legal system is the primary tool of incident-driven policing. Although people call the police to resolve a wide variety of difficulties, the responding officer's primary authority rests with the criminal law. If a solution to the difficulty is found, it is often because the presence of the officer is an explicit threat that the criminal law can be applied. Although efforts have been made to give officers non-criminal law alternatives for resolving certain types of disputes, the criminal law remains their single most important tool.

Finally, the performance of incident-driven police agencies is gauged primarily by aggregate statistics which group a variety of incidents over wide geographic areas. These performance measures include the FBI Index Crimes, clearance rates, arrest rates, drunk driving arrests, and many other sets of numbers. (2)

According to Eck and Spelman, incident-driven policing had several undesirable consequences. Generally, the calls that patrol officers responded to took up a great deal of time with few results. Thus, they caused frustration on the part of patrol officers, who often handled the same calls over and over again without noticing any progress. In addition, the public would become frustrated because they did not see any changes, even though officers had answered the same call on numerous occasions. Incident-driven policing had very limited objectives, and it could only have limited results if police agencies were not involved in problem solving (2).

THE PROBLEM-ORIENTED APPROACH

The new problem-oriented approach set forth by both Goldstein (1979, 1990) and Eck and Spelman (1988) was an attempt to go beyond the limtiations of the professional model by focusing on the ends of policing rather than the means; encouraging police initiative in addressing community problems; attempting to solve persistent problems by analyzing their causes; assessing the appropriateness of current strategies; developing new solutions and monitoring their effectiveness; and engaging community resources and input. Goldstein (1990) has described problem-oriented policing as resting on the following assumptions:

- Policing consists of dealing with a wide range of quite different problems, not just crime.
- These problems are interrelated, and the priority given them must be reassessed rather than ranked in traditional ways.
- Each problem requires a different response, not a generic response that is applied equally to all problems.

- Use of the criminal law is but one means of responding to a problem; it is not the only means.
- Police can accomplish much in working to prevent problems rather than simply responding efficiently to an endless number of incidents that are merely the manifestations of problems.
- Developing an effective response to a problem requires prior analysis rather than simply invoking traditional practices.
- The capacity of the police is extremely limited, despite the impression of omnipotence that the police cultivate and others believe.
- The police role is more akin to that of facilitators, enabling and encouraging the community to maintain its norm governing behavior, rather than the agency that assumes total responsibility for doing so. (179)

Exhibit 7–1 lists Eck and Spelman's guidelines for what a problem-oriented policing agency should do.

According to Eck and Spelman (1988), a problem can be solved in five ways:

1. *A problem can be solved by totally eliminating it.* In this case, effectiveness is measured by the absence of the type of incidents that this problem creates. It is unlikely that most problems can be totally eliminated, but a few can.
2. *A problem can be solved by reducing the number of incidents it creates.* A reduction of incidents stemming from the problem is a major measure of effectiveness.

Exhibit 7–1 What a Problem-Oriented Policing Agency Should Do

1. Focus on problems of concern to the public.
2. Zero in on effectiveness as the primary concern.
3. Be proactive.
4. Be committed to systematic inquiry as the first step in solving substantial problems.
5. Encourage use of rigorous methods in making inquiries.
6. Make full use of the data in police files and the experience of police personnel.
7. Group like incidents together so that they can be addressed as a common problem.
8. Avoid using overly broad labels in groupings.
9. Encourage a broad and uninhibited search for solutions.
10. Acknowledge the limits of the criminal justice system as a response to problems.
11. Identify multiple interests in any one problem and weigh them when analyzing the value of the different responses.
12. Be committed to taking some risks in responding to problems.

Source: Reprinted from J.E. Eck and W. Spelman, *Problem-Solving: Problem-Oriented Policing in Newport News,* p. 3, 1988, United States Department of Justice.

3. *A problem can be solved by reducing the seriousness of the incidents it creates.* Effectiveness of this type of solution is demonstrated by showing that the incidents are less harmful.

4. *A problem can be solved by designing methods for better handling the incidents* (treating participants more humanely, reducing costs, or increasing the effectiveness of incident handling). Improved victim satisfaction, reduced costs, and other measures can show that this type of solution is effective.

5. *A problem can be solved by removing it from police consideration.* The effectiveness of this type of solution can be measured by looking at why the police were handling the problem originally and the rationale for shifting the handling to others. (6)

Obviously, problem solving involves input and cooperation from various individuals, groups, organizations, and agencies. Eck and Spelman (1988) list five principles that should guide the development of a problem-solving process (Exhibit 7–2).

THE SARA MODEL

Eck and Spelman (1988) have outlined a problem-solving process for police agencies that they call the SARA model. Before problem-solving can be initiated, the problem or problems have to be defined. This process, which the authors tested in Newport News, Virginia, has been recommended ever since as *the* process to be used for solving problems in a problem-oriented policing agency. The acronym *SARA* refers to four stages of problem solving:

Exhibit 7–2 A Guide to the Development of the Problem-Solving Process

1. The final process must involve all department members—all ranks and units, all sworn officers, and non-sworn department members—in the identification, study, and resolution of the problems.

2. The final process must foster the use of a wide variety of data sources—from internal records and officers' knowledge to other government agencies and private individuals and organizations—to understand the causes and consequences of the problem.

3. The final process must encourage police departments to work with members of other public and private agencies to obtain diverse, effective, and long-lasting solutions to problems.

4. The final process must be capable of becoming an integral part of police decision making, without creating special units or requiring resources.

5. The final process must be capable of being applied to other law enforcement agencies.

Source: Reprinted from J.E. Eck and W. Spelman, *Problem-Solving: Problem-Oriented Policing in Newport News,* p. 6, 1988, United States Department of Justice.

1. *Scanning*—identifying the problem
2. *Analysis*—learning the problem's causes, scope, and effects
3. *Response*—acting to alleviate the problem
4. *Assessment*—determining whether the response worked (43)

Scanning

The problem-solving process begins with scanning, or identifying a problem. A problem can be defined as "a group of incidents occurring in a community, that are similar in one or more ways, and that are of concern to the police and the public" (Eck and Spelman 1988, 42). The Savannah (Georgia) Police Department (1994) requires that a problem have four attributes before it will be considered for problem-oriented policing:

1. The problem must involve a group of incidents.
2. The problem must involve two or more issues similar in nature.
3. The problem is causing harm or has the potential to cause harm.
4. The public expects the police agency to handle the problem. (5–6)

Herman Goldstein (1990) listed four characteristics that the police commonly use to cluster incidents and thereby identify a problem:

1. *Behavior.* Behavior is probably the most frequently used organizing theme for clustering incidents, especially when the focus is on a citywide problem: e.g., noise, theft of vehicles, sale of drugs, sexual assault, drinking drivers, spousal abuse, runaways.
2. *Territory.* A collection of different behavioral problems, concentrated in a given area, will typically result in the problems being identified by the place name of the area: e.g., Crown Heights neighborhood, Bryant Park, the intersection of Main Street and 2nd Avenue, Garden Village Apartments, Joe's Tavern, the residence at 2458 23rd Street. The problem at the intersection may involve merchants and pedestrians who are bothered by alcoholics, panhandlers, prostitutes, or individuals who are mentally ill and/or homeless. The most frequent problem identified in territorial terms is the low-income housing project, for the obvious reason that it is a convenient, umbrellalike way of describing collectively the wide range of interrelated behavioral problems (burglary, theft, vandalism, arson, truancy, child neglect, assault, liquor violations, drug use and sale) commonly found in such housing complexes. . . .
3. *Persons.* A problem may be identified in terms of offenders, complainants, or victims (e.g., the elderly); in terms of a large class of people (e.g., shoppers), a group (e.g., a gang), a family, or an individual (e.g., an identified repeat offender or a person who persistently harasses a neighborhood); or in terms of a condition that distinguishes the individuals, as would be true in describing a problem as the alcoholics, the homeless, the mentally ill, or the developmentally disabled.

4. *Time.* Both the public and the police tend to define some regularly recurring, predictable problems primarily in terms of an event or by the season, the day of the week, or the hour of the day in which they most often occur: e.g., an annual rock concert, Halloween party, or ethnic festival (any one of which may involve a subset of problems such as assaults, injuries, traffic congestion, sexual harassment, excessive consumption of alcohol or drugs, or potential racist conflict); thefts from parking areas in shopping malls prior to Christmas; burglaries in student housing when unattended in vacation periods; or drunk driving in the period immediately after the closing of bars on weekends. (67–68)

The problem should be described by everyone who is affected by it. The various viewpoints should be reviewed so that the problem description can be as complete as possible, and a collective (community) description should be developed.

Analysis

The next step is analyzing the problem. Data and information should be collected from a variety of sources other than just police records. An analysis asks the questions of who, what, how, when, where, and why. It identifies the people involved: the offenders, the victims, and the parties that should respond. It defines the scope of the problem and pinpoints when and in what locations the problem occurs. It attempts to trace the sequence of events that produces the problem and to identify factors that might precipitate or contribute to the problem: location, individual and group interests, agency and government policies and procedures. Finally, it examines the current community and institutional responses to the problem and the effects of those responses (Savannah Police Department 1994, 5-6–5-7).

Exhibit 7–3 provides a list of questions that can be used as a tool for the scanning and analysis phases of the problem-solving process.

Response

The third step in the problem-solving process is responding to the problem. The Savannah Police Department (1994) defined a problem response as "an action taken that directly and appropriately addresses the issue, event, or condition in question. The response takes into consideration resources, external support, and planning requirements. It seeks to eliminate the problem or reduce it to manageable size" (5–8).

The first step in responding is *conceiving the response.* This requires the cooperative labor of all those affected by the problem. Eck and Spelman (1988) emphasized that "it is extremely important that those who participated in the scanning and analysis of the problem be key collaborators in designing an appropriate response. After all, they know, in critical detail, the element and causes of the problem" (5–9). Community meetings should be held that give the residents or business owners the opportunity to provide input on what they think will work to solve the problem. The goal of this effort should be to develop a workable consensus about a solution.

Exhibit 7–3 Scanning and Analysis Checklist

The following questions will guide you through the scanning and analysis phase of the problem-solving process.

Scanning
- What actions or events constitute the problem?
- Has the problem been described by all affected persons or groups?
- Has the problem been scrutinized for unforeseen elements?
- Has the problem been identified by more than one person or group? How do others describe the problem?
- Does your description contain all the observed elements or components?
- After you have reviewed the problem from a variety of viewpoints, what are the common elements that can be used to develop a collective (community) description of the problem?

Analysis
- Who are the people involved or affected by the problem? Who are the victims? Who are the offenders? Who should respond to the problem?
- To what groups do the persons involved in the problem belong? What are the likely motivations of the action (committing crime, destroying property)? What activities of others contribute to the victims' vulnerability?
- What is the scope of the problem? How large of a problem is it?
- Why did the problem develop? How did the problem develop? What is the sequence of events that produce the problem?
- When and where does the problem occur? What is the physical setting in which the problem is taking place? Does the location contribute to the problem?
- What underlying conditions precipitate the problem? Have you examined the problem's causal relationship to agency or government policies, procedures, and practices? What is the current community and governmental response to the problem? How do persons and agencies react to the actions or problems? What have they done about the problem? What results have they had?
- Has all available information or data on the problem been collected and analyzed?

Source: Reprinted from Savannah Police Department Community Policing and Problem Solving Workshop, Participants' Handbook, pp. 5–8, 1994, United States Department of Justice.

Several factors should be considered when responding to a problem:

- the potential that the response has to reduce the problem
- the specific impact that the response will have on the most serious aspect of the problem (or those social interests deemed most important)
- the extent to which the response is preventive in nature, thereby reducing recurrence or more acute consequences that are more difficult to handle

- the degree to which the response intrudes into the lives of individuals and depends on legal sanctions and the potential use of force
- the attitude of the different communities most likely to be affected by adoption
- the financial costs
- the availability of police authority and resources
- the legality and civility of the response and the way in which it is likely to affect overall relationships with the police
- the ease with which the response can be implemented (Goldstein 1990, 143)

Response planning, or *action planning,* takes place following the conception of the response. This involves developing procedures for putting into place the precise response to solve the problem. It includes drafting goals and objectives and listing the steps needed to accomplish them. It also involves determining how the effectiveness of the plan will be measured so that assessment data can be collected on an ongoing basis.

Next, the plan will be implemented. Implementations of plans require flexibility.

The checklist in Exhibit 7–4 can be a useful tool for completing the response stage of the problem-solving process.

Assessment

The final stage in the problem-solving process is assessing the results and the effectiveness of the plan placed into action, identifying potential improvements, and revising the plan as necessary. According to Goldstein (1990), evaluating the police response to problems requires

- a clear understanding of the problem
- agreement over the specific interest(s) to be served in dealing with the problem and their order of importance
- agreement on the method to be used to determine the extent to which these interests (or goals) are reached
- a realistic assessment of what might be expected of the police (e.g., solving the problem versus improving the quality of the police management of it)
- determination of the relative importance of short-term versus long-term impact
- a clear understanding of the legality and fairness of the response (recognizing that reducing a problem through improper use of authority is not only wrong, but likely to be counterproductive because of its effects on other aspects of police operation) (145–146)

The assessment stage involves process evaluation, impact evaluation, and new problem assessment. The *process evaluation* examines how the response was implemented and whether it was carried out as planned. Also, it looks at how people and policies functioned in carrying out the plan. The next element, the *impact evaluation,* examines how effective the response to the problem was in obtaining the intended results. The last element of assessment, *new problem assessment,* should be viewed as

Exhibit 7–4 Response Checklist

The following questions will guide you through the response phase of the problem-solving process.

- Involve the persons involved in or affected by the problem (including residents) in a brainstorming session to identify potential responses.
- Has the response been tried before? What were the results?
- What action does the response include? What agencies need to be involved? Are representatives from those agencies present to discuss feasibility of action?
- What other resources are available for the response?
- What is the time frame for implementing the response?
- What barriers exist to implementing the response?
- What results can the stakeholders realistically expect?
- Using the collective description and the analysis of the problems and response (done above), obtain consensus on a response to be implemented.
- Set an action plan for implementing the response.
- Set response goals and objectives.
- Identify the action steps necessary to implement the response.
- Identify key stakeholders involved in or affected by each step.
- Assign individual responsibilities for carrying out each action step.
- Determine how the impact of the plan will be measured:
  - What will be measured and how? (e.g., Problem: Eliminate graffiti on school and recreational center buildings. Will you measure the number of artists arrested for vandalism or the amount of new graffiti that has appeared?)
  - What will the time period for assessment be? How long should the response remain operational to give it a fair opportunity to show success or failure?
  - How will you collect data? Will you use a citizen survey, examine police statistics, etc.?
  - How will you analyze data?

Source: Reprinted from Savannah Police Department Community Policing and Problem Solving Workshop, Participants' Handbook, pp. 5–11, 1994, United States Department of Justice.

an ongoing part of problem solving. The purpose of new problem assessment is to determine if the problem has changed or even if it has created a new problem. The checklist in Exhibit 7–5 can be a useful tool for evaluating the process and impact of a problem-solving response.

IMPLEMENTATION OF PROBLEM-ORIENTED POLICING

Successful implementation of problem-oriented policing requires the complete support of the head of the police organization—not just verbal support but also full comprehension and action. Every member of the police organization, regardless of his or her rank or specific job assignment, should be trained in the problem-oriented policing concept.

Exhibit 7–5 Assessment Checklist

The following questions will guide you through the assessment phase of the problem-solving process.

Assessment: Process Evaluation
- Was the response implemented as intended? What problems were encountered, and how were they overcome?
- What specific action steps were implemented? Were all actions steps implemented? Were steps implemented according to the timetable?
- Did these action steps lead to the attainment of specific objectives? Were objectives revised? Were any objectives added?
- To what extent were successes or failures a result of factors other than the response?
- What was the level of cooperation and participation among team members? Were the lines of communication between individuals and groups adequate?
- What changes in policies, procedures, or practices occurred or should have occurred during the implementation?

Assessment: Impact Evaluation
- Were the overall goals of the plan achieved?
- What were the expected results of the response?
- Were there any unintended effects?
- Was the response cost-effective?
- Did the response make a substantial difference?
- Will you replicate the response in other areas with the same problem?

Source: Reprinted from Savannah Police Department Community Policing and Problem Solving Workshop, Participants' Handbook, pp. 5–14, 1994, United States Department of Justice.

Performance evaluations of officers must be thoroughly revised. Numbers of arrests, traffic citations, and parking citations can no longer be the main criteria for determining whether an officer has been doing a good job. "Bean counting" must be replaced by evaluation of the officer's problem-solving role. If supervisors and managers claim to support problem-oriented policing but continue their traditional performance evaluation methods, police officers will continue to rack up high arrests and traffic citations at the expense of a problem-solving strategy in order to meet evaluators' expectations.

A study of the implementation of problem-oriented policing in the Newport News, Virginia, Police Department (Eck and Spelman 1988) discussed two sets of concerns, internal and external, that must be addressed in the implementation process. Internal concerns include leading, directing, supervision, decision making, cooperation, coordination, and communications within the police department, as well as the police role in the community and police administrative and operational procedures. External concerns include obtaining the support of local government agencies, private businesses and organizations, churches, and schools and universities so that they can assist the police in problem solving. See the problem analysis model in Exhibit 7–6.

As Goldstein (1990) and Vaughn (1992) have pointed out, problem-oriented policing requires major organizational changes in police agencies. For example, police administrators, instead of micromanaging their line officers, must trust them to make decisions pertaining to a problem in their sector. They must also give officers time to solve problems and arrange for crime analysis units to give officers the necessary information on trends and patterns of community problems. Further, the police must become more open to public input and criticism. For example, if a citizen calls the police chief to ask why the officer assigned to his neighborhood is being transferred, the chief, instead of getting angry and saying that it is his department and he does not want any citizen telling him how to run it, should be pleased that the citizen supports problem-oriented policing.

RESISTANCE TO PROBLEM-ORIENTED POLICING

Although Goldstein and Eck and Spelman have made many valid arguments for the implementation of problem-oriented policing and the abandonment of the professional model of policing, the professional model still has many supporters today. Some police

Exhibit 7–6 The Problem Analysis Model

Actors	**Incidents**	**Responses**
	Sequence of Events	
Victims	Events preceding act	Community
Lifestyle	Event itself	Neighborhood affected
Security measures taken	Events following	by problem
Victimization history	criminal acts	City as a whole
Offenders	Physical	People outside the
Identity and physical	Time	Institution
description	Location	Criminal justice
Lifestyle, education,	Access control and	agencies
employment history	surveillance	Other public agencies
Criminal History		
	Social context	Mass media
Third Parties	Likelihood and probable	Business sector
Personal data	action of witnesses	
Connection to	Apparent attitude of	
victimization	residents toward	
	neighborhood	

Source: Reprinted from W. Spelman and J. Eck, *Newport News Tests Problem-Oriented Policing,* p. 5, 1987, National Institute of Justice.

administrators consider problem-oriented policing, along with community-oriented policing, to be merely the latest buzzword or fad that will not survive the long haul. Others
believe that they support new approaches to policing such as problem-oriented policing,
but their actions indicate otherwise. Their management style is rigid, and they emphasize
"bean counting" over problem solving. Often, these administrators believe that a rigidly
managed police department will present a positive image of the police to the community.
Other factors as well have contributed to resistance to problem-oriented policing and the
persistence of a focus on means over ends:

- The diverse, poorly defined, and sometimes overwhelming character of the police job makes it difficult to establish what, precisely, is the end product of policing [i.e., the type of service that the public should receive]. Appeals to focus on
  the end product therefore understandably meet with some confusion and apprehension.
- Police are commonly viewed as palliators—as being concerned primarily with
  meeting immediate, emergency needs. It follows that greater rewards are attached to alleviating problems than to solving or curing them.
- Many of the problems that police deal with are unsolvable. This is the very reason they come to the attention of the police. The potential for doing anything
  about an age-old problem like prostitution or shoplifting is limited. Improving a
  communications system or establishing a new operating procedure, in contrast, is
  much more satisfying. Nonsubstantive matters are more self-contained within
  the agency, and the police are therefore less dependent on outside forces for their
  success in dealing with them.
- The constraints under which police operate in a democracy make police reluctant
  to take the initiative in addressing problems. Many officers view their function as
  simply doing what is formally required of them, even if it is widely recognized
  that this may be ineffective. (Goldstein 1990, 16–17)

FACTORS ENCOURAGING A SHIFT TO PROBLEM-ORIENTED POLICING

Goldstein argued in 1979 that various new trends would end up forcing the police
to reexamine their philosophy of policing:

1. *The Financial Crisis.* The growing cost of police services and the financial
   plight of most city governments, . . . are making municipal officials increasingly
   reluctant to appropriate still more money for police service without greater as-
   surance that their investment will have an impact on the problems that the police
   are expected to handle. . . .
2. *Research Findings.* Recently completed research questions the value of two
   major aspects of police operations—preventive patrol and investigations conducted by detectives. Some police administrators have challenged the findings;
   others are awaiting the results of the application. But those who concur with the

results have begun to search for alternatives, aware of the need to measure the effectiveness of a new response before making a substantial investment in it.

3. *Growth of a Consumer Orientation.* Policing has not yet felt the full impact of consumer advocacy. As citizens press for improvements in police service, improvements will increasingly be measured in terms of results. Those concerned about battered wives, for example, could not care less whether the police who respond to such calls operate with one or two officers in a car, whether the officers are short or tall, or whether they have a college education. Their attention is on what the police do for the battered wife.

4. *Questioning the Effectiveness of the Best-Managed Agencies.* A number of police departments have carried out most, if not all, of the numerous recommendations for strengthening a police organization, and enjoy a national reputation for their efficiency, their high standards of personnel selection and training, and their application of modern technology to their operations. Nevertheless, their communities apparently continue to have the same problems as do others with fewer advanced agencies.

5. *Increased Resistance to Organizational Change.* Intended improvements that are primarily in the form of organizational change, such as team policing, almost invariably run into resistance from rank-and-file personnel. Stronger and more militant unions have engaged some police administrators in bitter and prolonged fights over such changes. Because the costs in terms of disruption and discontent are so great, police administrators initiating change will be under increasing pressure to demonstrate in advance that the results of their efforts will make the struggle worthwhile. (239–241)

All of these developments are still affecting policing today. First, although police agencies today may not be in a financial crisis, public officials are increasingly holding police agencies accountable and are requiring that programs be effective in meeting their stated goals. Second, since the 1970s, when police research first received initiatives from governmental grants, research has increased substantially, and much of it is challenging the effectiveness of traditional practices. Third, consumer advocacy today is stronger than ever, as shown by the victims' rights movement and initiatives to change how police handle spousal abuse and child abuse cases. Fourth, massive infusions of money and effort into managerial and technological improvements have made it even clearer that these improvements will not, in themselves, solve community problems. Finally, endless organizational changes without results other than disruption have made both managers and line officers skeptical about changes and made it more necessary for initiators of change to enlist broad support and focus on practical projects that are likely to be successful.

PROBLEM-ORIENTED PROJECTS

Problem-oriented projects can range from fighting fear of crime, as Baltimore County did in the 1980s, to controlling prostitution, as Wichita, Kansas, did, to holding a community safety fair, as the Sedgwick County Sheriff's Department in Kansas did. They can focus on anything from minor community problems to very serious community problems: for example, gangs, curfew violations, drug dealing, crack houses, graffiti on buildings, burglaries, robberies, or malicious mischief. This section will review a number of problem-oriented projects that police officers in various departments have worked on and solved.

Citizen Oriented Police Enforcement

One department in the early 1980s that was willing to be innovative and use the problem-oriented approach to policing as advocated by Herman Goldstein was the Baltimore County Police Department. In 1982, the Citizen Oriented Police Enforcement (COPE) unit was created in response to several murders as an attempt to combat citizens' fear in the county of Baltimore, which includes the surrounding areas of the city of Baltimore. The COPE unit began attacking fear by confronting the social problems that were the cause of the fear. The police officers assigned to the COPE unit developed skills in social research, community organization, and problem solving (Taft 1986).

COPE officers conducted surveys to determine the major concerns of a neighborhood's residents. The emphasis was on what the community considered to be a problem and not on what the COPE officers considered to be problems. The officers then used the problem-solving approach to address these concerns. They solicited the community's input, they researched the problem through such techniques as crime analysis, brainstormed with other officers, and networked with other public and private agencies to come up with solutions.

In one neighborhood called Garden Village, the officers in surveying the community concluded that the citizens were troubled by the lack of recreational facilities for their children and by the physical deterioration of the neighborhood. The COPE officers got the electric company to repair broken street lights, convinced the county road department to repair potholes and cracked pavement, and got the county park and recreation officials to build a new park with basketball courts and playgrounds (Taft 1986, 14–15).

In addressing the issue that COPE officers were performing social work, one lieutenant responded by saying, "Yet COPE may be closer to 'real policing' than patrol itself—and more successful as well. For years people have called the police for any service they want, and the traditional response has been to send a car and solve that problem, whether it satisfied the community or not. . . . COPE's response makes much more sense: We're solving problems from the community perspective, not from the police perspective" (Taft 1986, 8).

Oaklawn/Sunview Community Fair

In July 1997, community policing officers of the Sedgwick County, Kansas, Sheriff's Department who were assigned to the Oaklawn/Sunview community conducted a brainstorming session with community residents to identify various methods that could improve the quality of life, enhance communication, and bring citizens of the Oaklawn/Sunview neighborhood together. Historically, the Oaklawn/Sunview community had been viewed negatively by other communities in the Sedgwick County/Wichita area because of its crime problem, particularly its reputation of having drug-dealing and gang activity. As a result of the meeting, the sheriff's department and the community decided together to hold a community safety fair in Oaklawn/ Sunview. The general public and residents from other areas in Sedgwick County and Wichita would be invited to attend the fair so that they would have a positive experience in the Oaklawn/Sunview community. The fair would feature games for children and refreshments and soft drinks sold at a nominal fee. The fair included information booths for parents, staffed by social service and governmental agencies, on such topics as substance abuse prevention. In addition to the sheriff's department and the Oaklawn Neighborhood Association, cosponsors for the fair included the Driving Under the Influence (DUI) Victim Center of Kansas and Big Brothers and Sisters of Kansas. Money made from the fair was donated to the DUI Victim Center (Lt. Michael Birzer, Community Policing Coordinator, Sedgwick County Sheriff's Department, personal communication, 1998).

The purpose of the fair was to foster a sense of community, pride, and security among residents of the Oaklawn/Sunview neighborhood. The event was a success, and the first step was made in getting Oaklawn/Sunview to function as a community. If Oaklawn/Sunview can work together on a fair, then it should be able to work together to solve some of its other problems, such as drug dealing and gang activity.

Sound Amplification Systems in Vehicle Violations

In 1996, an officer of the Wichita, Kansas, Police Department began a project to change the way that violations involving loud music in vehicles were handled.* At that time, such incidents required making a brief report of approximately two to three sentences. Officers who observed this violation being committed by a juvenile were required to complete an incident and arrest report, notify and wait for the juvenile's parent(s) to arrive, and enter the case through the case desk. This entire process, from initial contact with the juvenile to entering the case, took about 55 minutes. Given that most offenses were committed by juveniles, the arrangement took an exorbitant amount of an officer's time.

The proposal recommended that when officers observed this violation, they issue a moving citation to any violator who was 14 years of age or older. This would reduce the amount of time spent by officers on this type of violation by approximately 3 minutes for adult violators and 43 minutes for juvenile violators.

The benefits of this proposal included

1. The significant reduction in officers' time spent on this type of violation, thereby freeing officers to perform other tasks

---

*The project was completed by Sergeant Ken McMillian of the Wichita Police Department for a community policing class assignment for me.

2. The significant reduction in costs of paying officers to take action on this type of violation
3. The reduction of paperwork and cost of paperwork associated with processing this type of violation
4. The resolution of a community concern, predominantly involving juveniles, that had been only minimally addressed under departmental protocol

The author based his proposal on the following findings:

- A crime analysis showed that from September 1, 1995, to September 1, 1996, there were 147 adult violations involving sound amplification systems in vehicles.
- An attorney in the law department stated that the ordinance could be changed from Chapter 5 to Title 11, allowing violators 14 years of age and older to be issued a moving citation. The law department attorney stated that process could take place as soon as the department's command staff approved the proposed action and sent a Blue Letter to the law department requesting the change. The necessary change in the ordinance would then be made, and the new ordinance could be presented to the city council.
- The city attorney of Bel Aire (a suburb of Wichita) stated that it would be possible to change the ordinance to a traffic ordinance. He stated that Bel Aire had been issuing traffic citations for loud music in cars for a couple of months.
- Police departments of three suburbs of Wichita were contacted about their handling of the problem. All of them had changed their loud music ordinances, making it possible to issue moving citations to violators, and all had had good results.
- A Bel Aire judge stated that he had been presiding over cases in which moving citations were issued for loud music in vehicles and that he had been finding violators guilty.

On December 1, 1996, the chief of the Wichita Police Department signed a special order changing the departmental policy on sound amplification systems in vehicles. The city council changed the ordinance, and the new ordinance allowed officers to cite any operator of a vehicle who was 14 years of age or older on a moving violation.

This project shows how the initiative of one police officer was able to cut the paperwork for an entire department. It provides a good example of how to obtain information from other governmental units within the city and in areas adjacent to the city. Through his efforts, this officer was able to get the city council and police chief to change a city ordinance and police policy that would allow police officers to handle a minor situation quickly, thereby freeing them to perform other tasks.

Wichita Police Department South Central Prostitution Project

When the Wichita Police Department implemented community-oriented policing in specific high-crime target areas of the city, problem solving was initiated. Four Wichita police officers were assigned to the South Central Wichita area to implement problem-oriented policing. These police officers received two weeks of problem-solving training. They were also equipped with take-home vehicles, radios, pagers, and bicycles. The police chief and the city manager encouraged the officers to address the problems in the South Central neighborhood. The officers working on problem solving were allowed to work a flexible schedule so that they had more opportunity to meet the needs of the neighborhood.

In meeting with members of the South Central community, the officers quickly learned that the residents considered prostitution the biggest problem of their community. Therefore, the officers and neighborhood leaders developed a plan to combat prostitution. Support for the plan was obtained from the political leaders of Wichita, who included the mayor, the city council, the city manager, and police administrators.

The officers obtained the service of Wichita State University professors to conduct a study of the ethnic composition, employment, education, and general characteristics of the South Central neighborhood. The study found residents concerned with the adverse appearance of the areas, the high crime rate, and prostitution. Another study conducted by the Wichita Chamber of Commerce found four major concerns: illegal activity, poor communication with the police, poor access to government services, and deteriorated neighborhood appearance. Also, two surveys conducted by the community policing officers assigned to the area found that residents identified prostitution and drug dealing as major problems. More than 50 percent of area businesses attributed the loss of customers to the prostitution traffic, drug activity, and crime. The officers also surveyed the prostitutes working the area in order to obtain another perspective on the neighborhood's problems. The survey resulted in prostitutes' providing information about drug use, frequency of sexual contact, and the length of time they had been on the street.

A crime analysis for the period between January 1993 and December 1994 showed that the Wichita Police Department recorded 418 arrests for prostitution. Sixty-one percent of these arrests were made in the South Central neighborhood. Seventy-five percent of the arrests of johns occurred in the South Central neighborhood. A review of court records showed that the average fine for both prostitutes and johns was $200, with very few offenders receiving jail time. The officers also found a direct relationship between prostitution violations and other types of crimes. During the summer months, complaints about prostitution increased substantially.

Traditionally, the Wichita Police Department had emphasized enforcement when it received a complaint from the neighborhood, but it had not developed a strategic plan to control or eliminate prostitution, and there was no emphasis on crime prevention. Patrol officers indicated that there were insufficient ordinances to control prostitution and that the courts were too lenient in dealing with prostitution cases.

The problem-oriented trained officers worked with the South Central neighborhood association to solve the problems facing the neighborhood. Together, the police and the community established three goals: reduce prostitution and the crime associated with prostitution; improve police-community relations; and improve the neighborhood's appearance. Six recommendations were made to solve the prostitution problem in the South Central neighborhood: develop city ordinances, develop enforcement strategies, work with the courts, encourage community efforts, improve the environment, and work with the media. Each of these recommendations will be discussed.

City Ordinances

The officers assigned to the South Central neighborhood developed prostitution ordinances with assistance from the city's municipal court and law department. An ordinance was revised to allow police officers to arrest males who were loitering with the intent to solicit a prostitute. Prior to this ordinance, only females could be arrested with the intent to commit an act of prostitution. Another ordinance established an "Anti-Prostitution Emphasis Zone," which increased fines and jail time for anyone who was arrested within the major streets where prostitutes roamed. Arrest data on prostitution were used as the basis for creating boundaries for the zone.

Enforcement

The problem-solving officers met with supervisors, who approved the development of a 20-week strategic plan for weekly law enforcement actions. The plan was in place from May 1995 to August 1995, because prostitution increased during the summer months. The police actions taken included foot patrol, bicycle patrol, traffic enforcement, motel register checks, and sting operations. A "10 Most Wanted List" was created for the prostitution zone. It included prostitutes, drug dealers, pimps, and other criminal offenders who were active in the area. When those on the most-wanted list were arrested, the officers would request high bonds from prosecutors and judges. They also spoke to parole officers, probation officers, judges, and prosecutors about the offenders' actions in the neighborhood and requested longer jail sentences upon conviction. The problem-solving officers developed a "Reference Guide" that listed known offenders who frequented the area. The guide included photographs, which were placed in area patrol cars.

The officers provided other services as well. They left information brochures in every neighborhood business on how to document illegal activity. Communication channels were opened with the neighborhood so that, for example, officers were able to track down children abandoned by their prostitute mothers.

Courts

The problem-solving officers followed up every prostitution arrest and were able to get the prosecutors not to plea-bargain cases. Officers videotaped prostitutes and used the information to educate prosecutors on those involved in prostitution.

Community

Billboards were placed in the prostitution area warning potential offenders that prostitution would not be tolerated. The signs depicted a stop sign using the acronym *STOP,* which stood for "Stand Tough on Prostitution." A neighborhood action group was established that acted as a neighborhood steering committee. The problem-solving officers worked closely with the group.

In 1995, a citizen patrol group began patrolling the neighborhood. Citizen involvement was encouraged by police officers, who assisted in developing operational procedures, training curriculum, and program guidelines. Those involved in the citizen patrol groups were trained by the Wichita Police Department and received instruction on Neighborhood Watch programs and patrolling techniques.

Officers were able to hold a meeting with 30 motel owners. This meeting led to the development of a motel association that was formed with the intent to provide a networking system to warn motel owners of known prostitutes working the area. The goal would be to keep the prostitutes from moving from one motel to another. The local police substation also kept a list of convicted prostitutes that was made available to motel owners.

Environment

To improve the appearance of the neighborhood, the problem-oriented police officers organized cleanups of vacant lots, alleys, homes, and businesses. Improving the environment was an important aspect of the anticrime strategy for the South Central neighborhood, because prostitutes and drug dealers would make contact with customers on city streets and move to alleyways to conduct their business. Residents of the neighborhood, along with Boy Scouts and inmates from the local community corrections halfway house, were involved in the cleanups. City trucks were used to haul the debris away. Neighborhood business donated equipment and food for the workers involved in the cleanups. Areas that needed to be painted to improve the neighborhood's appearance were completed.

Housing inspectors trained a neighborhood residents group known as Neighbor to Neighbor to detect city housing code violations. This neighborhood group would identify housing code violations and tell the homeowner to correct the violations. If the home owner ignored the warning, the group informed the city housing inspector, who had the legal authority to take corrective action. Officers with problem-solving responsibilities had a close working relationship with the Neighbor to Neighbor volunteers and the Clean Team city code inspectors. These groups met to share information concerning housing code violations. The Neighbor to Neighbor volunteers would attend court trials to monitor cases pertaining to housing code violations. Clean Team inspectors were called by the police officers whenever a search warrant was executed in a criminal case. The Clean Team would perform an inspection and, where appropriate, would use the building code to close businesses that were used for criminal activities.

Media

The media highlighted the changes in the neighborhood. The city newspaper reported the improvements taking place, the electronics media reported ordinances changes, and one local television station ran a special news report on the cleanup of the neighborhood, and another aired reports on the effects of the community policing officers' problem-solving approach on the neighborhood.

Evaluation

The strategies for controlling the major problems of the South Central Wichita neighborhood were evaluated at the end of 20 weeks. Although prostitution was lower in the winter months, revisions could be made in the strategy for the summer months. The evaluation led to the maintenance program for the summer of 1996.

To determine the project's effectiveness, surveys were conducted of business, residents, court personnel, and patrol officers who patrolled the neighborhood. The surveys found that 75 percent of the businesses reported a decrease in prostitution and that 90 percent felt that community policing (problem solving) had had a direct effect on the prostitution problem. A residents' survey indicated that residents in general felt better about their neighborhood and its appearance. The improvement in the appearance translated to a feeling of safety for residents. Crime statistics were compiled and evaluated for the 20 weeks of the project. An examination of the statistics found an 11 percent decrease in crime. There was a 41 percent decrease in drug law violations, a 47 percent decrease in prostitution, and a 16 percent decrease in 911 calls to the neighborhood.

Court documents showed that 26 people were charged with the new Anti-Prostitution Area Ordinance and that 8 of those charged pled guilty. Seven prostitutes were mapped out under the new mapping ordinance along with two johns. Four prostitutes were arrested by patrol officers for violating the mapping ordinance. The mapping ordinance helped to reduce prostitution in South Central Wichita. Prostitutes learned the street boundaries for the enforcement zones and mapping areas and took their business out of the area. The physical appearance of the neighborhood improved as a result of the police and neighborhood working together (Sgt. K. McMillian, Wichita Police Department, personal communication, 1998).

The South Central Prostitution Project reduced crime and the fear of crime, improved the physical environment of the neighborhood, and improved police-neighborhood cooperation. The problem-solving efforts of the police officers in this project were a success. This award-winning project can be duplicated in other neighborhoods.

SUMMARY

Problem-oriented policing is an approach to policing that encourages and assists all members of a police agency to engage in proactive problem solving. For all practical purposes, the philosophy of community policing has as its foundaiton the problem-oriented policing design. The primary idea behind problem-oriented policing is police accountability to community and a focus on addressing community concerns. Community-oriented policing draws on this idea but takes it even further to emphasize full community participation in the problem-solving process.

Problem-oriented policing emerged in the late 1970s out of a dissatisfaction with the professional (traditional) model of policing. Herman Goldstein claimed that attempts to improve policing had for decades been focused primarily on administrative improvements, such as recruiting better personnel, modernizing equipment, and implementing business practices into police operations. Although these improvements were necessary, they had become ends in themselves, and police agencies had lost sight of the *effectiveness* of organizational changes in reducing crime and meeting community needs. Eck and Spelman (1988) similarly argued that the professional model of policing was incident driven, consisting of routine and limited responses to calls for service, instead of being problem driven. All these authors advocated an alternative style of policing that would focus on the ends of policing—what the public wanted—rather than the means; attempt to solve persistent problems by analyzing their causes; monitor the effectiveness of solutions; encourage police initiative; and engage community resources.

Before problem solving can be initiated, the problem or problems have to be defined. A problem can be defined as a group of incidents occurring in a community on several occasions that are similar in one or more ways and are of concern to the police and the public. The similarities could refer to people or types of people involved in incidents, whether they are victims (e.g., the elderly) or offenders (e.g., gangs or the homeless); behaviors (e.g., drunk driving, sexual assault); time (e.g., at an annual rock concert, before Christmas); or place (e.g., a public housing project).

The SARA model developed by Eck and Spelman is a model for solving police problems that has been widely used by departments that practice problem-oriented policing. *SARA* stands for the four stages of problem solving: scanning (i.e., identifying the problem), analysis, response, and assessment.

Successful implementation of problem-oriented policing requires the complete support and understanding of the head of the organization; training in problem-oriented policing for every member of the police organization, sworn personnel and civilians; and the replacement of "bean-counting" evaluations of officer performance (e.g., counting arrests) with evaluations that emphasize officers' problem-solving role.

KEY TERMS

analysis stage	incident-driven policing	response stage
assessment stage	problem-oriented policing	SARA model
Herman Goldstein	problem solving	scanning stage

REVIEW QUESTIONS

1. Describe the problem-oriented policing approach.
2. Who developed the problem-oriented policing approach, and what were his criticisms of professional-model policing?
3. What are five trends that have encouraged the adoption of problem-oriented policing?
4. What does the term *incident-driven* policing mean?
5. What are five ways that problems can be solved?
6. Describe the SARA model.
7. What are four ways that incidents in a community can be clustered to define a problem?
8. Describe some of the problems that can be encountered when implementing problem-oriented policing.

REFERENCES

Eck, J.E., and W. Spelman. 1988. *Problem solving: Problem-oriented policing in Newport News.* Washington, DC: National Institute of Justice.

Goldstein, H. 1979. Improving policing: A problem-oriented approach. *Crime and Delinquency* 25, no. 2:238–241.

Goldstein, H. 1990. *Problem-oriented policing.* New York: McGraw-Hill.

Moore, M.H., and R.C. Trojanowicz. 1988. *Corporate strategies for policing.* Washington, DC: National Institute of Justice.

Savannah Police Department. 1994. *Community Policing and Problem Solving Workshop: Participants' handbook.* Washington, DC: National Institute of Justice.

Taft, P.B., Jr. 1986. *Fighting fear: The Baltimore County C.O.P.E. Project.* Washington, DC: Police Executive Research Forum.

Vaughn, M.S. 1992. Problem-oriented policing: A philosophy of policing in the 21st century. *Crime and Justice* 19, no. 3:345.

Chapter 8

Community-Oriented Policing

CHAPTER OBJECTIVES

1. Have an understanding of community policing philosophy.
2. Have an understanding of the value of community policing.
3. Have an understanding of the potential pitfalls of community policing.
4. Have an understanding of obstacles to implementing community policing.
5. Be familiar with the Community Patrol Officer program in New York City.
6. Be familiar with how community policing works in Seattle.

INTRODUCTION

Public officials, from the president of the United States to local chiefs of police, herald community policing as the long-awaited solution to violent crime. In his 1994 State of the Union address, President Clinton prods the nation to "take steps to reduce violence and prevent crime, beginning with more police officers and more community policing." A local police chief unabashedly proclaims that community policing and precinct patrolling "have gone a long way toward preventing crime. . . . They have worked" (Holland 1994, 12). Both proponents and detractors of community policing agree that it constitutes a revision of the police role, away from that of crime fighter to that of planner, community organizer, problem solver, and broker of governmental services. Currently, scores of police departments not only in America but throughout the world claim to be implementing community policing. Although community policing is the newest policing model, this approach cannot be considered completely new. In 1900–1910, Arthur Woods, the New York City Police Commissioner, instituted a number of policies that could be described today as community policing. Woods believed that the rank-and-file police officer should be in a position of social importance and public value. He also was convinced that the public would benefit if they were informed of issues affecting the community and the local police precinct and would have more respect for the police once they realized the complexities and difficulties of police work (Skolnik and Bayley 1988, 37).

Woods placed into operation within the New York City Police Department what today would be termed community-oriented policing. Captains were directed to establish junior police leagues. Young people were provided with junior police badges and were trained to report violations of law that they observed in their neighborhoods. Sergeants were assigned to visit schools and to educate the students that policing was more than just making arrests. Officer's duties included working to improve neighborhoods and make them safer places for people to live. Woods emphasized that every police officer was responsible for the social conditions of the neighborhood he patrolled. Because unemployment, according to Woods, was a major cause of crime, police precincts were used to distribute employment information and beat officers were expected to help the unemployed in their neighborhood to find employment (Skolnick and Bayley 1988, 37–38).

Woods is credited with creating "play streets" for youngsters in neighborhoods that had no playgrounds, parks, or open places to play. The police would close a street for several hours by placing barricades that prevented traffic from entering the street. In addition, Woods communicated with ethnic newspapers, Greek, Italian, and Yiddish, and encouraged them to publish city ordinances and laws in the newspapers so that immigrants could read about the laws in their own language. Teenage males who were involved in acts of delinquency were placed in contact with such social service agencies as Big Brother and YMCA. Woods' policing philosophy was widely accepted in New York City and praised by the news media (Skolnick and Bayley 1988, 38). A journalist of Woods' time commented on Woods' policing philosophy that

> to many persons, and particularly the foreign-born population, . . . the law stands for a vast machine of menace. The new police idea is wholly different. It aims to do something that in America seems never to have been tried as an angle of police duty—to strive for the inculcation of the thought that the law is an engine of mutuality, of good will, of positive influence; that it is constructive. The new police idea is to present it as a protector. (MacCulloch 1915)

When Woods left his position as police commissioner, the philosophy he espoused fell by the wayside. As discussed in Chapter 1, police departments from the 1930s or 1940s through the 1960s or 1970s were largely dominated by a professional model of crime control through making arrests, expanding strategies of surveillance, and deploying technology and professional expertise. Close community relations were not considered important or were even discouraged.

The shift away from the crime control model can be traced back to the 1960s. The social, political, and civil turmoil of the 1960s created unprecedented changes in American policing. These changes affected the thinking, tactics, and resources of the police and forced police to focus on relations with the community (Skolnick and Bayley 1988). Two governmental reports of that era, *The Challenge of Crime in a Free Society* (President's Commission 1967) and the *Report of the National Advisory Commission on Civil Disorders* (National Advisory Commission 1968), underscored, in particular, the marked hostility of African Americans toward the police and the failure of police to cultivate and maintain constructive relationships with the African American community.

Subsequent research in the 1970s investigating the efficacy of policing as a formal mechanism of social control showed that the usual policing strategies were having little impact on crime or public safety. Skolnick and Bayley (1988) summarized this research.

> First, increasing the number of police does not necessarily reduce crime rates or raise the proportion of crimes solved. The same is true for budgetary expenditures on the police. The most that could be said was that if there were no cops there would be more crime; however, once a certain threshold of coverage has been reached, presumably long passed in the United States, increments of money and personnel are no longer efficacious. Variations in both crime and clearance rates are best predicted by social conditions such as income, unemployment, population, income distribution, and social heterogeneity. We have learned that you can't simply throw money at law enforcement and expect proportionate results.
>
> Second, random motorized patrolling neither reduces crime nor improves chances of catching suspects. Moreover, it does not reassure citizens enough to affect their fear of crime, nor does it engender greater trust in the police. Regular patrols by police officers on foot . . . [have been] shown to reduce citizens' fear of crime, [but] they have no demonstrable impact on the crime rate.
>
> Third, two-person patrol cars are no more effective than one-person cars in reducing crime or catching criminals. Furthermore, injuries to police officers are not more likely to occur in one-person cars.
>
> Fourth, saturation patrolling does reduce crime, but only temporarily, largely by displacing it to other areas.
>
> Fifth, the kind of crime that terrifies Americans most—mugging, robbery, burglary, rape, homicide—is rarely encountered by police on patrol. Only "Dirty Harry" has his lunch disturbed by a bank robbery in progress. Patrol officers individually make few important arrests. The "good collar" is a rare event. Cops spend most of their time passively patrolling and providing emergency services.
>
> Sixth, improving response time to emergency calls has no effect on the likelihood of arresting criminals or even on satisfying involved citizens. One recent and very large study showed that the chances of making an arrest on the spot drop below 10 percent if even one minute elapses from the time the crime is committed. Only instantaneous reaction, in other words, would be effective in catching criminals. Yet that can't reasonably be expected unless a cop is put on every corner. Speed of response makes so little difference because victims delay an average of four to five and one-half minutes before calling the police, even when they have been victims of confrontational crimes.
>
> Rapid response also doesn't satisfy citizens, surprisingly, because what most victims of crime want is predictable rather than quick response. Victims seem to recognize that in most crimes the criminal will be long gone

whenever the police arrive. What they want is a police response they can count on as they go about reordering their stricken lives. They would rather wait forty-five minutes, the research shows, for an assured reaction from the police than [experience] the uncertainty of waiting for an unpredictable response. They don't want to sit around waiting.

Seventh, crimes are not solved—in the sense of offenders arrested and prosecuted—through criminal investigations conducted by police departments. Generally, crimes are solved because offenders are immediately apprehended or someone identifies them specifically—a name, an address, a license plate number. If neither of those things happens, the studies show, the chances that any crime will be solved fall to less than one in ten. Despite what television has led us to think, detectives do not work from clues to criminals; they work from known suspects to corroborating evidence. Detectives are important for the prosecution of identified perpetrators and not for finding unknown offenders. (Skolnick and Bayley 1986, 4–5)

Klockars (1985) pointed out, in the same vein, that the patrol officer's typical tour of duty does not result in any arrests; the typical New York patrol officer, for instance, makes only three felony arrests per year. What makes a difference in apprehension rates is the cooperation of witnesses and neighborhood residents in providing information to the police.

This research resulted in a resurgent interest in experimenting with forms of policing that would improve police-community interaction. Goldstein's innovative problem-oriented policing article, published in *Crime and Delinquency* in 1979, and Wilson and Kelling's seminal article "Broken Windows," which appeared in the *Atlantic Monthly* in 1982, combined with police research published in the Perspectives on Policing Series sponsored by the National Institute of Justice and Harvard University, specifically the Moore et al. publication *Crime and Policing* in 1988, provided a strong impetus for policing strategies organized around the principles of service decentralization, neighborhood focus, and close interaction between citizens and police as coproducers of public safety. The stage was set for the development of community policing.

WHAT IS COMMUNITY POLICING?

A review of the community policing literature reveals that the concept has a variety of definitions and meanings. Some police administrators believe that community policing has always occurred in small towns where the police officer knows everyone on a first-name basis. Others equate problem-oriented policing with community policing. Some even believe that community policing is having specialized units work in a high-crime area. And many give lip service to community policing because they consider it to be the latest fad and claim that they have instituted community policing for public relations purposes. Community policing has become so fashionable that a majority of police departments claim to be doing it or, at least, planning to do it.

As one police scholar has pointed out, community policing "is one of those terms that simultaneously suggests so much that is general and so little that is specific that it risks being a barrier rather than a bridge to discourse about development in policing" (Wycoff 1988, 104). Although the term can cause confusion, several researchers have offered useful working definitions. A good brief definition was provided by Stipak (1994): "Community policing is a management strategy that promotes the joint responsibility of citizens and police for community safety, through working partnerships and interpersonal contacts" (115). Trojanowicz and Bucqueroux (1994) provide an expanded definition:

> Community policing is a philosophy and an organizational strategy that promotes a new partnership between people and their police. It is based on the premise that both the police and the community must work together to identify, prioritize, and solve contemporary problems such as crime, drugs, fear of crime, social and physical disorder, and overall neighborhood decay, with the goal of improving the overall quality of life in the area.
>
> Community policing requires a department-wide commitment from everyone, civilian and sworn, to the community policing philosophy. It also challenges all personnel to find ways to express this new philosophy in their jobs, thereby balancing the need to maintain an immediate and effective police response to individual crime incidents and emergencies with the goal of exploring new proactive initiatives aimed at solving problems before they occur or escalate.
>
> Community policing also rests on establishing community policing officers as decentralized "mini" chiefs in permanent beats, where they enjoy the freedom and autonomy to operate as community-based problem solvers who work directly with the community—making their neighborhoods better and safer places in which to live and work. (2–3)

Skolnick and Bayley (1988), who studied community policing in police departments throughout the world, concluded that there are some common elements to all the community policing approaches:

- A growing reliance on "community-based crime prevention" through the use of citizen education, neighborhood watches, and similar techniques, as opposed to relying entirely on police patrols to prevent crime
- The reorientation of patrol from being primarily an emergency-response force ("chasing calls") to using "proactive" techniques such as foot patrol
- Increased police accountability to the citizens they are supposed to serve
- Decentralization of command and police authority, with more discretion allowed to lower-ranking "generalist" officers, and more initiative expected of them

Brown (1989) came up with a similar list of components:

- a problem-solving, results-oriented approach to law enforcement
- articulation of police values that incorporate citizen involvement
- accountability of the police to each neighborhood

- decentralization of authority
- police-community partnership and sharing of power
- beat boundaries that correspond to neighborhood boundaries
- permanent assignment of patrol officers
- empowerment of patrol officers to show initiative
- coordination of investigations at both neighborhood and citywide levels
- new roles for supervisors and managers, as supporters of patrol rather than evaluators of patrol officers
- changes in the content of training at all levels
- new systems of performance evaluation, placing much less emphasis on "production" of quantified activities
- new approaches to "demand management," the response of the agency to calls for service (5–6)

Bayley (1994), who described community policing as "the most serious and sustained attempt to reformulate the purpose of and practice of policing since the development of the 'professional' model in the early twentieth century" (104), claimed that it rests on three insights. First, the police cannot solve society's crime problems alone. They need the public's assistance to deter crime and to apprehend criminal offenders. Second, the resources of the police have to be deployed proactively against crime and criminals. This approach addresses the circumstances that cause crime. Third, routine police patrol is too passive; it does not deter crime. Police patrol should create a moral order with a respectable public appearance, and behavior should reflect the standards of the local community (102–104).

To understand what community policing is, it pays to review what community policing is not. According to Trojanowicz and Bucqueroux (1990),

- *Community policing is not a technique.* . . . The entire department must be infused with the community policing philosophy, with CPOs [community police officers] as the department's direct link to average citizens. Community policing is not something to be used periodically, but it is a permanent commitment to a new kind of policing that provides decentralization and personalized community problem-solving.
- *Community policing is not public relations.* . . . Community policing instead enhances the department's image because it is a sincere change in the way the department interacts with people in the community. It treats law-abiding people as partners, a new relationship based on mutual trust and shared power. The traditional system often makes people feel that the police do not care about their wants and needs. In traditional departments, officers often see the people in the community as them, those nameless and faceless strangers whose reluctance to cooperate and share what they know makes them indistinguishable from the criminals. Community policing instead treats law-abiding people in the community as an extension of us.
- *Community policing is not anti-technology.* . . . Many people erroneously assume that community policing rejects technology because it refuses to lock the

officer into a patrol car and handcuff him to the police radio. CPOs can make tremendous use of new technologies, such as hand held computers, remote telephone answering machines, and cellular telephones. Yet community policing also recognizes that no technology can compete with a fully functioning, creative human being, whether the person is a police officer or a law-abiding citizen willing to help.

- *Community policing is not soft on crime.* Critics suggest that community policing's broad mandate and its focus on using tactics other than arrest to solve problems detract from a proper focus on serious crime. CPOs often face derision within the department from fellow officers who call them lollicops or the grin-and-wave squad. The reality is that CPOs make arrests just like any other officer does, but CPOs deal with a broader variety of community concerns in addition to crime, not as a substitute for addressing serious crime. . . .
- *Community policing is not flamboyant.* . . . The media reinforces the image of the macho police officer whose job is glamorous, tough, and often dangerous. That hero myth also appeals to police officers themselves, which is part of why many police officers refuse to volunteer for duty as a CPO. For community policing to succeed, however, the entire department must embrace the wisdom of this approach. All officers must focus on working together among themselves and with people in the community on solving community concerns.
- *Community policing is not paternalistic.* . . . Community policing threatens those who enjoy the traditional system, because it says that police superiors must stop treating line officers like children who must be watched constantly. It also says that police officers must stop acting like parents to people in the community and must begin treating them like respected partners. . . .
- *Community policing is not an independent entity within the department.* . . . Departments launching a new community policing effort can use war stories from elsewhere to educate everyone in the department about how community policing benefits the entire department. Police officials should also make a special effort to make sure that any initial successes are touted within the department in ways that do not seem to set CPOs apart from their fellow officers. Integrating the community policing philosophy into any day-to-day operation of the entire department is a challenge, one that requires care and feeding over time.
- *Community policing is not cosmetic.* . . . Community policing not only broadens the agenda to include the entire spectrum of community concerns relating to crime and disorder but offers greater continuity, follow-up, and accountability. Important as well is that community policing goes far beyond handing out brochures, making speeches, talking with community leaders, telling people how to guard against crime, and urging fellow officers to treat people with respect. As line officers directly involved in the community, CPOs have the opportunity to make far-reaching changes.
- *Community policing is not a top-down approach.* . . . Community policing decentralizes decision-making, opening up departments so that new ideas can surface. Community policing provides the department grass-roots input from both

community residents and line officers, which can help overcome a bureaucracy's inherent tendency to become more stodgy and stagnant over time.

- *Community policing is not just another name for social work. . . .* The fact is, social work has always been an important element of police work. When line officers talk about the use of police discretion, they mean that the job requires doing more than just sticking to rules and procedures. It also means allowing police officers the freedom to make immediate decisions on their own, including the freedom to solve problems in ways that have nothing to do with arresting bad guys. Though it conflicts with the superhero myth, good police officers have always tried to encourage youngsters to live within the law and help the elderly feel less vulnerable. (20–27)

Watson et al. (1998) have also contributed to our understanding of what community policing is not by describing four myths and misconceptions that opponents of community policing expound:

> The first myth is that community policing is so radically different from traditional law enforcement practices that it requires a complete restructuring of a law enforcement agency. . . . Organizational restructuring is not the first step in implementing community policing. It may not always be necessary even as the last step.
>
> The second myth is that community policing involves an enormous increase in officers' workload, especially for patrol officers, who are suddenly required to "interact with the community" in addition to all of their ordinary duties of preventive patrol and responding to calls for service.
>
> The truth is that many officers will spend part of their time in different activities, depending on their assignment and the nature of the community in which they are deployed. Patrol officers and their supervisors will be expected to take the initiative in establishing a partnership with the community. However, their ability to perform these new duties will depend on their being relieved of some of the burden of "chasing 9-1-1 calls," plodding through administrative paperwork, or laboring on tasks that either do not need to be done at all or can be done just as well by someone else.
>
> Ultimately, if community policing lives up to its promise of making neighborhoods safer, most patrol officers should be able to spend most of their time on productive, "proactive" duties without being in any way overburdened.
>
> The third myth is that community policing distracts police officers from their main responsibility of enforcing the law, and therefore leads to an increase in the incidence of crime. Evidence is sometimes offered to support this myth, in the form of crime rate statistics from some of the cities that have implemented one or another form of community policing. . . .
>
> The first duty of a police officer is to enforce the law by identifying and arresting offenders and gathering evidence to secure a conviction. That is

just as true in a community policing agency as in a traditional agency. The question is not what police officers should do, but why and how they do it.

Perhaps the most pernicious myth of all is that community policing turns patrol officers into "social workers." This myth is doubly damning. First, it implies that there is something demeaning about social work. . . . Second, the myth reflects a profound misconception about the nature of the criminal justice system and the role of police officers in the system. . . . The police cannot fulfill all of the roles by themselves. They do not have the resources to do so, and therefore must depend on other segments of the community (including social workers) to participate. (54–56)

Finally, it may be helpful to scan a list of questions devised by Trojanowicz (1994) in a study sponsored by the Department of Justice to determine if a police department has in reality implemented community policing. The following questions should be answered "Yes" if community policing is being practiced:

- Do citizens nominate the problem?
- Do citizens work with the police to help solve the basis of the problem, and not just react to the symptoms?
- Does the officer have a defined beat area?
- Is the officer full service, both reacting to and preventing crime?
- Does the officer make arrests?
- Does the officer have a permanent assignment of at least 18 months in the defined beat area?
- Is it possible for the officer to interact with most of the people in the particular area within a six- to eight-month period?
- Does the officer work out of a decentralized office?
- Does the officer actively work as a member of a team with non-law-enforcement agencies?
- Does the officer, with help, conduct a long-term evaluation to determine if problems are solved and not merely interrupted on a short-term basis? (38)

THE VALUE OF COMMUNITY POLICING

Benefits to the Community

The key question about community policing is whether it will produce safer communities. Although some anecdotal evidence suggests that it can, no definitive answer can be given at this point because community policing has not yet been sufficiently evaluated. Therefore, supporters of community policing have stressed other benefits to the community that are easier to demonstrate. Citizens' attitudes, particularly their fear of crime, their evaluation of the police, and their satisfaction with their neighborhood, have been surveyed before and after implementation of community policing strategies, such as distribution of police newsletters informing residents about crime prevention and police programs; community organizing to develop special projects (cleanup campaigns, property identification, "safe" houses for children); citizen contact programs in which police go door to door, introducing themselves to residents and businesspeople and asking if citizens have any problems they wish to bring to the attention of police; victim recontact programs in which patrol officers make telephone contact with victims to inform them of the status of their case, inquire whether they need assistance, offer to send crime prevention information, and ask whether victims can provide additional information; police community stations that provide a variety of services for the area; increased foot patrols; increased enforcement of disorderly conduct laws; and programs to reverse physical deterioration of neighborhoods (Wycoff 1988, 107–108). In a study conducted in Houston, Texas, and Newark, New Jersey, and sponsored by the National Institute of Justice (Wycoff 1988), such strategies were shown to reduce the fear of perceived social disorder in the area, reduce fear of personal victimization, reduce worry about property crime, improve evaluation of police, and increase satisfaction with the area (110). Other community benefits have been listed by Skolnick and Bayley (1988):

- *Public scrutiny.* Even when community policing is only rhetoric, an opportunity is created for legitimate public examination of police practices. If community policing minimally means greater involvement of the public in public safety, how can the police convincingly deflect public discussion of police strategies? Community policing is like a Trojan horse that attacks the pretensions of professional insulation from within. Even if community policing cannot do much in the way of preventing crime, it does offer the public a larger window into police activity.
- *Public accountability.* Community policing increases effective public control over the police. There are three primary ways in which the public can constrain what the police do: (1) by providing, or not providing, a framework of laws and money for police action; (2) by participating in policy-making with respect to the means of achieving desired objectives; and (3) by examining and possibly punishing errors in performance. . . . Community policing also makes civilian oversight of implementation more acceptable. . . . Community policing is a back door into comprehensible accountability. What could not be achieved by public demand through political channels may occur because the police believe wider community participation is essential to the achievement of organizational objectives. Accountability may occur under police auspices, where it could not under political ones.
- *Recruitment.* Community policing provides a double-barreled rationale for representative recruitment. On the one hand, it challenges the macho, martial model of policing. Traditional officers are being very acute when they distinguish "hard" from "soft" policing. The tactics of community policing are indeed soft, even though the goal of deterring criminality is not. Those of community policing are geared toward soliciting, enlisting, inviting, and encouraging, whereas those of traditional policing are geared toward warning, threatening, forcing, and hurting. Community policing is less direct than traditional policing. It is a kind of policing that can be done as well by

women as by men, by the short as by the tall, by the verbal as by the physical, by the sympathetic as by the authoritarian. (67–70)

Benefits to the Police Organization and Police Officers

Skolnick and Bayley (1988) have listed seven benefits to community policing to police organizations and individual officers:

1. *Political benefits.* Politically, community policing is a game the police can't lose. If coproduction through community participation leads to lower crime rates and higher arrest rates, the police can take credit as foresighted agents of change. If community policing fails to increase public security, the public is hardly likely to reduce support for policing because a new gambit doesn't work out. Moreover, even if the police cannot actually deliver on the large goal of crime reduction, a heightened police presence is reassuring. Thus, community policing reduces fear of crime—and, from the perspective of political benefits to the police, delivers the message that police care.
2. *Grassroots support.* Community policing offers a magnificent opportunity to build grassroots support for the police. It embeds the police in the community, giving them an opportunity to explain themselves, associate themselves with community initiatives, and become highly visible as concerned defenders with public safety. Community policing makes the population at large a "special interest group" supporting police-led programs. . . .
3. *Consensus building.* Community policing is a means for developing consensus between the police and the public about the appropriate use of law and force. The police have an obligation not only to catch criminals but to maintain order in public places. There are sound crime-prevention reasons for this, quite apart from enforcement of standards of decency and propriety. Research has shown that people are made fearful and insecure by disorder and incivility, not just by criminal activity. In fact, criminal victimization is rare; most people's knowledge of crime comes secondhand through the media. Inability to curb public disorder—loud music, vandalism, drunkenness, uncouth behavior—generates further disorder, more serious crime, and diffuse feelings of insecurity.

   In order to maintain public decorum, the police use the authority of law and sometimes the reality of constraint or the threat of force. This is a delicate balancing: if police under-enforce, the "signs of crime" multiply, encouraging further depredations; if they over-enforce, the public becomes mistrustful, worse yet hostile, perhaps even violent. Community policing is a vehicle for undergirding police action with support. Through community liaison the police can assimilate local standards of conduct and acceptable levels of performance. . . .
4. *Police morale.* Community policing probably raises the morale of police involved because it multiplies the positive contacts they have with a community's supportive people, those who welcome police presence and activity. Traditional deployment concentrates police contacts on "difficult" people—criminals and

incorrigibles as well as deserving but demanding claimants for police service, such as victims, incompetents, and mental cases. . . .

Community policing leads to unemotional, non-emergency interaction with citizens, increasing contact with people who want nothing more from the police than their reassuring presence. Community policing increases the likelihood that the public's quiet regard will be displayed to individual officers. . . .

5. *Satisfaction.* Because effective community policing requires that subordinate ranks take more initiative and responsibility, it makes the police job more challenging. Community policing cannot be managed in a quasi-military way, fulfilling easily measured norms and avoiding stipulated errors. Community policing may, in fact, be the operational strategy that is peculiarly fitted to the new breed of police recruit. Police managers report that today's more highly educated officers are less accepting of routine, more likely to question command, and more impatient with non-solutions to recurrent problems. Community policing may be the best program the police have devised for maintaining zest for the job.

6. *Professional stature.* Community policing raises the professional standing of the police by broadening the range of skills required. To be successful at community policing, police must be more than large, physical, and tough; they must be analytic, empathetic, flexible, and communicative. . . .

7. *Career development.* By enriching the strategic paradigm of policing, community policing creates more lines for career development. Because community policing encompasses and expands upon the traditional model, it provides more ways for personnel to be valuable. For community policing to work, police forces must reward a wider range of performance skills. This provides career opportunities to a more diversified group of officers. (71–73)

Several other benefits can be added to this list. Officers engaged in community problem solving are putting their energy into productive use and are less prone to carry out pranks on their fellow officers than officers with free time on their hands. If they develop a closer attachment to their assigned neighborhoods and more familiarity with residents, they are more likely to be concerned about the welfare of the residents. This concern can lead to better rapport with residents, fewer complaints about police, fewer police abuses, and greater conscientiousness on the part of police in performing their duties, resulting in fewer accidents and damages to equipment and perhaps even fewer disciplinary problems. Further, data from Newark indicate that foot patrol officers who are satisfied with their jobs have lower rates of absenteeism (Wycoff 1988, 115–116).

PROBLEMATIC ASPECTS OF COMMUNITY POLICING PRACTICE

Although community policing can produce many benefits, we need to recognize that it can have some problematic effects as well. Wycoff (1988) has discussed five possible pitfalls of community policing:

1. *Illegal policing.* This is possible in the event that community-based officers become more responsive to local norms than to legal constraints.
2. *Inequitable policing.* Many communities or neighborhoods are far from homogeneous, and it would be possible for some groups to benefit more than others from police service. We found this to be the case in Houston, where blacks and renters were much less likely to be positively affected by the fear-reduction strategies than were whites and homeowners.
3. *Politicization of the police.* Mayors and city managers have some reason to question the intent of community-oriented strategies. More than one politically ambitious chief has used the community link as a source of political clout—either in electoral or in bureaucratic battles. Officers might find it tempting to use their community organizing skills and good relationships with the neighborhood to mount a campaign to accomplish a political objective, whether it be a salary increase or the ouster of an unappreciated judge or chief.
4. *Corruption.* Close contact between police and businesspeople or residents always makes an anticorruption manager alert to possibilities for unacceptable behavior, especially in communities in which the civic culture does not dissuade corrupt practices.
5. *Police intrusion into private arenas.* Clearly, the better the police know us, the more they are able to know about us. In a democratic society, it is always appropriate to ask how effective we really want our police to be. (116)

It is unfortunate that advocates of community policing have not sufficiently addressed these five potential problem areas. Rather than waiting for a crisis to occur, they should be developing strategies to prevent inequities, abuses of power, and corruption from occurring under community policing systems.

PROBLEMATIC CONTRADICTIONS IN COMMUNITY POLICING PHILOSOPHY

Peter Manning (1997) has argued that "historical policing practices co-exist in variable amounts with 'community policing.' This suggests that it is not a definitive logical entity but a collection of ideas and concepts already present to some degree in many police departments" (13). Consequently, the community policing philosophy is full of ambiguities and contradictions:

1. The rhetoric of community policing positions it as customer oriented, yet this is patently impossible as a full role definition for police.
2. Community policing promises "co-production" with neighborhoods, but the police in various ways retain full control over every resource allocation decision, assignment, tactic and strategy, and type of "service" delivered to neighborhoods.
3. Despite emphasizing community integration, community policing has been used to rationalize street sweeps, harassing youths, drug crackdowns, and other "preventive" actions to reduce disorder.

4. "Neighborhood" and "community" are arbitrarily defined as isomorphic with police precincts or arbitrary political boundaries—not with natural social areas.
5. Community policing promises continuous area-based policing, but union contracts that privilege seniority in shift and beat assignments obviate this continuity.
6. Community policing promises a level of coordination of neighborhood service, but . . . [often] . . . dispatchers continue a division of labor that assigns crime and fast response calls to tactical units or special response teams and leaves routine and often boring work to the beat officer.
7. Community policing is generally symbolized as the first priority of the department, but officers are given no special training, rewards, perks, or promotional opportunities. Officers are given little or no training in the new roles and routines they are to adopt or the skills they are expected to display, such as leading meetings, organizing talks, and making maps and other data-based presentations. They are left to create strategies for serving the public as well as handling their calls.
8. Community police officers, or all officers in a community policing–oriented department, are to spend some time problem solving and working with the community, but they are given no recognized block of time to do so and continue to go where the radio leads. (13–14)

Police administrators who are adopting a community policing approach in their organization need to address all these other points to determine the extent to which each contradiction is manifested in their organization's practices and to reassess their own priorities. A department may proclaim that community policing is valuable, but does it back up its rhetoric with money and training, and does it structure patrol officers' work so that they have the time and discretion to perform their new duties? To what extent is the community really "empowered" in police-community collaborative projects, and to what extent do police keep the control? To what extent do police tactics disrupt a community instead of fostering unity? All these questions need to be openly discussed in every department that has incorporated community policing.

OBSTACLES TO THE IMPLEMENTATION OF COMMUNITY POLICING

Internal Obstacles

If community policing is to be incorporated into a police organization, changes in that organization and its culture must occur. But given that police personnel resist change, community policing is not always easy to implement. Skolnick and Bayley (1988) identified 10 obstacles to implementing community policing within police departments:

1. *The traditional culture of policing.* This includes the defensive attitude of suspicion that officers develop to deal with the public. Because citizens can be dangerous, police may not be inclined to develop closer relations with them and may prefer to socialize with one another. The solidarity among police reinforces distrust of civilians.

2. *Young police officers who exhibit macho attitudes.* They may have bought into the crime-fighting image sold to them while they attended the police academy. A community policing system requires officers to have qualities of maturity.
3. *Street cop culture.* Street cops are generally more apt to be cynical and skeptical about community policing and more resistant to innovative policing concepts and strategies in general. Management cops are more likely to welcome policing innovations such as community policing.
4. *The responsibility to answer emergency calls.* An emphasis on responding to 911 calls as the highest priority of police officers is antithetical to community policing.
5. *Limited resources.* This factor is closely related to the preceding one. When the emergency response system has top priority, many police administrators do not believe they have sufficient resources to implement community policing. More resources will be needed if community policing and an emergency response system are in operation at the same time.
6. *The inertia of police unions.* These may see community policing as a threat to the traditional model of policing. Unions are more concerned with police salaries and benefits than with supporting community policing.
7. *The two-officer patrol car.* This policy creates a shortage of personnel for many police departments and further constrains the development of community policing by distancing officers from the community and limiting their social interaction to each other.
8. *The reward structure of traditional policing.* A "bean-counting" approach to evaluation encourages officers to focus on making large numbers of arrests or giving large numbers of citations instead of on solving problems.
9. *The centralized, hierarchical, quasi-military command structure of traditional policing.* Community policing requires a decentralization of authority and a reorganization of the command structure to give lower-level officers more flexibility and initiative and to increase bottom-up communication.
10. *Lack of coordination with crime detection.* Often, when community policing activities are assigned to specialized units, such as crime prevention units or ministations, these units do their own thing and are not integrated into the crime detection activities of the department, so that their effectiveness is reduced. (49–66).

External Obstacles

The role of the community in community policing is crucial, but sometimes the cooperation and participation of the community can be difficult to obtain. One study, for example, reported "extreme difficulty in establishing a solid infrastructure on which to build their community programs" and concluded that "the most perplexing . . . [problem] was the inability of the police department to organize and maintain active community involvement in their projects" (Grinc 1994, 442, 437).

There are many reasons for difficulty in obtaining community cooperation:

1. Many families today have both parents working, and there are a substantial number of single parents. In addition to having full-time jobs, parents have to spend time with their children, run errands, cook, clean their home, and attend school functions with their children; no wonder they do not have time for neighborhood activities.
2. In many communities, the police and citizens have a long history of confrontation and alienation. Residents may fear, mistrust, and feel hostile toward the police. They may perceive the police as brutal, arrogant, and nasty and may be more interested in monitoring police misconduct than in developing a close relationship with the police.
3. Low-income high-crime areas generally lack the organizational infrastructure required for people to become involved.
4. Residents of high-crime communities may fear retaliation by neighborhood thugs and drug dealers.
5. After decades of traditional policing being sold to the general public, it is rather difficult to switch to community policing and to expect the public to understand the concept or to readily buy in to it. The police have for decades informed the public that they are the crime fighters, the specialists who will control crime. Now they say they want to develop a "partnership with the community," but they cannot expect attitudes formed by decades of traditional policing to change overnight. Nor can they expect a community to be interested in a partnership with them unless community residents are persuaded that the police and the community have similar interests and goals.
6. Residents of high-crime areas are all too familiar with police anticrime programs. They have seen such programs come and go and are likely to be skeptical.
7. Studies reveal that citizens who have been personally victimized by crime are usually more dissatisfied with the police than those who have not. Victims of crime want service, and community policing may not seem to have much to offer the crime victim.
8. Police programs imposed upon the community can often create problems because of the diversity of communities. Race, class, and lifestyle differences can all contribute to suspicion and fear among residents. Various groups contending for various governmental services, such as housing and public sector jobs, often find themselves in battle with one another over police priorities. Communities can become polarized if it is perceived that the police are favoring one group over another or treating a particular group unfairly. Further, to protect the rights of a minority, the police may have to act contrary to majority opinions. (Skogan 1996, 31–34)

GUIDELINES FOR SUCCESSFUL IMPLEMENTATION

Wesley Skogan (1996), in his evaluation of community policing in Chicago, offered several guidelines for police administrators who wish to implement community policing programs successfully.

First, the support must be won by the police and not assumed by the police. The police must respond to citizens' interests and deliver on community policing's strategy of problem solving. There must be an organizational framework for responding to citizens' concerns. There must be communication channels to allow the public to articulate their interests and priorities to the community policing officers. One mechanism that has been used in Chicago is beat meetings. Beat meetings are small groups of neighborhood residents and police officers who meet on a regular basis throughout the city to discuss and come up with solutions to community problems.

Second, citizens should be trained as well as the police. They need to know what they can expect from the police and what they can do to contribute to neighborhood problem-solving efforts. When citizens are trained along with the police, they are better able to make decisions about prioritizing problems. Trained citizens will have a better understanding of community policing and their importance to the community policing philosophy.

Third, community organizations have to be actively involved in community policing. Organizations have agendas that keep their members focused on issues even when their leaders change. They provide social benefits to members, such as a sense of solidarity, that can bring people out to meetings even in inclement weather. They also can provide political support for community policing.

EXAMPLES OF COMMUNITY POLICING PROGRAMS

The Community Patrol Officer Program in New York City

In 1982, the New York City Police Department began developing a demonstration project called the Community Patrol Officer Program (CPOP) in the 72 Precinct in Brooklyn. The Vera Institute of Justice, a New York City consultant firm, was invited to assist in developing the project and to study it to determine if it merited expansion to other precincts in the city. This project was designed to incorporate four elements of community policing—"(1) community based crime prevention, (2) proactive service as opposed to emergency responses, (3) public participation in the planning and supervision of police operations, and (4) shifting of command responsibility to lower ranks" (Bayley 1988, 266)—and to be guided by the following operational and organizational principles, which were considered to be common to all community policing programs:

- continuous assignment of police units to specific neighborhoods or beats
- insistence that the unit develop and maintain a knowledge base regarding the problems, cultural characteristics, and resources of the neighborhood
- emphasis on the importance of the unit's reaching out to neighborhood residents and businesspeople to assure them of the presence and concern of the police
- use of formal and informal mechanisms to involve community people in identifying, analyzing, and establishing priorities among local problems and in developing and implementing action plans for ameliorating them

- delegation of responsibility to the community police unit for addressing both crime and order maintenance problems of the neighborhood, and expansion of the unit's discretion in fashioning solutions to those problems
- emphasis on increasing information flow from the community to the police and on using that information to make important arrests and to develop important intelligence on illegal enterprises in the community
- sharing with representatives of the community accurate information on local crime problems and the results of ongoing efforts to address them (New York City Police Department 1987, 4)

In 1984, the project was initiated with the creation of a special unit within the 72 Precinct. The unit consisted of 10 CPOs and a sergeant who reported to the precinct commander. Upon acceptance as a CPO, each individual received 80 hours of training on the philosophy and aims of the CPOP program, tasks to be performed, ethical concerns, and city service agencies and community organizations in the precinct. Each CPO was permanently assigned to a beat ranging in size from 16 to 60 square blocks. Officers were expected to patrol their beats alone and on foot to enhance their accessibility to the people of their beat. They were relieved of the task of responding to 911 calls so that they would have sufficient time to proactively address crime, order maintenance, and serious quality-of-life problems within their beat. With the approval of their sergeant, they could decide the starting and ending times of their tour of duty so that their work schedule could coincide with the time frame in which the problems of their beat were occurring. CPOs were also allowed to determine what day they would have off, with the approval of their sergeant, to coincide with the time of fewest problems. They were expected to become knowledgeable about the community, identify with the community, and assist community residents and businesses. Each CPO was required to maintain a "beat book" that contained his or her monthly work plan and to record information on beat problems, priorities, collective strategies, implementation progress, community leaders/organizations, and businesses (McElroy et al. 1993, 8–9).

CPOs were described as having four major roles:

1. *Planner.* The first responsibility of the CPOs would be to identify the principal crime and order maintenance problems confronting the people within each beat area. Toward this end the CPOs would be expected to examine relevant statistical materials, record their own observations as they patrolled their beats, and solicit and secure input from residents, merchants, and service delivery agents in the community. The problems identified would then be prioritized and analyzed and corrective strategies designed. These strategies would be reviewed with the Unit Supervisor and incorporated in the CPO's Monthly Work Plan, which would form the focus of the officer's patrol for the coming month.
2. *Problem Solver.* CPOs would be encouraged to see themselves as problem solvers for the community. This would begin with the planning dimension of the role described above and proceed to the implementation of the action strategies. In the implementation phase, the officer would be encouraged to mobilize and

guide four types of resources against beat area problems: the CPO acting as a law enforcement officer; other police resources on the precinct and borough levels that could be brought to bear through the CPO sergeant and the precinct commander; other public and private service agencies operating, or available to operate, in the beat area; and individual citizens or organizations in that community. The strategies developed by the CPO would call for the application of any or all of these resources, and the CPO's success in resolving the problems identified would turn in large part on his or her success in marshaling them and in coordinating their application.

3. *Community Organizer.* Community resources cannot be brought to bear on crime and quality-of-life problems unless the community is willing to commit them for that purpose. Increasing the consciousness of the community about its problems, involving community people and organizations in developing strategies to address the problems, motivating the people to help in implementing the strategies, and coordinating their action so that they may contribute maximally to the solution are all aspects of the community organizing dimension of the CPO role. The CPOs would be required to identify potential resources and, where they were not adequate, to help in organizing and motivating the citizenry.

4. *Information Exchange Link.* Through his or her links to the community, the CPO would be in a position to provide the department with information about the problem conditions and locations, active criminals, developing gangs, illicit networks for trafficking in drugs and stolen property, information about citizens' fears, and insights into [citizens'] perceptions of police tactics. In turn, the CPO could provide citizens with information pertinent to their fears and problems, technical information and advice for preventing crimes and reducing the vulnerability of particular groups of citizens, information about the police view of conditions in the neighborhood and strategies for addressing them, and information about police operations in the community. This information exchange aspect of the CPO role was expected to result in arrests that might not occur otherwise, greater cooperation between the police and citizens in addressing crime and order maintenance problems, and a heightened sense in the citizenry that the police were a concerned and powerful resource for improving the quality of life in the community. (Farrell 1988, 78–79)

These four roles were the basis for a guide that instructed CPOs on how to carry out the activities connected with each role. Thus, the officers were not put out on their beats to fend for themselves without guidance. The project was well planned, and all CPOs were informed of what was expected of them.

CPOs' duties as planners included identifying the principal crimes and quality-of-life issues facing the people on their beat. To achieve this goal, the CPOs would review relevant statistical records, observe the conditions on their beat, and solicit information from beat residents and businesses. When a problem was identified, the officers were to give it top priority and solve it with the collective support and input for the people on their beat. The community was to play a key role in identifying problems and finding solutions to them. The vast majority of problems that the CPOs found were quality-of-life issues that were rarely addressed by traditional policing. Some of these issues, such as prostitutes doing business on a street without police interference, were not reported to the police by citizens. In a number of areas, CPOs found small-scale drug operations.

When the officers identified problems, the next step was problem solving. The CPOs had access to four types of resources that could assist in solving problems: (1) their own law enforcement role; (2) other police resources; (3) public or private service agencies; and (4) individual citizens or neighborhood organizations. Use of community organizing could involve enlisting existing community organizations to solve neighborhood problems or developing new community organizations, such as block watches. The CPOs were effective in using existing community resources, but they found it more difficult to establish new community organizations.

CPOs' role in information exchange involved (1) providing the precinct commander with information from citizens about crime, order maintenance, and citizens' attitudes regarding police strategies and tactics and (2) providing the citizens with information on police department activities and priorities. By working closely with citizens, CPOs were able to gain citizens' trust and thereby to gain access to information on criminal behavior in their beat areas that was unavailable to other patrol officers (Weisburd and McElroy 1988, 91–99).

The New York City Police Department implemented a number of preventive tactics to prevent misconduct by CPOs:

1. Careful screening and selection of personnel
2. Functional supervision by all patrol supervisors and the CPOP unit sergeant
3. Staff monitoring of the program by an inspectional unit of the chief of patrol's office
4. Weekly and monthly interviews of merchants and community residents within the CPO beat areas by the unit supervisor, precinct commander, and zone commander to determine the manner in which the CPOs were conducting themselves. (Farrell 1988, 86–87)

The CPOP project was expanded in January 1985, and in September 1988 it became operational in all 75 of the city's precincts. By that point, 750 police officers and 75 sergeants were involved (McElroy et al. 1993, 11).

Support of the CPOP project came from citizens who sent thousands of letters to the department praising the efforts of the CPOs assigned to the precincts. Precinct commanders also supported the CPOP program, because the CPOs filled a void that had existed in police operations: precinct commanders now had a resource to deal with order maintenance problems in neighborhoods (Farrell 1988, 82–83).

Seattle Public Safety Action Plan

In 1988, the National Institute of Justice provided funding for the Seattle Police Department to conduct a crime prevention project that led to community policing in that city (National Institute of Justice 1992). The southeast area of Seattle was the section of the city selected for the project. The southeast was economically behind the rest of the city, with a large proportion of low-income minorities. The crime rate was high, and there was a crack cocaine problem. Commercial burglaries were a major concern. In a meeting of the precinct police commander with businesspeople, the police commander was able to convince the business leaders that the prevalence of crack houses and drug use was tied to the high rate of commercial burglaries. From this meeting, business leaders made a number of recommendations to the mayor about crime. These recommendations led to the creation of a Crime Prevention League that established a variety of crime prevention programs, including a criminal trespass program, in which property owners gave police advance permission to enter their property to investigate and arrest loiterers, and an antigraffiti program in which volunteers would paint over graffiti. These crime prevention programs eventually spread throughout the city.

In 1989, the police department brought in a management consulting firm to recommend improvements in public safety. The city accepted the recommendations and placed them on the November ballot. The plan, known as the Public Safety Action Plan, was approved. It had several provisions to strengthen not only the police but a number of community organizations:

- *Crime Prevention Councils.* The plan authorized the city to allot $95,000 each year to increase citizen involvement in precinct work. This took the form of grants to South Seattle Crime Prevention Council (SSPC) and other crime prevention councils to pay for record keeping, mailing lists, board support, and other expenses.
- *Police Department Advisory Councils.* Funds were also earmarked for the development of citizen-based councils that would advise the precinct commanders on community issues. Precinct commanders would provide input on the selection of board members as well as agenda items.
- *Community-Police Teams.* A key recommendation of the consultants' study that was incorporated into Public Safety Action Plan was to introduce into each precinct of the city a community policing team composed of five officers and one sergeant. The team would give full-time attention to community policing and would be specifically excluded from the responsibility of answering 911 calls. The purpose of creating the specialty teams was to lock the concept of community policing into each precinct.
- *Joint Parks Department and Police Guide Program.* Funds were allocated to a program in which police union volunteers would work with older youth in an evening-hours recreational program.
- *Youth Intervention Program.* The plan called for a program to be jointly planned by the police, the department of human resources, the schools, and community agencies. The program's purpose was to prevent youth from getting involved in gangs and to intervene with youth who were already at high risk of involvement. (National Institute of Justice 1992, 7–8)

The Seattle Public Safety Action plan placed the city and police department in the middle of community policing, because all of the above programs required the cooperation of police and community organizations.

The Seattle policing philosophy was that traditional police methods were not incompatible with community policing. CPOs were expected not only to work proactively on community problems but to function as law enforcement officers by making arrests when appropriate.

SUMMARY

The community policing approach is not entirely new, but it represents a major break with the professional or crime control model of policing that dominated policing agencies for much of the twentieth century, and it is in large part a reaction to the failure of previous crime control methods to adequately prevent, solve, or reduce crime. The concept of community policing has a variety of definitions and meanings but in general can be said to include a problem-solving, results-oriented, consumer-oriented approach to law enforcement and an emphasis on partnership with the community in solving problems of crime, fear of crime, public disorder, and neighborhood decay. Although the effectiveness of community policing in preventing crime is not yet clearly proven, community policing can positively affect citizen attitudes such as fear of crime and satisfaction with the neighborhood. It also facilitates public examination of police practices, makes police more accountable to the community, and stimulates efforts to employ more women and minorities in police work. Benefits to police departments include a more positive image of police, more grassroots support for police, and raised police morale. Some potential problems with community policing are that police action to uphold local norms of order and respectability may go beyond legal bounds (e.g., in harassment of the homeless); that police strategies may selectively benefit the more well-to-do and powerful segments of the community at the expense of the others; that CPOs may use their close community ties for political ends or for their own gain (i.e., corruption); and that police may become too intrusive into residents' private lives.

The implementation of a community policing program may face several obstacles. Many aspects of traditional police organization and culture are resistant to the changes that community policing involves. Also, community involvement may be difficult to obtain, for reasons ranging from residents' lack of time, to residents' mistrust of police or fear of neighborhood criminals, to the lack of a neighborhood infrastructure in high-crime neighborhoods. Community support must be won by police and not assumed by police.

This chapter discusses two programs in detail: the Community Patrol Officer Program in New York City and the Public Safety Action Plan in Seattle. Both programs were initiated on a small scale and eventually expanded to the entire city. These programs were considered to be successful and provided many benefits to the community that traditional policing did not provide.

KEY TERMS

beat book	CPO	Arthur Woods
community policing	CPOP	Mary Ann Wycoff

REVIEW QUESTIONS

1. Define community policing philosophy.
2. Discuss the benefits of community policing.
3. Discuss the potential pitfalls of community policing.
4. Discuss obstacles to implementing community policing.

REFERENCES

Bayley, D.H. 1988. Community policing: A report from the devil's advocate. In *Community policing: Rhetoric or reality,* ed. J.R. Green and S.D. Mastrofski. New York: Praeger.

Bayley, D.H. 1994. *Police for the future.* New York: Oxford University Press.

Brown, L.P. 1989. *Community policing: A practical guide for police officials.* Perspectives on Policing, no. 12. Washington, DC: National Institute of Justice.

Farrell, M.J. 1988. The development of the Community Patrol Officer Program: Community-oriented policing in the New York City Police Department. In *Community policing: Rhetoric or reality,* ed. J.R. Green and S.D. Mastrofski. New York: Praeger.

Goldstein, H. 1979. Improving policing: A problem-oriented approach. *Crime and Delinquency* 25, no. 2:238–241.

Grinc, R. 1994. Angels in marble: Problems in stimulating community involvement in community policing. *Crime and Delinquency* 40, no. 3:437–442.

Holland, M. 1994. Gellatly: Two programs help cut city crime. *Savanna Evening Press,* January 26, p. 12.

Klockars, C.B. 1985. *The idea of police.* Newbury Park, CA: Sage.

MacCulloch, C. 1915. *Outlook,* February 10.

Manning, P.K. 1997. *Police work: The social organization of policing.* 2nd ed. Prospect Heights, IL: Waveland.

McElroy, J., et al. 1993. *Community policing: The CPOP in New York.* Newbury Park, CA: Sage.

Moore, M.H., et al. 1988. *Crime and policing.* Perspectives on Policing Series. Washington, DC: National Institute of Justice.

National Advisory Commission on Civil Disorders. 1968. *Report of the National Advisory Commission on Civil Disorders.* New York: E.P. Dutton.

National Institute of Justice. 1992. Community policing in Seattle: A model partnership between citizens and police. *National Institute of Justice Research in Brief,* August, pp. 7–9.

New York City Police Department. 1987. *The Community Patrol Officers Program: Orientation guide.* New York: NYPD.

President's Commission on Law Enforcement and Administration of Justice. 1967. *The challenge of crime in a free society.* Washington, DC: Government Printing Office.

Skogan, W.G. 1996. The community's role in community policing. *National Institute of Justice Journal,* August, pp. 31–34.

Skolnick, J., and D. Bayley. 1986. *The new blue line.* New York: Free Press.

Skolnick, J., and D. Bayley. 1988. *Community policing: Issues and practices throughout the world.* Washington, DC: National Institute of Justice.

Stipak, B. 1994. Are you really doing community policing? *Police Chief,* October, p. 115.

Trojanowicz, R. 1994. *Community policing: A survey of police departments in the United States.* Washington, DC: U.S. Department of Justice.

Trojanowicz, R., and B. Bucqueroux. 1990. *Community policing: A contemporary perspective.* Cincinnati, OH: Anderson.

Trojanowicz, R., and B. Bucqueroux. 1994. *Community policing: How to get started.* Cincinnati, OH: Anderson.

Watson, E.M., et al. 1998. *Strategies for community policing.* Upper Saddle River, NJ: Prentice Hall.

Weisburd, D., and J.E. McElroy. 1988. Enacting the CPO role: Findings from the New York City Pilot Program in Community Policing. In *Community policing: Rhetoric or reality,* ed. J.R. Green and S.D. Mastrofski. New York: Praeger.

Wilson, J.Q., and G.L. Kelling. 1982. Broken windows. *Atlantic Monthly* 249, no. 3:30.

Wycoff, M.A. 1988. The benefits of community policing: Evidence and conjecture. In *Community policing: Rhetoric or reality,* ed. J.R. Green and S.D. Mastrofski. New York: Praeger.

Chapter 9

Organizational Change and Community Policing

Michael L. Birzer

CHAPTER OBJECTIVES

1. Understand the need for organizational change as the police profession makes the transition to community policing.
2. Describe both traditional and new forms of police organization in terms of organizational structure, leadership, and command-and-control philosophies.
3. Understand how values play a role in defining the police organization's mission and objectives.
4. Identify the support mechanisms (internal and external) that are required for community policing.
5. Understand the concepts of the learning organization and total quality management and how they may be applied to police organizations.

INTRODUCTION

The American workforce is changing in many ways. Workers' self-efficacy and attitudes about work are becoming increasingly important to organizations. Like other fields, policing will need to go through organizational transformation to remain competitive and to attract and retain quality employees. The adoption of a community policing philosophy will also require this transformation.

To the casual observer, policing has made praiseworthy advances during the twentieth century. Police practitioners and academics alike have seen police agencies change from the largely corrupt, inefficient, and ineffective organizations that they frequently were throughout much of the political era, to the efficient, crime-fighting, paramilitary machines that police reformers August Vollmer and O.W. Wilson envisioned. Although we must acknowledge the positive changes that have taken place in policing, further changes will be necessary if police organizations hope to effectively implement community policing.

Implementing change within police organizations as they are traditionally run is not an easy task. As Victor Strecher (1997) pointed out, "The status quo is easier—it is easier to defend (people know it; they're more comfortable with a predictable, known quantity) and it is less risky" (12). Nevertheless, organizational transforma-

221

tion, also referred to as organization engineering, must be initiated and should be the first of many steps in the community policing implementation process.

TRADITIONAL POLICE ORGANIZATION

If we take a critical look at the history of American policing, we find that around the turn of the century there was a movement launched to professionalize the police and to eliminate the corruption and brutal methods that were plaguing policing at that time. This movement, which came to be known as the "reform movement," resulted in a "professional model" of policing that became dominant in police agencies throughout much of this century and in many cases remains in place today. Samuel Walker (1992) described the primary goals of the professional model of policing as

1. eliminating political influence
2. appointing qualified chief executives
3. establishing a mission of nonpartisan public service
4. raising personnel standards
5. introducing principles of scientific management
6. emphasizing military-style discipline
7. developing specialized units (13)

The professional model of policing is grounded in the classical theory of organization. This theory comprises three types of management models: scientific, administrative, and bureaucratic. Aspects of each of these models can be found in some form in police organizations.

Frederick W. Taylor, an industrial engineer who argued that scientific methods could create a more efficient workforce, developed *scientific management.* This system has had a profound effect on twentieth-century American workplaces, including police departments. Scientific management focuses on the basic physical activities involved in the production process (Palmiotto 1997, 111). Taylor used a very sophisticated process to study the motions of specific tasks in the factory production process and the time spent to perform them so that workers would be able to perform the most amount of work in the least amount of time (Koontz et al. 1986, 10). In other words, the aim of the system was to obtain maximum output (and maximum profit) from each worker by standardizing each work process according to "one best way" that would be scientifically determined and dictated from above.

Scientific management as applied to policing might take the form of studying the amount of time spent in processing a specific type of arrest case, studying police response time within a given jurisdiction, or studying the time spent to write traffic citations.

Luther Gulick and Lyndall Urwick developed the system of *administrative management* in the 1930s. They argued that within organizations there should be a unity of command (i.e., each employee answering to one supervisor) and a clearly defined division of labor. Most police organizations today follow administrative management: they are organized according to clearly definable divisions of labor, such as patrol, investigative, warrant service, and traffic functions, and there is also a clear unity of command, with each line police officer knowing the supervisor to whom he or she is accountable in any given situation.

Gulick (1969) coined the acronym POSDCORB, which has become well known in the management literature, to describe the functions of management:

- *P, Planning:* that is, working out in broad outline the things that need to be done and the methods for doing them to accomplish the purpose of the enterprise
- *O, Organizing:* that is, the establishment of the formal structure of authority through which work subdivisions are arranged, defined, and coordinated for the defined objective
- *S, Staffing:* that is, the whole personnel function of bringing in and training the staff and maintaining favorable conditions of work
- *D, Directing:* that is, the continuous task of making decisions and embodying them for specific and general orders and instructions and serving as the leader of the enterprise
- *CO, Coordinating:* that is, the all important duty of interrelating the various parts of the work
- *R, Reporting:* that is, keeping those to whom the executive is responsible informed as to what is going on, which thus includes keeping himself and his subordinates informed through records, research, and inspections
- *B, Budgeting:* including all that goes with budgeting in the form of fiscal planning, accounting, and control (13)

In most government agencies, a formal rational bureaucracy appears to be the norm, and policing is no exception. Most police departments are organized along bureaucratic lines. German sociologist Max Weber described *bureaucratic management* as having been introduced primarily as a reaction to managerial abuses of power. The emphasis of bureaucratic management was on eliminating managerial inconsistencies that would ultimately lead to ineffectiveness. Bennis (1966) described the basic principles of the bureaucratic organization as

1. a division of labor by functional specialization
2. a well-defined hierarchy of authority
3. a system of rules covering the rights and duties of employees
4. a system of procedures for dealing with work situations
5. impersonal relations between people
6. promotion and selection based on technical competence (5)

Typically, the term *bureaucracy* is used to refer to a specific set of structural arrangements and to specific patterns of behavior (Shefritz and Ott 1992, 31). If we take a critical look at police organizations, it is clear that many of the characteristics of bureaucratic management are deeply ingrained in the internal structure of the police organization. For example, most police organizations are shaped like a pyramid, with many layers separating the top command from the police officers at the bottom. Police organizations have a paramilitary rank system, a rigid chain of command, and standard operating procedures that dictate an officer's behavior in a given situation.

The bureaucratic model is often criticized on two grounds. First, it is said to be inflexible, inefficient, and unresponsive to changing needs and times. Second, it is believed to stifle the individual freedom, spontaneity, and self-realization of employees (Perrow 1972), as well as individual creativity, skills, and potentials. The bureaucratic model is also said to be resistant to change. The bureaucratic layers within most police organizations tend to isolate managers from the customer (i.e., the citizenry) and from employees, and this can be disastrous for an organization. Police managers need to be responsive and in touch with both the citizenry and the police employees who are delivering the service for the organization.

One other salient feature found in most police organizations is the adherence to a military authority structure and style of discipline. Characteristics of the paramilitary model of policing are

1. centralized command structure
2. rigid differences among ranks
3. military terminology
4. frequent use of commands and orders
5. rules, regulations, and discipline strongly enforced
6. individual creativity not encouraged
7. system resistant to challenge (Auten 1985, 122–123)

Although the paramilitary model facilitates close supervision of the officer's traditional role, it is inappropriate for the broader, more autonomous role that is necessary with community policing (Weisburd et al. 1989). Furthermore, the paramilitary policing model may not be appealing to the college-educated men and women that police agencies would most like to recruit.

Some police scholars posited that the paramilitary model of policing has created a warriorlike mentality on the part of the police. For example, according to McNeill (1982),

> The police constitute a quasi-military warrior class. In common with warriors generally, they exhibit bonds of solidarity [that] are fierce and strong. Indeed, [their] human propensities find fullest expression in having an enemy to hate, fear, and destroy and fellow fighters with whom to share the risks and triumphs of violent action. (viii)

Although such a warrior culture fosters police solidarity, it tends to discourage the development of close ties with the community of citizens that community policing seeks to create.

The traditional police organization's professional model of policing synthesized aspects of the classical theory of organization (scientific, administrative, and bureaucratic management) and military structure and discipline. The assumptions of the professional model of policing can best be described as follows:

1. The primary function of policing is crime control.
2. Police departments should be independent of politics.
3. Effectiveness and efficiency result from a highly centralized command structure and standardized operating procedures.
4. The police organization should be hierarchical and characterized by clear division of labor and task specialization.
5. Officers should be selected on the basis of established recruitment standards.
6. Officers should be well trained and disciplined.
7. Preventive random motorized patrol deters crime.
8. Policing should use modern technology.
9. Police officers should enforce laws impartially.
10. Crimes are solved by scientific investigative methods. (Fyfe 1997)

The classical theory of organization, based on a military structure of authority and style of discipline and a rigid centralized command, has continued to guide police agencies because it is a convenient way for supervisors to ensure that line-level police personnel follow orders dictated by central headquarters (Birzer 1996, 6–10). Although its harsh punishments for even the slightest infraction of departmental rules and regulations and its close supervision of rank-and-file officers corrected many of the problems that faced policing during an earlier time, it has, over time, created a host of new problems. The classical theory fostered an assembly-line mentality among rank-and-file police officers and a focus on technical efficiency and maximum outputs. As police departments move toward community policing, their current organization must change.

The classical organization of policing that developed with the professional model has had a profound effect on the service delivered to the citizenry. The classical organization nurtured an environment in which rewards and promotions were made on the basis of the number of outputs an officer achieved. In many cases, performance evaluations were weighted heavily on the number of cases, tickets, and arrests an officer obtained during a specific evaluation period. The message sent to line-level police officers was that maximizing the number of outputs (quantitative measures) was important to achieve recognition for a job well done. This was problematic inasmuch as it ignored the qualitative aspects of policing, which are important in a community policing approach. Herman Goldstein (1990) asserted that the police need to move from an efficiency model emphasizing numbers and response times to an effectiveness model in which qualitative issues such as eliminating the root causes to community problems become important.

The professional model of policing created an "us against them" attitude on the part of the police. The citizen's role was to stay out of the way and let the experts (police) do what they were paid to do, crime fighting. The main responsibility of the police under the professional model was to use the mechanisms of the criminal law and prosecutions to maintain order, and the effectiveness of police was judged in large part by their ability to do this.

The crime fighter mentality, over time, created a legacy that detached the police from the very communities they served. In essence, the police were a reactionary force that functioned autonomously from the community. In contrast, the community policing model requires the police and the community to work together to solve neighborhood problems, prevent crime, and improve the neighborhood's quality of life.

A comparison of traditional policing and community policing by Sparrow (1988) is shown in Table 9–1.

NEW MODELS OF ORGANIZATION

Total Quality Management

Total quality management (TQM) was a system initially devised by W. Edwards Deming to combat the stagnation of American-operated businesses and governmental institutions that had resulted from a bureaucratic management style. Since the early 1980s, TQM has been adopted by many companies, such as the Ford Motor Company and the Florida Power and Light Company, as well as colleges and universities, the military, and some police departments, such as the Madison, Wisconsin, Police Department. Deming management is based on what are called the Fourteen Points:

1. *Create constancy of purpose for improvement of product and service.* Deming suggests a radical new definition of a company's role: rather than to make money, it is to stay in business and provide jobs through innovation, research, and constant improvement and maintenance.
2. *Adopt the new philosophy.* Americans are too tolerant of poor workmanship and sullen service. We need a new religion in which mistakes and negativism are unacceptable.
3. *Cease dependence on mass inspection.* American firms typically inspect a product as it comes off the assembly line or at major stages along the way; defective products are either thrown out or reworked. Both practices are unnecessarily expensive. In effect, a company is paying workers to make defects and then to correct them. Quality comes not from inspections but from improvements of the process. With instruction, workers can be enlisted in this improvement.
4. *End the practice of awarding business on the price tag alone.* Purchasing departments customarily operate on orders to seek the lower-priced vendor. Frequently, this leads to supplies of low quality. Instead, buyers should seek the best quality in a long-term relationship with a single supplier for any one item.
5. *Improve constantly and forever the system of production and service.* Improvement is not a one-time effort. Management is obligated to continually look for ways to reduce waste and improve quality.
6. *Institute training.* Too often, workers have learned their job from another worker who was never trained properly. They are forced to follow unintelli-

Table 9–1 Traditional versus Community Policing: Questions and Answers

Question	Traditional	Community Policing
Who are the police?	A governmental agency principally responsible for law enforcement.	Police are the public and the public are the police: the police officers are those who are paid to give full-time attention to the duties of every citizen.
What is the relationship of the police force to other public service departments?	Priorities often conflict.	The police are one department among many responsible for improving the quality of life.
What is the role of the police?	Focusing on solving crimes.	A broader problem-solving approach.
How is police efficiency measured?	By detection and arrest rates.	By the absence of crime and disorder.
What are the highest priorities?	Crimes that are high value (e.g., bank robberies) and those involving violence.	Whatever problems disturb the community the most.
What, specifically, do police deal with?	Incidents.	Citizens' problems and concerns.
What determines the effectiveness of the police?	Response times.	Public cooperation.
What view do police take of service calls?	Deal with them only if there is no real police work to do.	Vital function and great opportunity.
What is police professionalism?	Swift, effective response to serious crime.	Keeping close to the community.
What kind of intelligence is important?	Crime intelligence (study of particular crimes or series of crimes).	Criminal intelligence (information about the activities of individuals or groups).
What is the essential nature of police accountability?	Highly centralized; governed by rules, regulations, and policy directives; accountable to the law.	To preach organizational values.
What is the role of the press liaison department?	To keep the "heat" off operational officers so they can get on with the job.	To coordinate an essential channel of communication with the community.
How do the police regard prosecutions?	As an important goal.	As one tool among many.

Source: Reprinted from M.K. Sparrow, Implementing Community Policing, *Perspectives on Policing,* pp. 8–9, 1988, United States Department of Justice.

gible instructions. They can't do their jobs well because no one tells them how to do them.

7. *Institute leadership.* The job of a supervisor is not to tell people what to do or to punish them but to lead. Leading consists of helping people do a better job of learning by objective methods of determining who is in need of individual help.

8. *Drive out fear.* Many employees are afraid to ask questions or to take a position, even when they do not understand what their job is or what is right or wrong. They will continue to do things the wrong way or not do them at all. The economic losses from fear are appalling. To assure better quality and productivity, it is necessary that people feel secure.

9. *Break down barriers between staff areas.* Often a company's departments or units are competing with each other or have goals that conflict. They do not work as a team so they can solve or foresee problems. Worse, one department's goals may cause trouble for another.

10. *Eliminate slogans, exhortations and targets for the work force.* They never help anybody do a good job. Let workers formulate their own slogans.

11. *Eliminate numerical quotas.* Quotas take into account only numbers, not quality or methods. They are usually a guarantee of inefficient and high cost. A person, to hold a job, meets a quota at any cost, without regard to damage to his company.

12. *Remove barriers to pride of workmanship.* People are eager to do a good job and distressed when they cannot. Too often, misguided supervisors, faulty equipment and defective materials stand in the way of a good performance. These barriers must be removed.

13. *Institute a vigorous program of education and retraining.* Both management and their work force will have to be educated in the new methods, including team work and statistical techniques.

14. *Take action to accomplish the transformation.* It will require a special top management team with a plan of action to carry out the quality mission. Workers cannot do it on their own, nor can managers. A critical mass of people must understand the Fourteen Points. (Walton 1990, 17–19)

Although Deming's Fourteen Points appear to be geared toward the private sector, the major premises can be applied by police administrators as well. The orientation to continually improving service, the replacement of quantitative goals (quotas) with qualitative goals, the emphasis on the crucial feedback of line personnel, the enlargement of the roles of lower-level employees through increased training, and the use of cross-unit, diverse teams to analyze production/service delivery processes fit in well with the new police strategies of problem-oriented policing and community-oriented policing.

The Learning Organization

One vision of organization in the contemporary management and organizational development literature that has shed light on many issues is the concept of the learning organization.

Peter M. Senge, in his landmark book *The Fifth Discipline: The Art and Practice of the Learning Organization* (1990), described the learning organization as one "where people continually expand their capacity to create the results they truly desire, where new and expansive patterns of thinking are nurtured, where collective aspiration is set free, and where people are continually learning how to learn together" (3). In essence, Senge proposed a systems approach to doing business—one that encourages a continuing flow of feedback from the external environment and through and among organizational units to promote the learning necessary for the organization to adapt to changing conditions.

William Geller (1997) offered several suggestions that may help to make police departments learning organizations. A few of these suggestions are creating a research and development division that would actually do empirical research (e.g., testing a possible correlation between foot patrol activities in specific areas of a community to decreases in crime in comparison to other areas, or measuring citizen satisfaction and fear of crime levels in various areas where different patrol methodologies are deployed); developing a structure or process to foster learning (e.g., crime analysis that spans work units); developing a process to foster learning involving senior police officials that could reduce battles between departmental units; taking a talent inventory of sworn and civilian personnel; organizing learning around problem solving; instituting a bottom-up performance appraisal of organizational performance; and expanding police-researcher partnerships.

A police agency that became a learning organization might well raise the morale of police employees. Employees could examine processes and recommend changes without fear of reprisal and could be given more scope for their problem-solving talents. Such developments would fit in well with the community policing model.

ORGANIZATIONAL CHANGE

Change in policing, as in any field, has been driven by a wide range of factors:

1. Increases in crime in a particular area and/or the fear of crime
2. Budgetary concerns—a need to do more with less
3. Citizens' greater expectations of the police
4. An increase in specific types of crimes (i.e., juvenile crimes and domestic violence)
5. Technological changes to the nature of police work

These trends require the police to critically rethink the ways in which they provide police service. Traditional policing has failed to keep pace with societal complexities.

As discussed in previous chapters, community policing is a response to these complexities. It requires that the police become proactive problem solvers in the communities they serve. Police are expected to become resource catalysts and direct links to the many public and private agencies that may foster the problem-solving process. Community policing is also *coactive* in that the police and the community work together to solve problems of the community. It requires the police to become street-level criminologists and to think more in terms of the root causes to problems—for example, "Why is this area of the community having a large amount of reported burglaries?" In this case, police would be expected to investigate a variety of potential causal factors that might include environmental characteristics, victim characteristics, and offender characteristics. They might link the burglaries to truancy rates at schools in the area, if for example, the burglaries were occurring during the day. Then they might address the problem of truant juveniles (root cause) to solve the burglary problem (result).

Because community policing represents a fundamental change in the way that police do business, it requires organizational change. Effective organizational change will involve virtually all aspects of the police organization. Four aspects are critical: (1) training, (2) promotions, (3) leadership, and (4) organizational structure.

Training

The nature of police training needs to be reexamined as organizations move toward community policing strategies. The Bureau of Justice Statistics found in 1983 that about 10 percent of police patrol activity is spent on criminal-related matters and the other 90 percent on dealing with a large variety of service-related calls. The problem here is that police agencies put a great deal of time and energy into training police for what they will be doing 10 percent of the time. Conflict resolution, communication skills, cultural diversity, resource identification, and problem solving are some of the many new subjects and skills that should be taught and fostered in the police training curriculum.

After recruit officers graduate from the academy, periodic in-service training should be offered to reinforce the community policing philosophy of the organization. In-service training can bring real problems faced in the community into the classroom for group problem-solving activities. Training in all aspects of community policing must be continual for everyone from the chief to the line-level police officer. Police training must be viewed as a dynamic process (Alpert and Dunham 1992, 52).

Promotions

Promotions are an important target for organizational change. One misfit or unqualified individual who is granted promotion may obstruct any prospect of change for many years. And what better way to foster the culture and direction the chief desires than to promote police officers who are committed to the process of change? The chief sends a strong message to the rank and file based on whom he or she promotes. According to Couper and Lobitz (1991),

> Promotions are the lifeblood of an organization and have a tremendous symbolic as well as actual impact. Who gets promoted sends a louder message than any words from management. It sends a strong message through-

out the organization as to the direction of the department and the values of top management. Promotion decisions can affect the department for years to come through the people who are selected. (67)

According to O.W. Wilson and R.C. McLaren (1977), "Selection for promotion presents greater difficulties than selection of recruits. It is also more important" (275). Promotions should be granted to those personnel who have the tenacity and willingness to effect substantial change from the traditional model to the community policing model. The days of promotions based on "whom one knows" or politics simply must end. The promotional process can be a tremendous opportunity to reinforce the department's direction and goals. As candidates prepare for the promotional test, they will be reminded of the important issues facing the department; as they take the test, they will be reminded that the organization supports and values those goals (Booth 1995).

Leadership

"By definition, leaders lead change" (Bardwick 1996, 131), and policing is at the crossroads of change. Thus, the profession is in desperate need of leaders. The traditional role of police leader, much like that of leader in the corporate world, has been largely one of cop, referee, devil's advocate, dispassionate analyst, professional, decision maker, naysayer, and pronouncer (Peters and Austin 1985, 311). But the new police leadership role, in the words of Peters and Austin (1985), will be more that of "cheerleader, enthusiast, nurturer of champions, hero finder, wanderer, dramatist, coach, facilitator, builder" (311). To effectively stimulate organizational change, leaders must actively engage their employees. They must pull their organizations forward and guide the organization through rough times. They must have a vision of where they want to take the organization and be able to articulate that vision to their subordinates. More important, they must articulate to their employees how they plan to accomplish the visionary journey and what is in it for the employees. And they must have the exuberance to encourage creativity and risk taking on the part of every police employee.

Total quality management, as discussed earlier in this chapter, has become a guiding force in business and industry in America and in many public agencies, including some police agencies. The Madison, Wisconsin, Police Department has developed, in the spirit of TQM, 12 principles of "quality leadership":*

1. Believe in, foster, and support **TEAMWORK.**
2. Be committed to the **PROBLEM-SOLVING** process; use it, and let **DATA,** not emotions, drive decisions.
3. Seek employees' **INPUT** before you make key decisions.

Source: Reprinted with permission from D.C. Couper and S.H. Lobitz, *Quality Policing in Madison: The Madison Experience,* p. 48, © 1991, Police Executive Research Forum.

4. Believe that the best way to improve the quality of work or service is to **ASK** and **LISTEN** to employees who are doing the work.
5. Strive to develop mutual **RESPECT** and **TRUST** among employees.
6. Have a **CUSTOMER** orientation and focus toward employees and citizens.
7. Manage the **BEHAVIOR** of 95 percent of the employees, and not the 5 percent who cause problems. Deal with the 5 percent **PROMPTLY** and **FAIRLY.**
8. Improve **SYSTEMS** and examine **PROCESSES** before placing blame on people.
9. Avoid "top-down," **POWER-ORIENTED** decision making whenever possible.
10. Encourage **CREATIVITY** through **RISK-TAKING,** and be **TOLERANT** of honest **MISTAKES.**
11. Be a **FACILITATOR** and **COACH. DEVELOP** an **OPEN** atmosphere that encourages providing and accepting **FEEDBACK.**
12. With **TEAMWORK,** develop with employees agreed upon **GOALS** and a **PLAN** of action to achieve them.

These 12 principles reflect a more democratic, flexible, and proactive leadership style that calls for continually monitoring of the environment and identifying better ways of doing business rather than merely reacting to breakdowns in the system or errors on the part of employees.

Under the traditional police command-and-control structure, leaders are not sufficiently challenged and stimulated by detailed and truthful feedback from lower levels. As Braiden (1992) commented, "We are in great need of people of character who will blow the whistle on some of the senseless things we do, as a matter of routine. Sadly, most chiefs have surrounded themselves with obsequious bureaucrats who routinely provide them with echoes of their own opinions on every issue" (91). Thus, leadership issues are intimately bound up with issues of organizational structure, which will be discussed next.

Organizational Structure

The traditional police organizational structure is pyramidal in design and has many layers of command separating the chief from the line-level personnel (Figure 9–1). Many scholars have concluded that this steep hierarchical structure often makes police departments dysfunctional and that police organizations have to be restructured to become more efficient, more effective, and more responsive to the community (Palmiotto 1997, 129). The problem is clear: America's police have been and are currently bound to an organizational structure designed to react to crime rather than to prevent crime or to address the host of issues that cause it (Wadman and Olson 1990, 18).

Some specific criticisms of the traditional police organization's structure are as follows:

1. The traditional structure cannot support new, more democratic leadership styles.
2. Today's police officers are very different from the officers of just 20 years ago. They want and need to ask questions about issues affecting them in the work

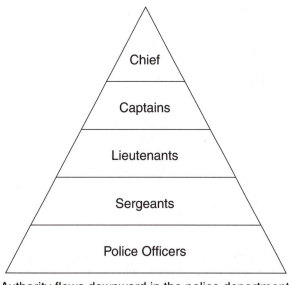

Authority flows downward in the police department.
This chart shows different levels of authority.

Figure 9–1 Pyramid/Hierarchy of Authority. *Source:* Reprinted with permission from M. Palmiotto, *Policing: Concepts, Strategies, and Current Issues in American Police Forces,* p. 118, © 1997, Carolina Academic Press.

environment. The traditional structure of police command and control is in direct conflict with this new mentality.

3. The traditional structure creates impediments to the adoption of new ideas. The traditional emphasis is on rules and regulations, which can mean that when ideas run counter to policy, accepted practice, or tradition, they are labeled "unworkable" and discarded before they have a chance (Wadman and Olson 1990, 57). Further, the requirement that line-level personnel's suggestions make their way up a steep chain of command means that the transmission of good ideas may be delayed, distorted, or prematurely stopped.

4. The traditional top-down structure is far too centralized. Resources are allocated according to generic plans drawn up at headquarters, and the officers who are in closest contact with operational problems have little opportunity to shape policy (Bayley 1994, 88).

A flattened, more horizontal and decentralized structure allows an organization to react to the changing environment more effectively. When the organization structure is flattened, employees at lower levels of the organization will make more decisions, including some decisions that have traditionally been made by middle management or, in some cases, upper command. The benefits of flattening the organizational structure are innumerable. They include

1. communication enhancement (fewer layers of the chain of command to send information through)
2. empowerment of employees at the bottom of the organization to make more decisions
3. increased participation in the problem-solving process of those employees closest to the problems
4. improved service delivery: the police officers at the bottom of the pyramid are closest to the customer (citizen) and are in an advantageous position to identify methods to improve service delivery
5. enhanced bottom-up input on policy development
6. improved coordination of services internally within the police department
7. reduced bureaucracy, red tape, and time delays

Structural change is the most difficult to effect within an organization. In many cases, accomplishing it is like the tedious and time-consuming task of turning an aircraft carrier. It involves everything from commander-officer relationships to communication patterns, decision-making procedures, accountability-commitment provisions, and reward systems (Gray et al. 1997, 43). Nevertheless, if organizations are to be successful with community policing, serious examination of the organizational structure must be undertaken.

ORGANIZATIONAL VALUES

Organizational change requires the clarification and articulation of the new values that the organization wishes to adopt. Values are the beliefs that guide an organization and the behavior of its employees. They are what the organization takes seriously and practices on a daily basis. The values of an organization can be seen in the actions and attitudes of employees, in jokes, in solemn understandings, and in internal explanations for actions taken. Values are what drive the men and women in the organization. The most important beliefs are those that set forth the ultimate purpose of the organization (Wasserman and Moore 1988, 270–276).

Every police organization maintains underlying values, whether articulated or not, that influence how it operates. Traditionally, the values that were deemed important within a police organization under the professional model of policing revolved around efficiency, law enforcement, the paramilitary chain of command, and strict observance of the procedure manual and/or rules and regulations. These values created a rule-driven and punitive environment and fostered a "Just the facts, ma'am" mentality on the part of the police. If community policing is to be successfully adopted, the values of police organizations must become partnership, problem solving, empowerment, and quality. Policing must abandon the rule-driven and law enforcement–driven value system that is still prevalent within many agencies.

Peters and Waterman, in their well-known book *In Search of Excellence* (1981), stated that excellent companies tended to be "hands on and value driven" (279) and to share certain core beliefs:

1. A belief in being the "best"
2. A belief in the importance of the details of execution, the nuts and bolts of doing the job well
3. A belief in the importance of people as individuals
4. A belief in superior quality and service
5. A belief that most members of the organization should be innovators, and its corollary, the willingness to support failure
6. A belief in the importance of informality to enhance communication
7. Explicit belief in and recognition of the importance of economic growth and profits (285)

These beliefs are consonant with a community policing model. (Belief no. 7 could be adapted to policing by the replacement of "economic growth and profit" with the nonmonetary "profit" of customer satisfaction, where "customers" are the citizens that police serve.)

The Value Statement

Value statements in policing may appear to be a new phenomenon, but a brief review of American police history reveals that many chiefs of the past established and espoused values for their organizations. Further, many chiefs of police throughout history have made their value statements public, thereby letting citizens know what their police department has deemed to be important. In this, they have acted like private industries, which often advertise that "they value their customers" and have devised value statements revolving around quality service to consumers.

O.W. Wilson, for example, while serving as chief of police in Wichita, Kansas (1928–1939), made public a set of values for the police department that he called the "Square Deal Code":

> TO SERVE ON THE SQUARE: To be a friend to MAN; to protect citizens and guests, safeguard lives, guarantee liberty and assist in the peaceful pursuit of happiness; to be honest, kind, strong and true, always proud of our department and City; to give friendly aid in distress; to be a gentleman, practicing courtesy and weaving a daily thread in the habit of politeness, always thoughtful of the comfort and welfare of others; to keep our private lives unsullied, an example for all; to be honorable that our character may be strengthened; to bear malice or ill will toward none; to guard our tongues lest they speak evil; to act with caution lest our motives be questioned; to be courageous and calm in the face of danger, scorn or ridicule; to ignore unjust criticism and profit by good; to be alert in mind, sound in body, courteous in demeanor, and soldierly in bearing, a comfort to the distressed, a protector of the weak, and a pride to our city; to practice self-restraint, using no unnecessary force or violence; to intimidate no one nor permit it to be done; to take appropriate action whenever an offense is committed; to have moral courage to enforce

the law; to be unofficious, firm but mindful of rights of others; never to permit personal feelings, animosities or friendships to influence decisions; never to act in the heat of passion; to assist the public in their compliance with regulations; to save unfortunate offenders from unnecessary humiliation, inconvenience and distress. With no compromise for crime, to be relentless toward the criminal, our judgment charitable toward the minor offender; never to arrest if a summon will suffice; never to summon if a warning would be better; never to scold or reprimand but inform and request. All this to make WICHITA a better place to live. For we stand for RIGHT, JUSTICE, and a SQUARE DEAL. (Bopp 1977, 138–138)

After examining the Square Deal Code, is there any question about what the Wichita Police Department stood for and what values they honored?

When police departments make the transition from traditional policing to community policing, they may find it helpful to publish a value statement expressing community policing values, both to educate the public and to guide the workforce of their own departments. Interestingly, the Newport News Police Department's value statement encompasses not only the external value system (service delivery) but also the internal value system (the way employees are treated). In general, value statements guided by the community policing philosophy include not only democratic participation of the community in policing matters but also democratic participation of rank-and-file police employees.

In 1982, the Houston, Texas, Police Department also developed a value statement to reflect its changing organizational strategies:

- The Houston Police Department will involve the community in all policing activities which directly impact the quality of community life.
- The Houston Police Department believes that policing strategies must preserve and advance democratic values.
- The Houston Police Department believes that it must structure service delivery in a way that will reinforce the strengths of the city's neighborhood life.
- The Houston Police Department believes that the public should have input into the development of policies which directly impact the quality of neighborhood life.
- The Houston Police Department will seek input of employees into matters which impact employee job satisfaction and effectiveness. (Wasserman and Moore 1988, 273)

Adherence to Values

Unfortunately, values in police organizations may be ambiguous at best. In many cases, police management sends out a contradictory message to the rank and file—for example, articulating a community policing culture but disciplining officers for not achieving a given number of cases, tickets, and arrests at the end of the month. This may result in role conflict on the part of police officers and ineffectiveness in the area of service delivery.

Sometimes, values are practiced by one entity of the department while being ignored by others. In many traditional police organizations, suboptimization—competition among individuals and units over scarce resources (Gaines et al. 1991, 37)—makes any realistic attainment and belief in common core values unlikely at best. It is crucial that values be shared throughout the organization and practiced by all.

OBTAINING OUTSIDE SUPPORT FOR COMMUNITY POLICING

For community policing to be successful, support must be obtained from a number of sources both outside and inside the police organization. This section will focus on the support from outside the organization, and the next will discuss support from inside the organization (management support).

According to Trojanowicz and Bucqueroux (1994), six groups must be identified and must work together to ensure the success of community policing efforts:

1. *The police department*—including all personnel, from the chief to the line officer, civilian and sworn
2. *The community*—including everyone, from formal and informal community leaders such as presidents of civic groups, ministers, and educators; to community organizers and activists; to average citizens on the street
3. *Elected civic officials*—including the mayor, city manager, city council, and any county, state, and federal officials whose support can affect community policing's future
4. *The business community*—including the full range of businesses, from major corporations to the "mom and pop" store on the corner
5. *Other agencies*—including public agencies (code enforcement, social services, public health, etc.) and nonprofit agencies, ranging from Boys and Girls Clubs to volunteer and charitable groups
6. *The media*—both electronic and print media (2)

Strategies for obtaining support from each of these segments will be discussed in detail below.

Community

It is critical that the police agency involve the community early on in the community policing implementation process in order to begin building public support. The first of many steps involving the community is usually educating the community about what community policing is and what it means for their neighborhoods. One of the most effective methods of educating community members is to hold several small community meetings. These should be open to any member of the public and should be publicized extensively. The police executive him- or herself should be present at each of these meetings, as well as precinct or sector commanders and as many lower ranking personnel as wish to attend. The presence of the chief or sheriff reinforces the importance of these meetings and assures the public that they are not merely "show-and-tell" sessions (Watson et al. 1988, 76). During these meetings, police executives and community police officers should provide the community with a general over-view of community policing and the roles of the police and community. At this time, leaders within specific neighborhoods will begin to emerge and be identified. Neighborhood leaders take on a crucial role later in the implementation process, assisting community policing officers in organizing groups and neighborhood teams and encouraging other citizens to become involved.

The location of meetings should be readily accessible to all community members. Schools, community recreation centers, and neighborhood police stations work well for meetings. The site for the meeting should comfortably accommodate a large number of people.

The Chicago Police Department held regular beat meetings during its transition to community policing in an effort to gain support of community residents. The officers who attended were those officers that patrolled the neighborhoods where the citizens in attendance resided. The meetings accomplished a number of objectives. There were many opportunities for the police and citizens to exchange information. Participants learned about police procedures, including when to call 911 rather than a list of alternate nonemergency numbers, and police learned something about residents' priorities (Skogan and Hartnett 1997, 113). Furthermore, the Chicago Police Department, by bringing residents and officers together, began to nurture community support for community policing and build relations of trust between the police and the community.

After neighborhood meetings are conducted, frequent meetings should be held for the purpose of identifying problems, cares, and concerns of the community. The police and affected community should be involved in tailoring solutions to identified problems. When citizens become engaged with the police in the problem-solving process, their support will become even stronger.

Elected Officials

The support of politicians (e.g., mayor, city/county commission, city/county manager) in any new police venture is advantageous for the police organization. In fact, it is invaluable. In obtaining support from politicians for community policing, the kind of information that police executives furnish to politicians is important. Traditionally, police executives have presented numbers to the politicians in order to secure additional personnel and/or a larger share of the budget. But if politicians are to support the community policing philosophy, they must be educated in all its aspects. For instance, police executives should inform politicians that rapid response to calls has been shown to be less effective at catching criminals than educating the public to call the police more quickly after a crime is committed (Peak and Glensor 1996, 189; Petersilia 1993). Also, politicians must understand that an increase in crime-related reports may indeed occur under community policing because the increased personal contact between the police and the citizenry will make citizens feel more comfortable in reporting crime to the police.

Politicians must be actively engaged in the community policing implementation process. Many police executives make the unfortunate mistake of assuming that politicians are well versed on police operations and matters. This is normally not the case. Politicians should be invited to observe and actively participate in community policing training with community members and police department personnel. They must understand the community policing philosophy in order to provide the support necessary.

Recently, when I provided community policing transitional training for the Waterford, Michigan, Police Department, I brought local politicians and community leaders into the training by designing separate training sessions specifically for them. These sessions provided a basic overview of community policing and what was required from politicians. The Waterford Chief of Police addressed the participants of the training sessions and articulated the new direction in which he was going to take the police department with community policing. This training proved to be very effective, and politicians came away from the training sessions as staunch supporters of community policing.

News Media

The news media, both print and electronic, play a crucial role in gaining support for community policing. Historically, the police have used the media only for limited purposes and have overlooked the many opportunities that the media can provide. Within the framework of community policing, the media become coactive partners with law enforcement. They can keep the community informed on community policing developments, particularly during the implementation process, publicize community meetings, and report on police community problem-solving endeavors. Media sources are an excellent avenue for the chief executive to market the community policing philosophy. Media accounts of community policing and other activities can assist the police department in obtaining additional resources (Peak and Glensor 1996, 190).

Business Leaders

Business leaders can provide invaluable support for community policing. They should be invited and encouraged to attend neighborhood meetings, particularly if their business is in close proximity to areas where community policing has been or will be implemented. Their influence can be beneficial for many reasons. First, they can provide financial support and donate equipment for community policing projects. Second, they can become allies in the problem-solving process. Third, they can donate time, provide valuable information, and become a major factor in determining, for example, whether a downtown area receives funds for obtaining more officers or the department receives additional equipment (Peak and Glensor 1996, 189).

One especially valuable resource of the business community is landlords. Landlords can assist in situations where the police are continually responding to repeat calls—for example, at the same apartment complex. Landlords can work with police in tailoring solutions to problems affecting apartments and other rental housing. They should be informed of the wide array of services that community policing officers can provide them under community policing, such as screening potential tenants and keeping the area free from abandoned vehicles and trash (Peak and Glensor 1996, 189). The police should educate landlords about the important roles they can play in community policing. Effective community policing may increase property values by reducing crime and improving the quality of life, thereby encouraging more ownership in the area (Trojanowicz and Bucqueroux 1994, 23).

Other Organizations

Several other organizations deserve mention for their support of community policing. The Boys and Girls Club of America is one such organization. It provides many activities for young people, thereby keeping them involved in constructive activity. At the time of this writing, the one-time dues for membership in the Boys and Girls Club of South Central Kansas was $5.

The Wichita, Kansas, community policing program provides one example of the partnership and support that the police and Boys and Girls Clubs can offer one another. Community policing deputies from the Sedgwick County, Kansas, Sheriff's Department, along with the Boys and Girls Club, a local neighborhood association, and Community Housing Services of Wichita/Sedgwick County, planned and successfully conducted a "Hoops for Youth" basketball tournament that targeted area youth between the ages of 14 and 17. The purpose of the tournament was to allow these agencies to come together to work with and recruit teenaged youth into the club. The tournament also provided an opportunity for young people to see deputies in a different role than that of traditional law enforcement and to form a bond with them.

Other agencies that can offer support to the police are United Way, the Salvation Army, neighborhood associations, food banks, homeless shelters, community housing services, and the various street outreach programs that operate in many communities. It is imperative that key leaders of these organizations be given a general overview of community policing. The police should actively solicit and form coalitions with these and many other private and public agencies.

OBTAINING MANAGEMENT SUPPORT FOR COMMUNITY POLICING

For a successful transition to community policing, internal support from police management is as important as or perhaps more important than external support. Executives and first-line supervisory support for community policing is especially critical. The police executive must not only articulate the community policing philosophy at every opportunity but also incorporate it into everything from selection of new officers to promotions.

Within the hierarchical structure of most police organizations, it makes sense that the most ardent supporter of the community policing philosophy should be the chief executive (chief or sheriff). The chief executive must be a change agent in several key areas: (1) decentralization, (2) empowerment, (3) communication, and (4) work environment.

Decentralization

The chief executive must allow for the decentralization of police operations. A common criticism of most police organizations is that the command-and-control structure is too centralized. At one time, a centralized command-and-control structure was assiduously promoted by reformers like August Vollmer and later O.W. Wilson as the best prescription for the corruption that plagued many police organizations. This structure remained in place for many years and is still the status quo in many organizations. But for management to support the changes required for community policing, it must now change.

For community police officers to work out of a central location removed from the community would defeat the purpose of community policing, which is responsiveness to community needs.

The chief executive and his or her staff must allow operations to decentralize and move into the community. Many agencies have accomplished this by opening up neighborhood stations, storefronts, satellite offices, and substations.

Decentralization is such a popular idea that forces of all sizes now say they are doing it (Bayley 1994, 88). The benefits of decentralization are many. Commanders and supervisors working out of a decentralized location will be closer to community problems and will be in a position to have input into decisions involving resource allocation, strategies, and allocations—decisions that in the past were often reserved for central headquarters. Furthermore, personnel working out of decentralized locations in the community will become more in tune to the problems, cares, and concerns of that specific community and thus will be in a more advantageous position to tailor and implement solutions.

Rosabeth Moss Kanter, addressing organizational change in her book *The Change Masters* (1983), wrote that "the last broad condition for power circulation is local access to resources" (169). If we apply Kanter's underlying thesis to policing, the implication is clear. The decentralized police facility allows the community to have better access to police services, thereby empowering them and may, ultimately, lead to more satisfaction on the part of both the police and the community.

Empowerment

The police executive and top command must empower those at the bottom to make more decisions and to develop innovative solutions to community problems. Empowerment includes encouraging risk taking at all levels of the organization. The police executive should ensure that the rules and regulations manual contains enough discretion to allow the rank and file to experiment and take risks. Empowerment is critical in sustaining long-term organizational change. Organizations that are change oriented will have a large number of integrative mechanisms encouraging fluidity of boundaries, the free flow of ideas, and the empowerment of people to act on new information (Kanter 1983, 32).

The police executive has a responsibility to ensure that top command is wholeheartedly supportive of empowerment. Police commanders must be made to understand that empowering subordinates does not necessarily mean giving up their power and that it is a means of providing more tools for employees to participate in the organization. An empowered organization will potentially respond and adapt to change more easily then an organization constrained in the traditional command-and-control system.

Communication

To maintain support for community policing, communication must be continual between management and rank-and-file police employees. Police employees must be kept apprised of the change process and what exactly is occurring within the department. In many organizations, this is difficult due to the many layers that separate the command staff from the line-level police officer.

Within police organizations, there are basically three types of formal communication channels: downward, upward, and lateral. In most police organizations, the downward and upward communication channels appear to be used the most.

Downward communication transmits information from the top of the organization to the bottom. Most of this information is in the form of standard operating procedures, directives, special orders, and general orders. In essence, downward communication is the flow of information down the chain of command.

Upward communication transmits information from the bottom of the organization to the top. This is the communication channel used by subordinates in the organization. Upward communication usually focuses on making management aware of problems, presenting ideas, suggesting improvements in performing certain tasks, and expressing feelings about certain aspects of the job. Upward communication can be somewhat frustrating for line-level police officers because of the chain of command that the information that must flow up through. For example, a police officer who wishes to communicate with a major must first go through his or her sergeant and then the lieutenant, the captain, and finally the major. This can be frustrating in and of itself, not to mention the confusion that may be created in the process. In some cases, the information may be filtered out by gatekeepers or simply stopped before it reaches the designated commander at the top of the organization. In some cases, it may reach top command after a delay of several weeks. Or, it may reach the top and be ignored, so that the subordinate never receives a reply or an answer to a request or proposal. This can be debilitating for any true form of community policing. Yet another problem with upward communication is that often employees communicate only what the boss wants to hear.

The lateral communication channel is the channel used by peers. For example, a police officer working on the day watch may report an activity to a detective or an officer working the night watch. This type of activity is critical in policing for the coordination of activities and events. Lateral communication can become ineffective when one functional unit withholds information from another unit or when one unit sees itself in competition with another unit. For example, day watch officers may be competing with night watch officers to get additional officers for that specific shift.

Nevertheless, lateral communication within police organizations is important so that functional units can be continually apprised of what the others are doing.

To foster the change to community policing, police management must correct obstacles to the organizational communication process. This may involve deviating from the traditional chain of command when upward or downward communication is desired. The traditional style of formal communication must be relaxed, and information from the top must be articulated directly to the subordinates at the bottom of the pyramid.

The traditional information system must be altered and perfected both inside and outside the organization to sustain effective change. In traditional policing, communication is organized around a need-to-know basis, and information is distributed according to position and rank. In contrast, community policing demands open communication with the community, frequent exchange among units within the police department, and ongoing communication and networking with other public and private agencies (Gray et al. 1997, 41–48).

The recent influx of technology within policing is altering many of the traditional communication channels in policing. Computers, with their e-mail and information networking capacity, allow for a diagonal form of communication that jumps across the traditional chain of command. This new communication channel is increasing the responsiveness of today's organizations. The traditional chain of command often takes too long to negotiate (Buhler 1997).

Computer communication technology will correct some of the problems that past communication channels created. Further, e-mail will allow police employees to communicate with members of other agencies and community members with ease and may improve on the dissemination of information both in and outside the police organization. Management must support and implement new styles of communication systems and encourage subordinates to suggest ways of providing better service and not just say what the boss wants to hear. An empowered, engaged, and high-trust police organization has a free-flowing and relaxed system of communication.

Work Environment

One final area in which management must be willing to support in the process of implementing community policing is that of the work environment. Police employees want to work in a meaningful environment where their ideas and suggestions are taken seriously. Police management must be committed to continually improving the working environment of the organization. Successful organizational improvement can only come about through a three-stage process:

1. Change yourself.
2. Improve the workplace (primarily by improving the organizational leadership style).
3. Ask your employees to treat customers the same way you are trying to treat them: i.e., improve services to citizens. (Couper and Lobitz 1993, 18–19)

Changing oneself means examining issues and reflecting in ways that one has not engaged in previously. It means being receptive to all viewpoints and becoming more democratic in thought. Supporting a democratic work environment entails changing to a more proactive leadership style.

Improving the workplace involves thinking in terms of a systems approach and in terms of creating a learning organization where processes are continually examined and alternate methods explored. It involves being willing to listen to subordinates and encouraging innovation. Improving organizational leadership involves leading by example and living the core values and vision that have been articulated to the rank and file.

Managers who support a customer service environment where employees are asked to treat others as they would want to be treated are well on their way to creating a community policing organization. Police officers must treat with dignity and respect all who require and request police service and treat every citizen concern as if it is the most important. Here are some suggestions of how a meaningful work environment may be fostered:

1. Share the company vision in detail with employees.
2. Show how individual jobs contribute to achieving that vision.
3. Include employees in decisions that affect their jobs.
4. Foster a spirit of collaboration.
5. Celebrate diversity.
6. Promote open and honest communication.
7. Don't punish employees for taking risks.
8. Empower employees to share equally in decisions that affect the shape and direction of the organization.
9. Provide ongoing growth and development opportunities.
10. Encourage regular feedback. (Caudron 1997, 24–27)

SUMMARY

Organizational change is a crucial prerequisite for the implementation of community policing. The police organization has traditionally been modeled after the principles of the classical theory of organization (i.e., scientific management, administrative management, bureaucratic management) and has been based on a paramilitary command-and-control system. This style of organization corrected many of the problems of policing in the past but ended up creating a host of new ones.

The traditional command-and-control, rule-driven environment of policing must change and evolve into a more democratic model where police employees at all levels are involved in the decision-making process. Recently, there has been increasing interest in the learning organization and total quality management approaches. The concept of the learning organization implies that an organization takes a systems approach to learning by encouraging a continuing flow of feedback from the external environment and through and among organizational units. Total quality management is a system for meeting and exceeding customer demands by using cross-unit, diverse teams that analyze and modify production/service delivery processes and programs to increase their effectiveness.

Organizational change includes changes in personnel training and recruitment, the criteria by which promotions are granted, leadership style, and organizational structure. There is an emerging trend within policing to flatten the traditional hierarchical structure of police organizations. Many agencies are eliminating layers of the command structure, and this has proven to be beneficial to the implementation of community policing.

Organizational change requires the clarification and articulation of the new values that the organization wishes to adopt. Values are the beliefs that guide an organization and the behavior of its employees. The values of the community policing philosophy represent a major break with those of traditional policing, and police leaders of agencies in transition may find it helpful to publish a value statement that can both educate the public and guide their own workforce.

Community policing requires both external and internal support. Sources of external support include elected civic officials, the business community, community leaders, organizers, residents, and public and nonprofit agencies.

Internal support is the support that comes from within the police organization. To support community policing, police executives must become change agents in four key areas: decentralization, empowerment of employees, communication, and the work environment.

Police executives must allow police operations to decentralize and move into the community. They must allow supervisors and commanders working in neighborhood stations to have more input into matters affecting police operations.

Empowerment involves encouraging employees to commit themselves to the values, goals, and mission of the police organization as a free and rational choice and to take on responsibilities themselves as opposed to being directed and controlled. Effective police leaders empower subordinates to create, experiment, innovate, and take risks. They overlook good-faith mistakes and give employees the information and training they need to make their own decisions.

In traditional police organizations, communication has often been ineffective due to the many layers of command that information flows up and down. Traditional methods of communication along the chain of command will have to give way to more effective methods, such as diagonal communication, or communication across the chain of command—for example, via e-mail. Upward communication from subordinates and lateral communication between units should be encouraged, and downward communication from the chief executive should be direct and candid.

Chief executives should encourage their organizations to become true learning organizations that research the effectiveness of their innovations. They should foster changes that make the work environment more democratic, more focused on customer satisfaction, and more meaningful to employees.

KEY TERMS

administrative
  management
bureaucratic management
diagonal communication
decentralization
downward
  communication

empowerment
lateral communication
leadership
learning organization
organizational change
paramilitary model
scientific management

suboptimization
total quality management
upward communication
value statement
values
vision

REVIEW QUESTIONS

1. Discuss the classical theory of organization.
2. Discuss the differences between the traditional police organization and the evolving community policing organization.
3. Define *organizational core values.*
4. Define the concept of empowerment.
5. Discuss the emerging police leadership style.
6. Discuss why the traditional organizational structure will not support the community policing culture.
7. Discuss the support mechanisms for community policing that are needed from the community and from police management.
8. Discuss the advantages of making a police agency a learning organization.

REFERENCES

Alpert, G.P., and R.G. Dunham. 1992. *Policing urban America.* 2nd ed. Prospect Heights, IL: Waveland.

Auten, J.H. 1985. The paramilitary model of police and police professionalism. In *The ambivalent force,* 3rd ed., ed. A.S. Blumberg and E. Niederhoffer. New York: Holt, Rinehart & Winston.

Bardwick, J.M. 1996. Peacetime management and wartime leadership. In *The leader of the future: New visions, strategies, and practices for the next era,* ed. F. Hesselbein et al. San Francisco: Jossey-Bass.

Bayley, D.H. 1994. *Police for the future.* New York: Oxford University Press.

Bennis, W. 1966. *Beyond bureaucracy.* New York: McGraw-Hill.

Birzer, M.L. 1996. Police supervision in the 21st century. *F.B.I. Law Enforcement Bulletin* 65, no. 6:6–10.

Booth, W.S. 1995. Integrating COP into selection and promotional systems. *Police Chief* 62, March, pp. 10–24.

Bopp, W.J. 1977. *O.W. Wilson and the search for a police profession.* Port Washington, NY: Kennikat.

Braiden, C.R. 1992. Enriching traditional roles. In *Police management: Issues and perspectives,* ed. L.T. Hoover. Washington, DC: Police Executive Research Forum.

Buhler, P. 1997. Managing in the 90s. *Supervision,* September, pp. 23–26.

Bureau of Justice Statistics. 1983. *Report to the nation on crime and justice.* Washington, DC: Government Printing Office.

Caudron, S. 1997. The search for meaning at work. *Training and Development,* September, pp. 24–27.

Couper, D.C., and S. Lobitz. 1991. *Quality policing: The Madison experience.* Washington, DC: Police Executive Research Forum.

Couper, D.C., and S. Lobitz. 1993. Leadership for change: A national agenda. *Police Chief,* December, pp. 15–19.

Fyfe, J.J., et al. 1997. Police administration, 5th ed. New York: McGraw-Hill, p. 15.

Gaines, L.K., et al. 1991. *Police administration.* New York: McGraw-Hill.

Geller, M.A. 1997. Suppose we were really serious about police departments becoming learning organizations. *National Institute of Justice Journal,* no. 234, pp. 2–8.

Goldstein, H. 1990. *Problem oriented policing.* New York: McGraw-Hill.

Gray, K., et al. 1997. Community policing and organizational change dynamics. In *Community policing in a rural setting,* ed. Q.C. Thurman and E.F. McGarrel. Cincinnati, OH: Anderson.

Gulick, L. 1969. Notes on the theory of organization. In *Papers on the science of administration,* ed. L. Gulick and L. Urwick. 1937. Reprint, New York: August M. Kelly.

Kanter, R.M. 1983. *The change masters.* New York: Simon & Schuster.

Koontz, H., et al. 1986. *Essentials of management.* 4th ed. New York: McGraw-Hill.

McNeill, W.H. 1982. *The pursuit of power: Technology, armed forces, and society since A.D. 1000.* Chicago: University of Chicago Press.

Palmiotto, M.J. 1997. *Policing: Concepts, strategies, and current issues in American police forces.* Durham, NC: Carolina Academic Press.

Peak, K.J., and R.W. Glensor. 1996. *Community policing and problem solving: Strategies and practices.* Upper Saddle River, NJ: Prentice Hall.

Perrow, C. 1972. *Complex organizations: A critical review.* Glenview, IL: Scott Foresman.

Peters, T.J., and N. Austin. 1985. *A passion for excellence: The leadership difference.* New York: Warner.

Peters, T.J., and R.H. Waterman, Jr. 1981. *In search of excellence.* New York: Harper & Row.

Petersilia, J. 1993. The influence of research on policing. In *Critical issues in policing: Contemporary readings,* 2nd ed., ed. R.C. Dunham and G.P. Alpert. Prospect Heights, IL: Waveland.

Senge, P.M. 1990. *The fifth discipline: The art and practice of the learning organization.* New York: Doubleday.

Shefritz, J.M., and J.S. Ott. 1992. *Classics of organization theory.* 3rd ed. Pacific Grove, CA: Brooks/Cole.

Skogan, W.G., and S.M. Hartnett. 1997. *Community policing: Chicago style.* New York: Oxford University Press.

Sparrow, M. 1988. *Implementing community policing.* Perspectives on Policing. Washington, DC: U.S. Department of Justice.

Strecher, V.G. 1997. *Planning community policing: Goal specific cases and exercises.* Prospect Heights, IL: Waveland.

Trojanowicz, R., and B. Bucqueroux. 1994. *Community policing: How to get started.* Cincinnati, OH: Anderson.

Wadman, R.C., and R.K. Olson. 1990. *Community wellness: A new theory of policing.* Washington, DC: Police Executive Research Forum.

Walker, S. 1992. *The police in America: An introduction.* New York: McGraw-Hill.

Walton, M. 1990. *Deming management at work.* New York: Perigee.

Wasserman, R., and M.H. Moore. 1988. *Values in policing.* Perspectives on Policing, no. 8. Washington, DC: National Institute of Justice.

Watson, E.M., et al. 1988. *Strategies for community policing.* Saddle River, NJ: Prentice Hall.

Weisburd, D., et al. 1989. Maintaining control in community oriented policing. In *Police and policing: Contemporary issues,* ed. D.J. Kenney. New York: Praeger.

Wilson, O.W., and R.C. McLaren. 1977. *Police administration.* 4th ed. New York: McGraw-Hill.

Chapter 10

Planning the Implementation of Community Policing

CHAPTER OBJECTIVES

1. Be familiar with the planning process.
2. Be able to define the term *planning*.
3. Be familiar with the value of strategic planning in implementing community policing.
4. Be familiar with the value of continuous assessment in implementing community policing.

INTRODUCTION

As Malcolm Sparrow (1982) remarked,

> Those who accept the desirability of introducing community policing confront a host of difficult issues: What structural changes are necessary, if any? How do we get the people on the beat to behave differently? Can the people we have now be forced into the new mold, or do we need to recruit a new kind of person? What should we tell the public, and when? How fast can we bring about this change? Do we have enough external support? (2)

Before community policing can be implemented to any degree in a police organization, a plan must be established to address such questions and many others. A plan often will depend upon the size of the police organization and the number of qualified personnel who are familiar with the planning process. Large police organizations may have a planning section with trained planners, whereas small ones may not have any trained planning personnel or planning unit. Depending on whether there are planning personnel, a police organization may have either a very detailed strategic plan or a "plan as you go" approach. A "plan as you go" approach may have goals, objectives, and a mission for implementing community policing, but goals and objectives are not detailed, and there may not be any consideration of potential problems, issues, or impediments to implementing community policing. A strategic plan includes mission, vision, and values statements, goals and objectives, analysis of community and department needs and resources, analysis of potential obstacles, action steps for accomplishing goals, timeframes and sometimes personnel assignments for action steps, and mechanisms of monitoring and assessment.

Police departments that have the capabilities to develop a strategic plan are advised to do so. A well-developed strategic plan can assist in knocking down barriers to the implementation of community policing. Both police personnel and members of the community should be involved in the planning process. A well-thought-out strategic plan should make community policing easier to implement and increase chances of obtaining the support of police personnel and community members.

The transition from traditional policing to community policing is not easy, and this must be taken into consideration by the chief executive of the police organization. Patience and understanding are important in implementing community policing.

WHAT IS PLANNING?

Planning has been defined as "setting objectives and deciding how to achieve them" (Tansik and Elliott 1981, 33), or, in more detail, as

> a management function concerned with visualizing future situations, making estimates concerning them, identifying the issues, needs and potential points, analyzing and evaluating the alternative ways and means for reaching desired goals according to a certain schedule, estimating the necessary funds and resources to do the work, and initiating action in time to prepare what may be needed to cope with changing conditions and contingent events. (Mottley 1972, 127)

Planning entails decisions making; it involves selecting a route that an organization will follow. There are various types of plans, ranging from the detailed to the broadly defined plan for getting organizational goals on track. Planning asks the basic questions, What to do? How to do it? When to do it? Who will do it? (Koontz et al. 1986, 35).

Plans to implement community policing vary a great deal in scope. Some are plans to set up a special community policing unit within a department that is free from responding to radio calls and is thus able to work on community problems full time. Others are plans to turn every officer in the department into a community policing officer. Some plans cover a specific neighborhood, others an entire city. The scope of the plan will generally depend upon the size of the agency, the number of police personnel who can be assigned to the plan, the ability of the police organization to accept innovation, the extent of changes required, and the expectations of the community concerning the police. But even agencies with ambitious plans tend to start out small and initiate community policing in a single neighborhood.

WHY PLAN?

Planning brings an organization's present situation into line with its goals for the future and lets members of an organization know what these goals are so that they can more effectively accomplish the organization's mission. By focusing on organizational goals, it allows the organization to concentrate its resources on goal-oriented priorities. It provides a blueprint and a process for achieving goals and makes clear who is responsible for what. It allows administrators to anticipate problems and to think about how to sidestep them or lessen their impact. It encourages long-range thinking. It cuts down on haphazard actions and makes possible better coordination of action so that different actions do not undercut each other or cancel each other out. Finally, it increases the chances of obtaining the external and internal support necessary for the plan to succeed (Whisenand and Ferguson 1989, 157–159).

A plan should be considered as a specified outline of the format to be executed. A plan follows the following points:

1. The need for the plan must be recognized. An apparent need must be verified by a more intensive investigation and analysis.
2. The objective must be stated, and the general method of operation (the manner in which the objective is to be attained) must be determined.
3. Data necessary in the development of the plan must be gathered and analyzed. Included will be answers to the questions of "what," "where," "when," "who," and "how."
4. The details of the plan must be developed: personnel and equipment must be provided and organized, procedures developed or applied, schedules drawn up, and assignments made.
5. Planning reports must be prepared.
6. Planners should participate in a staff capacity during implementation, if this is requested by persons carrying out the plan.
7. Plans must be reviewed, and modified if necessary, to accommodate changes in need and technology. (Fyfe et al. 1997)

Silverman (1995) has vividly described the chaos and confusion that result from insufficient planning:

1. *We are not sure what the problem is.* Definitions of the problem are vague or competing, and any given problem is intertwined with other messy problems.
2. *We are not sure what is really happening.* Information is incomplete, ambiguous, and unreliable, and people disagree on how to interpret the information that is available.
3. *We are not sure what we want.* We have multiple goals that are unclear or conflicting or both. Different people want different things, leading to political and emotional conflict.
4. *We do not have the resources we need.* Shortages of time, attention, or money make a difficult situation even more chaotic.
5. *We are not sure who is supposed to do what.* Roles are unclear, there is disagreement about who is responsible for what, and things keep shifting as players come and go.
6. *We are not sure how to get what we want.* Even if we agree on what we want, we are not sure what causes what.

7. *We are not sure how to determine if we have succeeded.* We are not sure what criteria to use to evaluate success. If we do know the criteria, we are not sure how to measure them. (37–45)

In making the transition to community policing, planning is particularly important because community policing involves new concepts and strategies that may be unfamiliar to many people and that may conflict with many established police practices and organizational structures (see Chapter 9).

HOW MUCH PLANNING BEFORE IMPLEMENTATION?

A key question facing police agencies is how far ahead and how extensively they should plan a community policing program before putting it into operation. The Bureau of Justice Assistance (1994b) outlined three possible approaches, along with their strengths and weaknesses:

1. *Plan, then implement.* This method entails developing a detailed long-range plan, with tasks and timelines, and assigning officers to execute the plan. This approach clearly delineates a set of strategies and actions that impart a sense of direction to implementation efforts; however, the initial planning stage for a large agency can take months or even years, and even a very detailed plan will be unable to predict the obstacles that will arise. In the absence of experienced-based feedback, some part of the implementation process may be miscalculated. Planning can also be complicated by the size of the staff involved. Keeping the planning staff relatively small may prevent the process from becoming unwieldy; however, it may not adequately represent all levels of command, function, and experience within the organization, thus creating the risk that the plan will not be implemented. Planning can also become excessive and may stifle enthusiasm.
2. *Plan and implement.* In this approach, planning and action occur simultaneously. While the planning process continues, the agency, to get started quickly, involves more personnel at the outset and permits further planning to benefit from feedback. However, the agency risks false starts, confusion, and major blunders unless effective, rapid, and regular communication takes place between planners and implementers.
3. *Implement with little planning.* The third option is for an agency with little preparation or knowledge of the nature of community policing to quickly launch into the action phase and then, on the basis of feedback, to retool the effort and begin the cycle again. This process is continuous, with each reevaluation cycle advancing the idea of community policing a bit further within the organization. This approach assumes that a limited knowledge of community policing may prevent agencies from initially planning in a meaningful way. Advocates note that the almost immediate action will catch officers' attention at all organizational levels and will harness the existing enthusiasm to help mobilize support.

However, the constant shifts in goals and actions can be highly unsettling to the organization and the community it serves. (29–30)

According to the Bureau, none of these approaches is the "one right way." Police administrators will have to choose among them on the basis of agency priorities, resources, and support for change.

WHO PLANS?

All personnel in a police department should be involved in planning, but the expectations from personnel will vary. Kuykendall and Unsinger (1979) describe general planning responsibilities of each of three levels of administration:

1. Top management is concerned with long-range considerations and general planning, and considerable time is invested in planning activities. In large police departments, top management is placing increasingly greater reliance on specialized planning and research units.
2. Middle managers participate in development of all ranges of plans in the agency, oversee the short- and intermediate-range plans, develop many of the details in long- and intermediate-range plans, and usually play a key role in making adjustments in the execution phases of plans.
3. Supervisors participate in the planning process by overseeing operational plans, and in many small police departments may perform the functions of middle managers as well; they also assist in developing the specifics or details for translation of plans into action. (105)

But as discussed in Chapter 9, community policing requires a more democratic management style, and this should begin in the planning for the transition to community policing. Therefore, all community policing plans should include not only the three levels of management but also line, staff, and civilian personnel. Bottom-up thinking should be employed rather than top-down thinking. Command staff should have limited involvement in strategic planning. A police officer could be the leader of the committee, and it would not be a bad idea for the chief *not* to be on the committee.

Community policing especially needs the understanding and support of line personnel if it is to have an impact on their community. The gap between "management cops" and "street cops" can be closed if management shows a willingness to make the street cop an important component of community policing planning, and this can make a big difference in the plan's success. A study sponsored by the National Institute of Justice of neighborhood-oriented policing of eight cities found that gaining police acceptance was a principal obstacle to implementing community policing and that officers who had limited or no knowledge about community policing or who felt that they had no voice in planning it did not support its implementation (Sadd and Grinc 1996, 7–11).

The participation of a cross-section of personnel—administrators, middle managers, supervisors, line and staff personnel, civilian personnel, and representatives from all department units and divisions—makes it possible to deal with and overcome resistance to community policing at an early stage—for example, claims that too much work will be required, that the plan will be used against employees at a later date, or that the plan limits officers' law enforcement powers. The more people see that their input is solicited and incorporated, the more support the plan will have.

Planning for community policing should also involve the community.* A strategic plan created without soliciting the cooperation of the community is doomed to failure. The community must buy in to the plan for it to be implemented successfully. When a community policing strategic plan involves the community, the process includes bringing together a community as a whole to plan the policing philosophy and practices in a specific geographic location and developing community consensus and political momentum to ensure long-term support and commitment to change (National Center 1997, 8).

Active community and neighborhood members, along with persons who reflect the diversity of the community, should be made members of the strategic planning team. Community team members should represent the interest of homeowners, apartment dwellers, and business owners. Planning should also involve other governmental and service agencies who are involved in communities and neighborhoods.

HOW TO PLAN

A precursor to making a strategic plan is developing a vision, a values statement, and a mission statement (see Chapter 9):

- *Vision:* a picture or image of where people want the organization to be
- *Values statement:* statement of the principles, standards, or qualities that are regarded as worthwhile or desirable by the organization
- *Mission statement:* a statement of the organization's purpose or what it is striving for (National Center 1997, 17)

This is to ensure that whatever the organization plans will be in keeping with its values and will contribute to accomplishing its mission and fulfilling its vision.

Then, constructing the plan involves answering three major questions:

1. *Where do we want to be?* This is the process of strategically setting goals.
2. *Where are we now?* Relative to the strategically set goals, what is the current position? What resources do we have? What limitations or barriers are we facing?
3. *How do we get there from here?* What programs, projects, and policies will allow us to achieve the goals? (Tansik and Elliott 1981, 39; see also Fyfe et al. 1997, 214–215)

Source: The next two paragraphs are adapted from National Center for State, Local, and International Law Enforcement Center on Strategic Planning and Implementation, *STAR Manual,* Federal Law Enforcement Training Center.

The first step, defining realistic goals, is not easy particularly because goals of organizations are often contradictory. For example, the community policing goal of proactive problem solving in the community might come into conflict with a goal of rapid response time to every call because there might not be enough officers or enough time for officers to accomplish both goals. In this case, it might be necessary to relinquish or modify the goal considered to be less important.

Goal setting, like every other stage of planning, requires community participation. Goals should be what both the police department *and* the community want to accomplish. Also, goal setting requires input, from both internal and external surveys, to gauge departmental and community views, priorities, needs, and concerns.

Step 2 involves analyzing how far the department is from achieving its goals, whether the goals are realistically achievable, and the current circumstances that could help or obstruct goal attainment. Here, resources are a major issue, in the form of assistance and expertise (e.g., from community organizations) and in the form of manpower and material resources such as equipment and money. Are there a sufficient number of officers to establish community policing? Does the chief have to obtain new funding or take resources from or even eliminate another policing activity?

Attitudes of support or opposition are also crucial to recognize. They include

- *Leadership attitudes.* Police administrators are often autocratic. They often feel personal ownership of the department and run it rigidly with a philosophy of "My way or the highway." The democratic style of community policing may threaten them.
- *Rank-and-file attitudes.* Adherence to the crime-fighting image of policing and a view of the citizen as the enemy are contradictory to the community policing philosophy. Also, concerns about changed workloads and responsibilities will have to be addressed. For example, if some officers are freed from responding to calls for service in order to do community policing, how will positive working relationships be maintained with officers who must take on their workload?
- *Attitudes of elected officials.* Policing is embedded in politics. Governmental officials provide the budget for the police department to hire more personnel, obtain more equipment and institute pay raises. Therefore, these officials must buy in to the community policing philosophy for it to be successfully implemented (Rush 1992, 44–45; Bureau of Justice Assistance 1994a, 57).

Finally, aspects of traditional organizational structure may interfere with the implementation of community policing. These are discussed extensively in Chapter 9.

The third step involves developing a sequence of specific actions to achieve the goals. Alternative actions should be designed in order to give community policing the best chance of succeeding in a constantly changing environment (Tansik and Elliott 1981, 40).

FINE-TUNING THE PLAN

Once planners have specified goals, assessed needs, resources, and barriers, and specified a sequence of actions to attain goals, they must work out more details of how the plan will be implemented. First, people must be assigned to work on each action and see it through to completion. Then realistic time frames for completing specific tasks should be established. It should be taken into consideration that if implementation moves too slowly, enthusiasm may be checked and any momentum that existed may be lost, but that moving too quickly may create confusion and chaos.

Training is an important aspect of implementing any community policing program and will be discussed in detail in the next chapter. Planning for training involves four steps, each of which must be assigned an estimated time frame:

1. developing a curriculum
2. identifying and possibly training the trainers
3. acquiring training materials (books, handouts, etc.)
4. developing a training schedule and making arrangements for officers to attend (Bureau of Justice Assistance 1994a, 53)

The community policing implementation plan should also describe how community policing activities will be monitored. The monitoring function should be made as much a part of the normal activities of the police department as possible. For example, if command meetings are regularly scheduled, then the progress on the community policing plan should be placed on the agenda and discussed. If department heads are required to provide activity reports, then community policing–related information should be included in the report. Units or individuals responsible for the specific community policing objectives should provide feedback about the progress being made toward achieving these objectives. The primary purpose of monitoring the progress of goals and objectives of the community policing plan is to inform personnel responsible for specific programs if they are achieving their objectives. If they are not, the monitoring process will allow them to make corrections in their activities in order to achieve the objectives they were assigned (Meese 1991, 57).

Finally, mechanisms for assessing whether the objectives of the plan are being achieved must be established from the beginning, as part of the planning process, and assessment must be a continuing process. Continuous assessment can determine whether the plan is on track and, if not, can prompt the necessary decisions to get it back on track. The assessment gives the police personnel direction and helps to determine what procedures should be changed, eliminated, or altered.

Assessment can help answer two questions: whether the project is proceeding as planned and whether it is having the desired impact. The first question can be answered through regular meetings of the chief executive of the police organization with those responsible for implementing the community policing plan to find out about problems and how to correct them, and by regular reports submitted to the police agency's head on the progress in implementing community policing and any problems pertaining to the attainment of specific community policing objectives and goals. In addition, the implementation team can periodically survey police personnel to obtain their input as to which community objectives and goals are working. This input can be used to correct any problems where appropriate.

The second question, concerning the project's impact on the organization and the community, cannot be answered by using traditional methods of assessment. Measures such as the number of 911 calls served, number of arrests made, response time to calls, and number of arrests made and traffic citations issued are not useful for assessing the impact of community policing. Crime statistics can be a measure but should be only one of many used to assess community policing. Because the community policing philosophy has as its goals eliminating the fear of crime, improving the quality of life in the community, and developing a partnership between the police department and the community, these must all be assessed as well.

Meese (1991) has suggested using criteria of effectiveness, efficiency, and equity to assess community policing (44–51). *Effectiveness* is important to measure because one outcome expected from the community policing strategy is that it will be more effective than the traditional approach to policing. Dimensions of effectiveness here would include reducing neighborhood crime, decreasing citizens' fear of crime, and improving communities' quality of life. Also assessed is the success of the strategies used to solve neighborhood problems.

Efficiency involves whether the community policing strategy has been obtaining the most results from the available resources. Are resources being used to their fullest possible extent to solve neighborhood and community problems? Use of technologies, use of community organizations, prioritizing of service calls, redefining of job descriptions, and coordination of problem-solving activities all fall under this head.

Equity, or fairness, criteria should be used to assess every community policing program and its results:

- *Equal access to police services.* All citizens, regardless of race, religion, personal characteristics, or group affiliation, must have equal access to police services for a full and productive partnership with a community. The paramount commitment of community policing should be respect for all citizens and sensitivity to their needs. . . . Supervisors should help to ensure that police services are readily available throughout the community.
- *Equal treatment under the Constitution.* Police must treat all individuals according to the constitutional rights that officers are sworn to protect and enforce. Careful attention to the constitutional rights of citizens, victims, or perpetrators will help to engender bonds of trust between the police and community. Police must treat all persons with respect and impartially—including the homeless, the poor, and the mentally or physically handicapped. They must reject stereotypes, ignore skin color, and use reason and persuasion rather than coercion whenever possible because inequitable or harsh treatment can lead to frustration, hostility, and even violence within a community. Such unethical behavior will imperil the trust so necessary to community policing.
- *Equal distribution of police services and resources among communities.* Because community policing customizes policing services to the needs of each community, services should be distributed equitably among poor and minority communities. Care must be taken, however, to ensure that this is the case. For

equitable distribution of resources among communities, each community must articulate its needs and be willing to work with the police to ensure its share of police services. Each neighborhood officer must listen to the community members and be willing to work with the community members to meet those needs. Poor and minority neighborhoods can present particular challenges for some patrol officers, who may have to bridge differences of race and class before a level of trust and cooperation can be established. (50–51)

Community policing needs to be continuously assessed and reevaluated to determine its successes and failures. Adjustments to ongoing programs should be expected and made where appropriate.

THE FINAL PLAN DOCUMENT

Generally, the final plan document will contain a chief executive summary, an introduction to the plan, an overview of the community and police department, a vision statement, a values statement, a mission statement, an organizational table, internal and external analysis, goals, objectives, action steps, time frames and sometimes job assignments, and mechanisms of monitoring and assessment. Initially, a draft of the plan is developed, and a committee of team members is selected to review it. The team reviews the goals and objectives and works to eliminate redundancy and maintain the plan's focus. After review, the plan is finalized and presented to the police department, local government, and the community. It is recommended that all personnel receive a copy of the strategic plan.

A strategic plan is a living document that is flexible and open to change and adjustments. It provides a road map for the police department to use as a guide for operations and encourages strategic thinking and organizational learning. It should be noted that a strategic plan will not be etched in concrete and that it in no way substitutes for leadership.

SUMMARY

Before community policing can be implemented by a police agency, the implementation must be planned. Planning is setting objectives and deciding how to achieve them. A complete strategic plan includes mission, vision, and values statements, goals and objectives, analysis of community and department needs and resources, analysis of potential obstacles, action steps for accomplishing goals, time frames and personnel assignments for action steps, and mechanisms of monitoring and assessment.

Police departments that have the capabilities to develop a strategic plan are advised to do so. A well-thought-out plan gives a department direction, focuses its resources on agreed-upon priorities, allows administrators to anticipate and circumvent problems, encourages long-range thinking, cuts down any haphazard actions, coordinates agency actions, and increases the chances of obtaining internal and external support.

Planning should involve the community at all stages and should enlist the participation of all police personnel at all levels. The input of lower-level "street cops" should be especially encouraged.

The planning process begins with statements of the agency mission, vision, and values, because any goals must be in keeping with them and help to fulfill them. Planners should then determine where they want to be (goals), where they are now (how far they are from achieving the goals and current circumstances that could help or obstruct goal attainment), and how they can get there from here (specific action steps to reach goals).

The finished plan should not be regarded as unalterable or as a substitute for leadership. Its workability and its impact must be continually assessed as it unfolds so that it can be modified as changing circumstances require.

KEY TERMS

assessment	mission statement	time frame
effectiveness	monitoring	values
efficiency	planning	vision
equity	strategic planning	

REVIEW QUESTIONS

1. Describe the planning process.
2. Discuss the importance of strategic planning to community policing.
3. Discuss who should be involved in planning and why.
4. Discuss the various approaches to how much planning should be done before implementation.
5. Discuss why assessment is important to the implementation of community policing.

REFERENCES

Bureau of Justice Assistance. 1994a. *Neighborhood-oriented policing in rural communities: A program planning guide.* Washington, DC: U.S. Justice Department.

Bureau of Justice Assistance. 1994b. *Understanding community policing: A framework for action.* Washington, DC: U.S. Justice Department.

Fyfe, J.J., et al. 1997. *Police administration.* 5th ed. New York: McGraw-Hill.

Koontz, H., et al. 1986. *Essentials of management.* New York: McGraw-Hill.

Kuykendall, J.L., and P.C. Unsinger. 1979. *Community policing administration.* Chicago: Nelson-Hall.

Meese, E., III. 1991. *Community policing and the police officer.* Perspectives on Policing. Washington, DC: National Institute of Justice.

Mottley, C. 1972. Strategy in Planning. In *Planning, programming, budgeting: A systems approach to management,* 2nd ed., ed. J.F. Lyden and E.S. Miller. Chicago: Markham.

National Center for State, Local and International Law Enforcement. 1997. *Strategic planning and implementation.* Glynco, GA: Federal Law Enforcement Training Center.

Rush, G. 1992. Community policing: Overcoming the obstacles. *Police Chief,* October, pp. 50–55.

Sadd, S., and R.M. Grinc. 1996. *Implementation challenges in community policing: Innovative neighborhood-oriented policing in eight cities.* Washington, DC: National Institute of Justice.

Silverman, E. 1995. Community policing: The implementation gap. In *Issues in community policing,* ed. P.C. Kratcoski and D. Dukes. Cincinnati, OH: Anderson.

Sparrow, M. 1982. *Implementing community policing.* Perspectives on Policing. Washington, DC: National Institute of Justice.

Tansik, D.A., and J. Elliott. 1981. *Managing police organizations.* Belmont, CA: Duxbury.

Whisenand, P.M., and F. Ferguson. 1989. *The managing of police organizations.* Englewood Cliffs, NJ: Prentice Hall.

Chapter 11

Selected Approaches to Training and Planning

CHAPTER OBJECTIVES

1. Be familiar with new developments in community policing training.
2. Be familiar with the Royal Canadian Mounted Police approach to training.
3. Be familiar with one small police agency's approach to community policing.
4. Be familiar with one midsized police agency's approach to community policing.
5. Be familiar with the Illinois State Police approach to community policing.
6. Be familiar with the elements of one strategic community policing plan.

INTRODUCTION

The training of police officers in the community policing philosophy is a key element for the successful transition from traditional policing to community policing. Police officers must understand the community policing concept if they are expected to use it.

There are a variety of training programs and methods to teach the community policing philosophy. Training can range from the traditional lecture method, with the instructor feeding the police student information, to computer-based instruction, role playing, and adult-based learning. In the last several years, police training academies have been approaching training police officers using adult-based learning. It is believed that adult-based learning gives police officers a better opportunity to learn and retain information and holds them responsible for their learning. The first part of this chapter presents selected police agencies' approaches to training in community policing.

When police departments make a transition from traditional policing to community policing, they use different approaches to implement it depending on the demographic characteristics of the community being served, the major problems and needs of the community, the support of government officials, the availability of resources, the willingness of police command staff and officers to accept the community policing philosophy, police administrators' managerial and policing philosophy, the progressiveness of the community, and the educational and training level of police personnel. The second part of this chapter showcases selected police departments' approaches to implementation.

NEW APPROACHES TO POLICE TRAINING

The training of police officers is going through a transition. Community policing will revolutionize both the content of police training and how police are being trained. According to Edwin Meese (1991), a former Attorney General of the United States, "The content of police training must go beyond merely preparing officers for the mechanical aspects of police work. . . . Training should help them understand their communities, the police role, police history, and even imperfections of the criminal justice system" (6). Meese's comments are especially true for training in community policing. Palmiotto et al. (1998) have advanced three guidelines for the content of such training:

1. *Police officers should possess a sense of social history.* An understanding of police history can contribute to an understanding of policing's current mission and objectives. For practitioners to prepare for and create the future of policing, it is imperative that they examine where policing has been. They need to know what has been tried, what has succeeded, and what has failed in police practice and why. They should know the characteristics of each era in policing history (prepolitical era, political era, and reform era). Only by understanding the past will police personnel be in a position to understand and implement needed change in their organization and to improve interaction with the community.
2. *Police officers should have a sense of how societies and communities function.* Responding to community disorder is at the heart of community policing. Instruction should include social indicators of crime and disorder (e.g., crime rates, crime trends, jail and prison populations, victimization numbers, juvenile crime rates, poverty indicators, unemployment figures, and other economic antecedents and consequences of crime) as well as related indicators of "quality of life" and neighborhood wellness.
3. *Police officers should be equipped with the skills and knowledge for incorporating community policing into their work.* Before instructors teach any police training class, they should attend a training session on the value of incorporating the community policing philosophy into their courses. (9–11)

A shift in training *methods* is also being recommended. The traditional method, referred to as the *pedagogical technique,* emphasizes information-based learning, primarily through the lecture-driven class. In this approach, learning is passive: the instructor dictates to students what they should learn, and students play no role in deciding what is important; they merely digest the information and give it back in the form of written examinations. In this model, students assume no responsibility for their learning.

The alternative to the pedagogical model is the andragogical model. Its supporters believe that the pedagogical approach is best fitted to children, whereas the andragogical approach is best fitted to adult learners. According to Malcolm Knowles, the leading proponent of andragogy,

Andragogy is premised on at least four crucial assumptions about the characteristics of adult learners that are different from the assumptions about child learners on which traditional pedagogy is premised. These assumptions are that, as a person matures, (1) his self-concept moves from one of being a dependent personality toward one of being a self-directing human being, (2) he accumulates a growing reservoir of experience that becomes an increasing resource for learning, (3) his readiness to learn becomes oriented increasingly to the developmental tasks of his social roles, and (4) his time perspective changes from one of postponed application [to one of immediate application], and accordingly his orientation toward learning shifts from one of subject centeredness to one of problem centeredness. (quoted in Cross 1981, 222–223)

According to this approach, adults learn differently from children or people with no experiences or knowledge base to build upon.

Knowles (1990b) described five characteristics of adults as learners:

1. *The need to know.* Adults learn more effectively if they understand why they need to know or be able to do something. . . .
2. *The need to be self-directing.* Adults have a deep psychological need to take responsibility for their own lives—to be self-directing. . . .
3. *Greater volume and quality of experience.* Adults, by virtue of having lived longer, accumulate a greater volume and a different quality of experience than children and youth. The greater volume is self-evident—they have done more things, and the longer they have lived, the more things they have done. . . .
4. *Readiness to learn.* Whereas youth have been well conditioned to be ready to learn what they are told or have to learn, adults become ready to learn those things that they perceive will bring them greater satisfaction or success of life.
5. *Orientation to learning.* Whereas children and youth have been conditioned to enter into a learning activity with a subject-centered orientation to learning, adults have a life-centered, task-centered or problem-centered orientation. (6.8–6.12)

With an andragogical approach, the instructor or teacher functions as a facilitator and resource: he or she guides and assists the students but does not necessarily make learning decisions for them. The learning environment created by the facilitator is characterized by mutual trust and respect between students and facilitator, the encouragement of student inquiry, and openness to learning (6.13–6.14).

According to Knowles (1990a), the primary purpose of schooling is to help individuals develop the skill of learning, with the final objective of becoming involved in self-directed inquiry in self-actualizing directions. Learning, in his view, should involve the following seven abilities:

1. The ability to develop and be in touch with curiosities. Perhaps another way to describe this skill would be "the ability to engage in divergent thinking."

2. The ability to perceive's one's self objectively and accept feedback about one's performance nondefensively.
3. The ability to diagnose one's learning needs in the light of models of competencies required for performing life roles.
4. The ability to identify human, material, and experiential resources for accomplishing various kinds of learning objectives.
5. The ability to design a plan of strategies for making use of appropriate learning resources effectively.
6. The ability to carry out a learning plan systematically and sequentially. This skill is the beginning of the ability to engage in convergent thinking.
7. The ability to collect evidence of the accomplishment of learning objectives and have it validated through performance. (174)

Royal Canadian Mounted Police

In 1990, the Royal Canadian Mounted Police (RCMP) reviewed their techniques of traditional training by interviewing graduates and trainers, interviewing top management, and reviewing evaluations, audits, and training practices of other police organizations. A major finding from this review was that the skills needed for policing were changing and that the process for learning police skills would consequently have to change as well. The RCMP were at the time making a transition to community policing that involved flattening the organizational structure, encouraging the use of problem-solving techniques, and developing partnerships with the community. They recognized that police officers, to fit into this new approach, would need to be innovative problem solvers with excellent people skills in such areas as negotiation and alliance development. The RCMP therefore introduced a new approach to training that incorporated many aspects of the andragogy model. Trainees were held responsible for their own training, with the instructor-trainer functioning as a facilitator (Himelfarb 1997, 33–34).

The RCMP developed a community policing problem-solving model known as CAPRA and restructured their police training so that all officers began their careers committed to it. The purpose of CAPRA is to promote discussions in small groups of employees, clients, and community partners. The CAPRA model emphasizes the importance of

- developing and maintaining partnerships and trust within communities/the workforce to establish priorities for service delivery and preventive problem solving
- understanding clients' perspective on work-related matters for establishing priorities and potential partnerships in service delivery
- encouraging ongoing feedback for continuous improvement (RCMP, n.d., 1)

The acronym *CAPRA* stands for

- *C = Clients.* Police assess and define problems through understanding the needs and expectations of diverse clients. Starting with the client means that the police must step outside their own professional understanding, their own culture, and learn to see the world through the eyes of those they serve. . . . Effective policing requires an understanding of the diverse and changing needs of the full range of clients in any particular situation and the ability to integrate or balance compet- ing interests.

- *A = Acquiring and Analyzing Information.* Key to preventing, defining, and resolv- ing problems, especially in our information-driven society, is the ability to collect, organize, analyze, and document information. . . . Effective community policing requires information beyond a specific case or incident. It requires information that helps police to understand their clients' concerns. New technologies provide unprec- edented access to information on patterns of crime, community profiles, client/com- munity perceptions and expectations, services available, etc. . . .
- *P = Partnership.* . . . At a minimum, increasingly complex police problems re- quire multidisciplinary teams that bring together various skills, often both police and non-police agencies. Additionally, partnerships now include more specific clients, community groups, and their representatives, for example, in the form of advisory committees. . . .
- *R = Response.* . . . Enforcement is one possible response. Service, crime prevention, and protection are also possible responses depending on clients' needs, and usually a combination of responses will be required. All police responses, including enforce- ment, fall within the CAPRA model. Thus, planning and implementing the appropri- ate response will require an understanding not only of police duties, responsibilities, and powers, but also [of] the principles that guide the use of discretion and, increas- ingly, the role of police in supporting non-enforcement responses.
- *A = Assessment for Continuous Improvement.* Increasingly, effective policing demands approaches and techniques for ongoing assessments that promote con- tinuous improvement and learning. Complementing traditional approaches to audit, evaluation, and management review will be the commitment of each mem- ber to continually review and learn from each problem-solving process. Policing in modern society must build in processes of adaptation, promote flexibility, and instill continuous learning. (Himelfarb 1997, 34–37)

The RCMP approach to police training is learner centered rather than instructor centered. The teaching of recruits involves problem solving through research, group problem-solving exercises, information gathering, case studies, and the use of sce- narios. This kind of application of the andragogical model to teach police officers may be the wave of the future. Several police departments and academies in the United States have adopted the RCMP training approach. However, because this training approach is relatively new to policing, several decades may pass before we can deter- mine whether it is an improvement over the traditional approach to police training.

Savannah Police Department Training

The Savannah Police Department, like most police departments, recognized that before they could implement community policing they had to train police personnel in its philosophy and principles. The department's training director and the deputy chief responsible for implementing community policing developed seven training modules to bring police personnel up to date on the philosophy of community policing.

The first module was for command staff. It was divided into two parts: the first previewed the next six, thereby giving administrators an overview of community policing and the content of instruction for personnel, and the second introduced total quality management (TQM) and the essential role of a TQM leadership style in the successful implementation of community policing.

The second module of training provided an overview of community policing. It began with an in-depth analysis of crime in Savannah and the conditions found in high-crime areas. Community policing principles were reviewed, and the goals of the Savannah Police Department were outlined. The specific roles and responsibilities of managers, supervisors, and police officers were addressed.

In module three, problem-oriented policing was examined. Because the Savannah Police Department considered problem-oriented policing to be a major component of community policing, officers were instructed on how to identify problems and go about solving them. The officers were informed that community resources could be focused on solving recurring problems.

The fourth module dealt with referrals and the specific agencies available in the city of Savannah. Officers were given city codes that could assist them in eradicating problems endemic to Savannah. In the fifth module, "Developing Sources of Human Information," the focus was on communicating with Savannah's residents in order to develop their trust. This module reviewed field interviews and investigative detentions in detail.

The sixth module discussed how to conduct neighborhood meetings, surveys of citizens' needs, and tactical crime analysis. The common thread in these subject areas was that each requires face-to-face contact between the police and citizens. The seventh module provided an understanding of crime prevention within the context of community policing. The emphasis of this module was that crime prevention is the responsibility of all police officers.

Los Angeles Sheriff's Department

As mentioned in previous chapters, supervisors and managers should receive training in community policing. Just as the Savannah Police Department developed a specific module to train their command staff on the community policing concept, the Los Angeles Sheriff's Department developed a 6-hour course specifically for supervisors and managers. This course defined community policing, provided a brief history of policing, discussed problem-oriented policing and the "broken windows theory," and covered various community policing strategies, such as combining investigative and patrol functions, allowing lower-level flexibility in policy making, and permanently assigning patrol officers to specific neighborhoods. There were sections on "Why We Need Community Policing," "Problem Solving," "Surveying the Community," and "Solutions for Criminal Nuisance Activity." Additional sections covered strategic planning, managing organizational change, and supervising community policing.

Regional Community Policing Institute Training

The Regional Community Policing Institute of Florida at St. Petersburg Community College operates under a cooperative agreement with the Department of Justice's Office of Community Oriented Policing Services (COPS). The institute offers community policing training without cost to law enforcement officers, community residents, city employees, social service agencies, and private sector representatives throughout Florida. This training is available to government employees, business employees, community leaders and citizens, police personnel, probation officers, police officers, and campus police. Training provided by the institute can be adapted to the specific needs of an agency, business, or community. Course materials are also provided without charge. A list of courses available from the institute follows:

- Introduction to Community Policing (16 hours)
- Survival Skills for Community Policing Officers (16 hours)
- Problem Solving for the Police Officer and Citizen (16 hours)
- Bridging the Gap: Police-Community Partnership (16 hours)
- Ethical Considerations in Community Policing (8 hours)
- Reaching Your Goals Through Code Compliance (16 hours)
- Adult Education Principles/Training for the Trainer (8 hours)
- Three-Part Series:
  1. Managerial Buy-In (16 hours)
  2. Managerial Advantage (16 hours)
  3. Tool Kit for Managing Organizational Change (16 hours)

Building Bridges

"Building Bridges: Police/Community Partnership" was a two-day conference developed and coordinated by the Regional Community Policing Institute at St. Petersburg Junior College in St. Petersburg, Florida, and attended by both police and 250 community residents. The chiefs of police and sheriffs of the St. Petersburg region participated in a panel discussion on the vision, philosophy, and future of community policing; residents were also afforded the opportunity to meet and network with their chiefs and sheriffs; and workshops were held on problem solving and community policing partnerships that were cofacilitated by residents and community police officers.

P3 Probation-Police Partnerships

P3 Probation-Police Partnerships is CD-ROM training developed by the Regional Community Policing Institute of St. Petersburg Community College. The multimedia CD-ROM is geared specifically toward bridging the information gap between probation officers and community policing officers throughout Florida. It outlines the steps that police and probation agencies must take to form a partnership. By sharing information on an offender, both officers better serve the community, as well as develop an understanding of each agency's mission, role, and scope as they relate to one another.

The CD-ROM contains five instructional units, each with an interactive quiz. Topics covered include

- importance and purpose of partnering community policing officers with probation officers
- foundations for integrating community police officers and probation officers
- importance of using a zone approach to supervision with a probation officer
- methodology for implementing and managing zone supervision
- definitions and examples of community oriented policing and SARA problem-oriented policing model
- how to administer and manage a partnership between probation and police officers

The training program also includes video simulations that depict real-life situations in which probation-police partnerships are used and that provide opportunities for user interaction.

Citizen Police Academies

As a facet of community policing, many police agencies in the 1990s have established citizen police academies. Citizen police academies originated in England. Shortly afterward, the Orlando, Florida, Police Department conducted the first citizen police academy in the United States. The academies are open to citizens who volunteer to attend classes for approximately 12 weeks, usually meeting one evening a week for three hours. The training program generally consists of classroom and "hands-on" instruction. The police departments provide comprehensive instruction that covers a different area of policing each week. Training topics include the responsibilities of patrol officers, criminal investigative skills, crime scene search, gang activities, communications, drugs, tactical operations, and community involvement. Citizens may spend time on patrol with officers and take a tour of police headquarters or a district police station. They may spend time on the firing range firing the exact weapons that officers are assigned. Classes may include lectures by the district attorney, judges, and other criminal justice personnel. Some police departments have conducted citizen police academies in Spanish for their Hispanic citizens.

SELECTED APPROACHES TO IMPLEMENTING COMMUNITY POLICING

Astoria Police Department

Generally, medium-sized and big police departments receive publicity on their community policing approaches, but small police departments are neglected by the media and researchers. Small departments may not have the resources to do extensive planning to implement community policing. One such department is the City of Astoria Police Department, located in the state of Oregon.*

---

*The information in this section was obtained from a letter from Chief Robert E. Deu Pree and from copies of memoranda of the Astoria Police Department.

The Astoria Police Department is a small agency with 16 officers serving 10,000 people in a 5–square-mile area. The department has a chief who directly supervises a captain, an emergency communication manager, and a support services manager. The captain supervises four sergeants, who perform all the functions of patrol officers, and a detective and a community policing officer. The emergency communication manager dispatches as well as supervises seven full- and part-time dispatchers, who provide regional 911 call-answering and dispatching services for 17 police and fire agencies. The support services manager is the department's administrative assistant and supervises two clerical personnel who provide records maintenance and evidence custody services.

In 1992, the department developed and published a plan to implement community policing, but because the city had limited financial resources, very little was done to put the plan into action. In 1993, the plan was reviewed and revitalized. It called for one senior officer to be trained in community policing and to put into practice the philosophy of community policing in a particular neighborhood from one to two years. This officer would not be assigned radio calls. When it was determined that community policing was working, two additional officers were selected for community policing activities. They were assigned radio calls but were expected to be involved in community policing when freed from service calls and other assignments. Two more officers will be assigned to community policing activities once they have sufficient experience. Eventually, all police officers and sergeants will be community policing officers and will be assigned to specific neighborhoods. It will take approximately 10 years to implement community policing in Astoria if no additional police officers are hired. Astoria has one detective, who rotates from patrol every two years, and patrol officers perform most follow-up investigations. Therefore, incorporating the detective into community policing is not an issue. The department uses volunteers to conduct some community policing activities, such as citizen surveys.

The Astoria Police Department has a simple plan for a small police agency that for all practical purposes should work. With city officials, business owners, neighborhood residents, and police personnel behind community policing, the plan should make for better and closer relations between the police and citizens.

Bellingham Police Department*

One department that developed a detailed strategic plan was the Bellingham, Washington, Police Department. The information here was obtained from their strategic plan.

The City of Bellingham is the county seat and major city of Whatcom County, which is located between the cities of Seattle, Washington, and Vancouver, British Columbia. Bellingham covers an area of approximately 26 miles. The department has 159 employees with 97 police officers. In 1997, the Bellingham Police Department

---

Source: Adapted from *The Bellingham Police Department Strategic Plan: January 1997–December 2001,* 1997, Bellingham Police Department.

published a strategic plan for adopting community policing as its organizational philosophy. The plan stated that community policing is based on the four principles of partnership, problem solving, participatory management, and progressive visionary leadership, which it described as follows:

1. *Partnership.* Partners enlighten and empower the community to solve their problems, concerns, and fear of crime. Partnerships bring together people from all parts of the community that are willing to work together to problem solve for the betterment of the entire community.
2. *Problem Solving.* Problem solving occurs in a community when there is a discernible combination of a cooperative effort, a comprehensive systematic model, and an environment that lends itself to equal input. Issues in a community can, and will, be dealt with in a more complete and successful manner when the problem-solving model is used as the structure for resolution.
3. *Participatory Management.* Ideas that are embodied in the concept of participatory management are the following: the integrity of the department will be maintained with as few levels of management as possible, all levels of the organization will participate in developing and organizing the efforts of the department, decision-making responsibilities will exist at the lowest levels possible, and, finally, the department will work to be as inclusive and supportive of one another as possible.
4. *Visionary Leadership.* Community policing employs the organizational structure of the inverted pyramid rather than the classic bureaucratic structure. Leaders within the department are committed to the future development of the department and personnel of the department. Leaders within the department will allow, and enhance, the change process.

The Bellingham plan is founded on the department's mission statement and values statement. According to the mission statement,

> The primary mission of the Bellingham Police Department is to coordinate and lead the efforts with the community to preserve the public peace, protect the rights of persons and property, prevent crime, and generally provide assistance to citizens in urgent situations. The department is responsible for the enforcement of all federal laws, Washington State laws, and city ordinances within the boundaries of the City of Bellingham. The department must enforce the law in a fair and impartial manner, recognizing both the statutory and judicial limitations of police authority and constitutional rights of all persons. It is not the role of the department to legislate, render legal judgement, or punish. The department serves the people of Bellingham by providing law enforcement service in a professional and courteous manner, and it is to these people that the department is ultimately responsible.

The values of the department are as follows:

- Individual members of the Bellingham Police Department are valued for their unique contributions. The department is strengthened through the creativity and participation of each member.

- The Bellingham Police Department is committed to close and constant contact with our community members to help maintain the quality of life enjoyed and expected.
- The Bellingham Police Department is committed to furthering democratic values. Every action of the department reflects the importance of protecting constitutional rights and ensuring basic personal freedom of all citizens.
- Honesty and integrity are essential qualities of all members of the Bellingham Police Department. These qualities foster an environment where members are supportive of each other and a "family-like" trust exists.
- The Bellingham Police Department is committed to maintaining our high level of professionalism.
- The Bellingham Police Department is committed to a problem-solving partnership with the community as the means to maintaining public peace and order.

As part of their implementation plan, the Bellingham Police Department conducted a random survey of citizens. The patrol officers who distributed it described it as a method for the department to evaluate the services provided by the department and encouraged citizens to give their honest opinions. Along with the survey instrument, a stamped addressed envelope was provided. Fifty-six percent of the citizens who received the surveys returned them. The survey covered citizens' opinions of the Bellingham Police Department, the amount of contact citizens had with the police, and the major crime and disorder issues of concern. This survey helped a 40-member strategic planning team to establish department goals and objectives. The strategic planning team included business people, school officials, citizens, and police personnel from all levels within the organization. Goals and objectives were developed for three different divisions—administration, operations, and services—under the headings of empowerment, partnership, organizational development, and organizational effectiveness. As an example, the goals and objectives of the operations division are outlined here:

1. *Organizational Development*
   - *Goal:* Assist employees in their efforts to perform problem-solving techniques while working in cooperation with community members.
   - *Objectives:* Develop and implement community policing resource/procedure manual for employees.
2. *Organizational Effectiveness*
   - *Goal:* Provide previous shift information to oncoming shift.
   - *Objectives:* Disseminate previous shift's activities to create broader base of knowledge for oncoming shift.
3. *Empowerment*
   - *Goal:* Empower employees to work within the philosophy of community policing.
   - *Objectives:* Allow employees the ability to have more input and affect their community policing neighborhoods.
4. *Partnership*
   - *Goal:* Integrate downtown foot beat/transit officer position with existing COP program.

- *Objectives:* Recruit and assign officers to downtown foot beat/transit duty with the intention of fully meshing these responsibilities with the establishment goals of the COP program; enhance the COP attention to growing downtown business community.

The strategic plan also devised strategies on how the goals and objectives were to be established.

Eugene Police Department*

The City of Eugene, Oregon, has a population of approximately 180,000 within 47 square miles. In early 1990, a study done by the City of Eugene found that the city's population was growing and that the number of calls for service was increasing much more rapidly than the resources available to meet this need. The department's incapacity to handle service calls was diminishing the quality of life in the city and was a direct threat to life and property. Major root causes of public safety problems were determined to be drug abuse and unemployment. Because of the city's problems and the limited resources to deal with them, the city's planning team determined that fundamental changes in the community's approach to its public service needs were required—specifically, increased cooperation of human service agencies, neighborhood groups, and individuals with public safety agencies. The City of Eugene therefore adopted a new approach called the partnership model, which stressed the need for outreach and networking to address underlying social problems that contributed to criminal behavior and unsafe neighborhood conditions. It was believed that this approach would enable the department to have a more positive and consistent presence in the community and would allow for more productive use of personnel, equipment, and public facilities. The primary goal of the partnership model was to promote a modern, competent, and service-oriented department accessible and answerable to the community it served. The planning team developed a 20-year, long-range plan that allowed for regular review, amendments, and refinements. It included policies, goals, and strategies that addressed public safety needs. To allow for a timely transition, the partnership model established eight key strategies:

1. *Community-Oriented Policing*—The existing program, concentrating the combined resources of the Department of Public Safety, the Parks, Recreation, and Cultural Services Department, the White Bird Clinic, and other community agencies on quality-of-life concerns—particularly street-level crisis intervention and referral as opposed to traditional enforcement measures—needs to be continued and expanded.
2. *Prevention and Education*—In addition to the department's current efforts in this area, resources should be added allowing fire crews to participate in presen-

---

Source: Adapted with permission from *Partnership for Public Safety: Long Range Public Safety Plan,* 1991, City of Eugene.

tations to local schools and community groups, citizens to receive training in organizing and conflict resolution, more crime prevention presentations [to be held] and [opportunities for] positive, regular interaction with the community's young people [to be provided].

3. *Community Service Specialist*—Greater utilization of this noncertified paraprofessional classification will provide support needed to maintain existing service levels by supplementing the work of certified police officers and fire fighters and will help to address some of the community policing and prevention/education needs noted above. Cost saving also will be realized as these employees assume some of the duties now being performed by certified personnel.

4. *Fire/EMS Redeployment*—To maintain an adequate level of fire protection citywide, while recognizing changes in the nature of the fire and emergency medical service (decreasing fire suppression activity, increasing requirements related to hazardous materials, fire prevention, and medical emergencies), the plan recommends redistribution of fire stations in the community. Another key advantage of this redeployment is that it will bring the Eugene community closer to citywide availability of a four-minute emergency response capability.

5. *Public Safety Stations*—The plan proposes establishing a presence in neighborhoods with facilities that would include community meeting rooms, note taking and referral services, public safety resource materials, and fine and bail collection capabilities. This could be done, at least in part, in conjunction with the Fire/EMS redeployment plan.

6. *Metro Coordination*—In view of the success of regional public safety programs such as 9-1-1, Emergency Medical Services, etc., the city should participate in examining the potential for creation of a Metro Public Safety Board, as well as in the formulation of a regional plan for public safety programs, possibly to include fire prevention and suppression, purchasing, training, and other programs and services.

7. *Training and Employee Development*—New strategies will be required to attract and retain high quality public safety employees to perform effectively in a community-oriented public safety environment. Changing community demographics will require a more culturally diverse work force.

8. *Planning, Evaluation, and Amendments*—The long-range plan should be reviewed annually, with a full evaluation and amendment procedure every five years, by the Public Safety Advisory Committee (PSAC), supported by the department's planning, research, and development staff. This process will include abundant opportunities for citizens' participation, in accordance with the fundamental principles of the partnership model.

This plan, at the time it was developed and implemented, might well be considered innovative. It provided an opportunity for a variety of governmental and community agencies to become involved in public safety issues, and it recognized that public safety involves development of a partnership among fire fighters, emergency teams, and police.

In 1997, the Eugene Police Department developed what they referred to as a Community Involved Policing (CIP) Action Plan* for their community policing philosophy. The CIP Action Plan was based on the 1991 partnership model plan. It defined community involved policing as

> officers who know their beats. Residents and business who know their beat officers. No graffiti, gang presence, or broken street lights and windows. No blighted neighborhoods. Streets that are cared for, clean and safe. Quick response to emergencies, thorough follow-ups, complete, high-quality, tough-on-crime investigations that lead to arrests. Police are there to help through close partnerships with social services, victims' services, neighborhood groups, business associations, other city departments, charitable associations. Or just [to help citizens] feel safer. (3)

The new action plan had seven components:

1. *Interdepartmental Cooperative Action Team (ICAT).* The purpose of the Interdepartmental Cooperative Action Team (ICAT) is to identify, prioritize, and refer crime and neighborhood safety issues within the city organization and to our community partners. The ultimate objective of this activity is to make neighborhoods look safe, feel safe, and be safe.
2. *Differential response.* The objective of this action item is to redirect the need for officers to respond to certain categories of incidents and investigations, so that more time is available for police to provide neighborhood interaction, problem solving, and other activities that are being requested.
3. *Area coordinators, crime analysis, and beat health.* The concepts of beat health and area coordinators are integral to a successful long-range community involved policing philosophy. . . . Beat health and the area coordinator are just two components to reduce crime and increase the livability of Eugene's diverse neighborhoods.
4. *Youth strategies.* The objective of this action is to collaborate with community agencies to develop and implement a regional crime prevention strategy. Public safety personnel will provide intervention, alternatives actions, skill-building, and role models for youth at risk.
5. *Community involvement.* To accomplish this objective, meaningful citizen input will be encouraged and considered in policy development and evaluation of police performance.
6. *Recruitment, diversity, training, and promotion.* The objectives of this action item are to select diverse recruits interested in community policing; to increase the quality of training provided to our recruit officers; to reduce the training time necessary; to make the training more community oriented; to provide depart-

---

Source: The next three paragraphs (and lists) are adapted with permission from *Community Involved Policing (CIP) Action Plan: October 1, 1997–December 31, 1997,* 1997, Eugene Police Department.

ment supervisors and managers with the training necessary to assist them in implementing and managing CIP; and to foster success in community policing into promotion for supervisors and managers.

7. *Regional coordination and automation.* This action item will develop and implement strategies to improve coordination and cooperation among regional criminal justice stakeholders, including social services and the private sector, while minimizing the possibility that efforts in one part of the criminal justice system will have unanticipated adverse impacts on other parts. (14–31)

A review of CIP Action Plan in 1998 made recommendations to enhance the community-police partnership:

- *Beat-level citizen/police groups:* As much as possible, use preexisting groups in each beat rather than creating new ones. Where necessary, set up new structures consistent with city plans for neighborhood group revitalization that allow people in the beat to communicate with the beat officer and the police department on an on-going basis about issues and concerns in that beat. For some beats, a neighborhood group will be the answer. For others, a neighborhood advisory board or a preexisting business group will be the key. Many beats should coalesce into one group, both to recognize shared issues, to cut down on the number of groups and meetings, and to save staff time. The exact structure involved needs to remain flexible so that people in each beat are best able to define what they need and change these structures over time. Beat officers will attend all meetings of the beat groups and will be able to call upon whatever resources are relevant to helping people in the beats to resolve public safety concerns through the development of partnerships, proactive approaches, and collective efficacy. . . .
- *Police Forum:* The focus of the Police Forum will be to work on community-wide issues, with a special focus on community policing efforts. The forum will help the department by reviewing its plans; acting as advocates for revenue measures; and providing a forum for people to bring their issues. The forum can refer issues to beat groups, the Police Advisory Committee, the Human Rights Commission, or the complaint/commendation process, as appropriate. . . .
- *Police Advisory Committee:* Create a standing Police Advisory Committee by ordinance. This group will be advisory to the city council and the chief of police regarding police policy issues and will replace the Council Public Safety Committee. Again, details will need to be worked out with the group itself, but the broad charge will be to review police policies, practices and priorities and mesh them with community values; to review revenue measures and alternative strategies; to take on council projects; and to provide a sounding board for public concerns. Membership will need to include representatives from education, social services, seniors, youth, minority committees, businesses, the city council, and at-large public seats. . . .
- *Complaints/commendation process:* If the voters so choose, an External Review Board will be added to this process. (Eugene Police Department 1998)

The Eugene Police Department is continuously revising and updating its approach to community policing. The beat-level citizen/police groups, Police Forum, Police Advisory Committee, and complaints/commendation process are all indicators that a concentrated and committed effort has been made by the Eugene Police Department to implement community policing. The department is continuously working toward developing a police-community partnership and empowering its officers to deal with issues on their beats by making the beats small enough so that the officer assigned to a particular beat can not only function as a community police officer but be the crime prevention officer and investigate minor crimes.

Illinois State Police

Generally, state police agencies are not involved in community policing, because they have statewide police jurisdiction and are not assigned to populated areas that would have a police agency. However, several state police agencies are involved in community policing either as community policing trainers or as resource centers. Students of community policing should be aware that several state police agencies are involved in community policing. The Illinois State Police were selected for this chapter because they are one of the more advanced state agencies involved in community policing.

The Illinois State Police implemented community policing in 1993 by deploying six state police officers in a pilot Statewide Community Oriented Policing Effort (SCOPE). State, municipal, and county police agencies usually have relied on the state police for support services such as crime scene services and data system information, but by adopting community policing, the Illinois State Police expanded this service role to include the full range of policing responsibilities, including patrol, traffic, and investigation. From six state police officers involved in community policing in 1993, the number has increased to 350 in 1999. These officers have been involved in over 800 problem-oriented projects throughout the state (Illinois State Police n.d.).

The Illinois State Police community policing philosophy led to the development of Resource Support Centers. These centers collect, process, and analyze information about major crimes and incidents. Their databases are accessible to all state law enforcement agencies and are backed up by a centralized mapping unit that provides valuable analytical and mapping assistance to police departments' patrol and investigative units. The Resource Support Centers allow police departments to deal efficiently with crime trends (Illinois State Police n.d.).

The Illinois State Police refer to their community policing strategy as geographically oriented community oriented policing (GEOCOM). According to the Illinois State Police, GEOCOM is a breakthrough in community policing in that it broadens the definition of community beyond neighborhood-oriented, beat-walking, and bike-riding strategies. One Illinois state police GEOCOM officer is assigned to each Illinois county. GEOCOM state police officers are required to reside in their GEOCOM area (county) and become part of the community. They act as liaisons between the state police and local police agencies. They patrol the state and local highways and network with citizen advisory councils and state, county, and municipal agencies to target and identify specific crime problem areas. The goal of GEOCOM is to decrease duplication efforts and direct resources to problem areas (Illinois State Police n.d.).

The Illinois State Police initiative in establishing community policing on a state-wide basis is exemplary. Although community policing on a statewide level is in the early stages in Illinois, the attempt at a minimum indicates that the Illinois State Police are innovative risk takers.

Madison Police Department

The Madison, Wisconsin, Police Department can trace its involvement in community policing to the mid-1980s. Community policing in Madison is referred to as quality policing and is based on the system of total quality management (TQM), which was discussed in Chapter 9.

The Madison Police Department, when establishing "quality policing," drew on Deming's Fourteen Points to create five points:

1. A clear and understandable mission statement
2. A commitment to engaging in teamwork and problem-solving in each and every unit
3. The creation of a community orientation and customer focus on the part of each and every employee
4. The full utilization of the important resources of the organization by recognizing employee worth and the untapped resources and potential of employees
5. An overall focus on quality (Couper and Lobitz 1991, 64)

The police department rewrote its mission statement to reflect the principles of "quality policing" and drew up "Principles of Quality Leadership" as well (Chapter 9). Thus, the concepts of "quality policing" have been fully integrated into the department. The experience of the Madison Police Department illustrates several key ingredients in organizational change:

- a strong vision of the future
- a strong unyielding commitment from the chief executive
- a commitment to developing the skills and abilities of leaders as well as employees in the organization and continually training them
- a patient and persistent focus on the long-term operations of the organization (Couper and Lobitz 1991, 87)

Portland Police Bureau

In the late 1980s, the City of Portland, Oregon, had a sharp increase in crime, gangs, and drug problems that severely affected the quality of life for citizens, neighborhoods, institutions, and businesses and placed a severe strain on limited police resources. At that time, the police bureau made the traditional response to an increase in crime when resources are limited. Because the city government decided that police resources would not or could not be increased, the bureau removed police officers from support functions to patrol activities, demoted detectives, eliminated crime analysis, scaled back crime prevention and planning and research units to skeleton crews, and eliminated tactical units. Civilian positions were reduced so that more funding would be freed up to put more patrol officers on the street. But even with more patrol officers on the street, radio service calls continued to increase, and response time to priority calls averaged over 9 minutes. The numerous service calls allowed the patrol officers little time to develop community relations with citizens. The gap between the community and the police was not closing; rather, it was widening. Citizens viewed police merely as responders to problems, and police viewed citizens as complainants who wanted the police to fight crime without their help.

During this period, there was a growing awareness of community policing. The mayor, upon hearing about the success of the community policing concept in other cities, directed the police chief to adopt the community policing philosophy. The police chief placed in action the events that led to the development of the Community Policing Transition Plan (Portland Police Bureau 1990, 4).

The Portland Police Bureau has periodically revised its strategic plan over the years to meet the needs of the community. When the plan was last revised in 1998, strategies were given with detailed descriptions, listings of the police units involved, applicable work plans, and partner agencies used to carry out the specific strategy. All the versions of the strategic plan have included a vision statement, mission statement, values statement, priority issues, goals, objectives, and work plans. Since the implementation of community policing, the bureau has continued to look for feedback from its officers and the community to improve the department's approach to community policing.

The history of the bureau's strategic planning process is as follows:

- 1988
  - A policy is drafted proposing that the bureau make a transition from being a traditional policing agency focusing on arrests and crime rate to being a community policing agency focusing on community problem solving and results.
  - The mayor directs the police bureau to prepare a strategic management plan incorporating a community policing philosophy.
  - The community policing planning process is initiated; the Community Policing Working Group is created from police, the Office of Neighborhood Associations, the community, and the mayor's office.
- 1989
  - A community policing concept paper is created with a vision statement definition of community policing, an outline of the strategic planning process, and five-year goals; it is distributed to community members for comment.
  - A critical path chart (with timelines and benchmarks) is created. The critical path has four stages: *definition phase* to define community policing; *design phase* to create recommendations for realignment of services and functions; *planning phase* from various committees and their action; and *implementation phase* to create an operational strategy/action plan for FY 1990–91.

- Five committee meetings, with surveys, are conducted by the Office of Neighborhood Associations. It is attended by the mayor, the chief, precinct captains, and members of the community.
- Community policing transition committees are formed with community members, representatives from other agencies, and bureau employees. The committees are Menu (to respond to issues raised in the five community meetings); Media/Education; Evaluation; Productivity/Workload Analysis; Information and Referral; Legal/Legislative; Training and Recruitment; Grants/Finance; and Criminal Justice.
- Resolution No. 34627 is passed by unanimous vote of the city council. This resolution includes a definition of community policing that was created from community meetings.

- 1990
  - Resolution No. 34670 is passed by the city council, adopting the Community Policing Transition Plan. The plan outlines a mission statement and one-, two-, and five-year goals and objectives. Each strategy is analyzed to determine if its implementation would require additional resources.
  - Three demonstration projects are selected, one in each precinct: Iris Court in North; Central Eastside in East; and Old Town/Chinatown in Central.
  - The Citizens' Crime Commission funds an analysis by the Institute of Law and Justice on law enforcement climate; organizational structure and resource deployment; staffing levels; management practices; service demand and workload; human resources management; management; communication; and budget process.
  - The Chief's Forum, a policy advisory group, is created.

- 1991
  - A Citywide Community Policing Workshop and survey is conducted.
  - The Year 1 Report on Community Policing Implementation is adopted by the city council. The report contains Year 1 strategies employed, highlights of activities, and Year 2 strategies. Adopted with this report are the bureau's attributes and success factors for community policing. The attributes are a set of qualities desired in a bureau that has fully implemented community policing. Success factors are a set of factors that measure how the qualities have been achieved.

- 1992
  - The Portland city auditor surveys Portland residents to gauge the performance of city government, which includes the police, and publishes the Service, Efforts and Accomplishments (SEA) baseline report.
  - The Human Goals Statement is adopted as a bureau general order. The statement defines the human resources and work environment goals for the bureau as a whole.
  - A National Institute of Justice $366,000 evaluation grant to create community policing performance measures is awarded to the police bureau.

- 1993
  - The second SEA report is published by the city auditor. It compares response on police services and overall perception of safety from 1992 to 1993. Data

collected are used by the police bureau as performance measurements in its 1994–96 budget and are reported on three times a year in budget monitoring reports.

– Work begins on creating the second strategic plan. Initial research is conducted by the bureau on the transition plan to determine what has been working and what has not been working, what obstacles exist, etc. The bureau mission statement is revised to include the aspect of maintaining and improving community livability. Bureau goals of the previous strategic plan are restated as values. New goals are adopted: reduce crime and fear of crime; empower the community; develop and empower personnel; and strengthen planning, evaluation, and fiscal support.

– A draft of the 1994–96 strategic plan is reviewed by RU managers, the budget advisory committee, the precinct advisory committees, the Chief's Forum, and internal advisory committees. The draft is distributed to neighborhood associations, community groups, and individuals; more than 600 are distributed for review. The draft is reviewed by the mayor and the city council. Employee job satisfaction survey results are released. The survey measures bureau employees' assessment of job satisfaction, supervision support, autonomy, recognition, teamwork, fairness, and problem-solving support. Surveys were distributed to all 1,200 employees, with a 46 percent rate of return.

• 1994

– The city council adopts the bureau's 1994–96 strategic plan, with an updated mission statement, five values, and the goals of reducing crime, empowering the community, empowering personnel, and strengthening planning and fiscal support.

– The first integrated work plan reports, incorporating progress reports and performance measurements on the 1994–96 strategic plan, are released to the city council and citizen advisory council.

– The Portland Police Bureau 1994 Community Assessment Survey, based on interviewing from July to August 1994, is released. The analysis is based on questions regarding crime, livability, victimization, satisfaction with police services, familiarity with police and crime prevention specialists, and recommendations on how to improve services.

• 1995

– The results of the second employee job satisfaction survey results are released. The survey measures bureau employees' assessment of job satisfaction, supervisor support, autonomy, teamwork, recognition, fairness, and organizational culture. The report released compares the 1995 with the 1993 results. Surveys were distributed to all 1,250 employees, with a 43 percent rate of return.

– The final report on Community Policing Performance Measures, supporting the National Institute of Justice grant to the City of Portland, is released. The report contains findings from a partnership agreement survey and assessment, interagency focus groups, an implementation profile analysis of bureau manag-

ers, a disaffected youth survey, a youth in school survey, and a domestic violence reduction unit evaluation.
- Work begins on the 1996–98 strategic plan, with a review of the assessment
  information and recommendations gathered to date and a review of the current
  integrated work plan process. The 1996–98 strategic plan is connected to individual division work plans and contains performance measurements for the
  first time.
- The Police Bureau conducts its second National Community Policing Conference, focusing on the area of organizational development to support the transition to community policing and other contemporary policing topics. More than
  600 attendees from 35 states and five countries participate.
- 1996
  - Performance measurements corresponding to the four program areas in the
    budget are developed and included in the 1996–98 budget submission. Reports
    on the city council benchmarks are also included in the budget submission.
  - The second Portland Police Bureau Community Assessment Survey is con-
    ducted and released in 1996. The analysis is based on questions regarding
    crime, livability, victimization, satisfaction with police services, familiarity
    with police officers and crime prevention specialists, and recommendations on
    how to improve services.
  - For the City's Comprehensive Organizational Review and Evaluation (CORE)
    efforts, the bureau produces a report outlining recommendations made to the
    bureau from outside agencies or major task forces for the last 10 years, along
    with the changes made as a result of those recommendations.
  - An extensive review of all bureau programs and services is performed to assess
    the response to requests for budget cut packages due to the passage of property
    tax limitation Measure 47.
  - The bureau applies for, and receives, federal grant support from the U.S. De-
    partment of Justice Office of Community Oriented Policing Services (COPS)
    to fund 60 officer positions for three years, with a commitment from the city
    council to fund the positions after the grant period ends. This assists in bringing
    the bureau up to strength after 47 officer positions were cut after Measure 47.
- 1997
  - The results of the third employee job satisfaction survey are released. The sur-
    vey measures bureau employees' assessment of job satisfaction, supervisor
    support, autonomy, teamwork, recognition, fairness, and organizational cul-
    ture. The report compares the 1997 with the 1995 and 1993 results. Surveys
    were distributed to all 1,250 employees, with a 59 percent rate of return.
- 1998
  - Work begins on the 1998–2000 strategic plan, with a review of the 1996–98
    plan and two public input opportunities. The 1998–2000 strategic plan is con-
    nected to individual division work plans and contains a national and regional
    trends analysis section for the first time.

Savannah Police Department

In 1991, Savannah, Georgia had a hotly contested race for mayor in which crime, especially violent crime, was a major campaign issue. In 1991, violent crime was up 17 percent over the previous year and had risen by over 66 percent since 1989. The city had experienced a large increase in homicides: from 20 in 1989 to 35 in 1990 to 59 in 1991. The increase in violence was contributed to the crack cocaine epidemic. Because of the hotly contested race, the incumbent city manager directed the police chief to develop a strategy to combat crime. In response, the police department developed a community policing strategy called the Comprehensive Community Crime Control Strategy. The executive summary for the strategy identified six major problems with the deployment* practices of the department:

1. *Lack of Geographic Accountability.* Currently, officers are only accountable for crime reduction efforts during their eight hour shift. The current zone and beat boundaries are not conducive to officer accountability because officers are not consistently assigned or deployed to the same areas. Captains supervise and are responsible for an eight hour shift. In essence no single individual is accountable for a particular problem; thus accountability is lost. In addition, current beat configurations do not square with the geographic distribution of known crime problems, that is, current boundaries do not group problem areas together.

2. *Unequal Distribution of Patrol Time.* Unequal distribution of patrol time is a problem because it does not permit officers sufficient time to engage in preventive patrol or problem-solving activities in areas where it is needed most. As presently structured the patrol deployment plan has an unequal distribution of the patrol time. The problem lies with the distribution of manpower by shift. The shift that operates from 4 p.m. to midnight accounts for 48.7% of the crimes but spends only 24% of available patrol time on preventive patrol. In contrast, the shift that operates between 12 midnight and 8 a.m. accounts for only 26.8% of the city's crimes yet spends more than 53% of its patrol time on preventive patrol. . . .

3. *Need for Better Interaction with Residents.* Research consistently shows a strong relationship between the levels of police-citizen interaction and citizen satisfaction with police services. The unequal distribution of patrol time in Savannah shows that officers spend a disproportionate amount of time on answering calls for service, leaving little time for the kinds of activities that would enhance positive interaction between the police and neighborhood residents. . . .

4. *Need for More Proactive Patrol.* While responding to calls for service is the most essential service provided by patrol officers, responding to calls, by definition, is reactive. Current police services are incident-oriented rather than problem-oriented. The effect is that police are neither solving problems nor preventing crime. If patrol is limited to satisfying demands on calls for services, then patrol becomes a reactive service with little impact on crime.

Source: Adapted from *Executive Summary: Comprehensive Community Crime Control Strategy,* 1991, Savannah Police Department.

5. *Inadequate Police Response System.* At present the police call response system does not make optimal use of available resources. . . . In essence, a mobile police response was sent to answer almost all calls for service, even though only 19% were calls in which immediate mobile response was needed. Nor does a rapid response affect crime rates. . . .

Tying up officers on calls that could be delayed or handled by telephone detracts from an officer's time and attention to proactive activities such as problem-solving and crime prevention. More important, however, it reduces the department's capacity to quickly and adequately respond to calls requiring an immediate police response. Consequently, a new method for prioritizing and managing calls-for-service workload is needed.

6. *Inadequate Data Management.* Accurate data are needed to adequately manage effective patrol deployment. Current data availability limits continued detailed analysis of crime and deployment conditions.

The six major problems identified by the Savannah Police Department can all be related to traditional policing. Although many aspects of traditional policing are valuable and ingrained into the police culture and our society's culture, aspects of traditional policing can be improved upon. The Savannah Police Department was a good traditional police department, but it was not responding to the crime problem of its city. The department's officers were responding to service calls, many of them trivial, when their officers could have been working on solving the problems that led to the city's high crime rate.

The same study made several keen observations concerning the limitations of the criminal justice system in the Savannah area. First, it found a lack of cooperation among the criminal justice components. For example, the community lacked a process for coordinating the flow of criminal cases. The criminal justice system consequently suffered from a case overload. Arrests, indictments, criminal trials, and probation revocations were all increasing substantially, and most offenders were repeaters. Another criminal justice problem was the lack of an integrated database for the various criminal justice components so that the criminal justice agencies could not coordinate their activities and operate effectively and efficiently.

In their transition from a traditional police department to a community policing agency, the Savannah Police Department developed a number of strategies, both internal and external. The internal strategies included the following:

1. *Command Accountability.* Accountability will be geographically distributed. Each of the twelve areas will be placed in one of four zones (District Police Stations). A captain will be assigned to and be held accountable for all police services in his zone 24 hours a day. He will direct police strategies for his zone and will specify objectives which are tied to each of the goals. To accomplish this, each captain will assign manpower for each zone and each area in the zone, specifically how this manpower will be used and when it will be used. The captain will freely schedule officers according to activity in the area. This may include staggered shifts, overlapping shifts, or hour-by-hour scheduling. In addition, the captain will maintain continuing contact with the people living in his

zone. The service areas are aggregated into four zones to group crime problems into manageable sections of the city and to facilitate patrol management. . . .

2. *Manpower Deployment.* The new manpower deployment plan will resolve the problem of unequal distribution of patrol time. Manpower will be distributed to each of the service areas and zones on the basis of calls for service, time to provide for preventive patrol, citizen participation, and other activities. . . .

3. *Community-Oriented Policing.* Community-oriented policing directly addresses the need for patrol officers to interact more with residents in the neighborhoods they patrol. . . .

4. *Problem-Oriented Policing.* Problem-oriented policing (POP) is a proactive approach to patrol operations that identifies, analyzes, and responds to specific programs.

5. *Differential Police Response.* To resolve the problem of inadequate police response and to allow patrol officers more time for problem solving, a revised differential police response system (DPR) is recommended. DPR is a deliberate stacking of certain non-emergency calls so as to not occupy too many officers at once, and to allow the dispatch to go to the neighborhood officer when he/she becomes available. . . .

6. *Improved Analytical Capabilities.* A planning and research director has recently been staffed to oversee data management efforts, since effective patrol deployment requires accurate and timely information.

The Savannah Police Department took an acceptable approach in their transition from traditional policing to community policing. They recognized the importance of decentralization. District police captains would now be accountable for specific areas of the city with an established number of officers under their command. They were held responsible for the crime rate in their district along with crime prevention. The patrol officers worked for one captain in the same district of the city. The department accepted problem-oriented policing and community-oriented policing as its policing strategy. In addition, it addressed differential police response and improved the department's analytical capabilities.

The Savannah Police Department, like other community-oriented police departments, recognized that crime was a community problem, not just a police problem. The department developed a plan to build trust between the police and the community. All police officers were encouraged to increase contact with citizens, share information about crime with citizens, be visible in the community, and engage in cooperative problem solving with citizens. In addition, the department made a concentrated effort to coordinate criminal justice activities with other criminal justice agencies. For example, it worked with other criminal justice agencies on jail overcrowding, recidivism, and escalating caseloads.

Wichita Police Department

In the early 1990s, the City of Wichita Police Department, with the encouragement of the mayor, established community policing in two high-crime neighborhoods. A lieutenant and several community policing officers were assigned to perform problem-solving activities. Under the command of the two lieutenants, problem-solving activities and cooperation from the neighborhoods were successful. In 1995, the police chief resigned, and a new police chief was appointed who was committed to community policing and its implementation citywide. The new police chief, observing the success of community policing in the pilot neighborhoods, directed that a strategic plan be developed. The strategic plan, referred to as the Public Safety Initiative (Wichita Police Department 1996) was implemented in 1996. It should be noted that the newly appointed police chief had the support of the city council, the newly elected mayor, and the city manager.

The Wichita Police Department's primary aim was to improve the department's ability to deliver police services by using the community policing philosophy. A transition from traditional policing to community policing involving the entire department was to take place. Under the Public Safety Initiative, the department was combining community officers' duties with the mission of the patrol officers and other departmental units in a policy that integrated all departmental operations.

The policy established 36 patrol beats, each of which had seven police officers assigned to it. One beat police officer would function as a community policing officer, and the remaining officers would provide police services to the beat. Two patrol officers were assigned to each shift. The responsibilities of the community policing officer included collecting crime analysis and staffing information and meeting with beat officers, neighborhood associations, community groups, and others to coordinate police services unique to the beat's needs. The six other beat officers were to use the information provided by the community policing officer to implement tailored police services and assist in implementing community policing activities. All the officers assigned to a specific beat were expected to assess problems within the beat and to work with the community to develop and implement solutions to beat problems. The beat officers and the community policing officer were held accountable for solving problems on their beats.

The Public Safety Initiative of the Wichita Police Department (January 30, 1996) had the following goals:

- *Make Wichita a safer community through a reduction in crime.* By developing partnerships throughout the community, the root of crime problems will be determined, and solutions to those problems will be designed and implemented in conjunction with the community. By addressing the community's public safety problems through partnerships, existing crime and the prospects for future crime will be reduced.
- *Provide high-quality customer-service.* By providing the community with a better staffed and equipped police force that is specially trained in the community policing philosophy, the Police Department will improve its ability to deliver police services in a more timely, customer-oriented fashion.
- *Ensure timely, effective follow-up investigations.* One of the major improvements provided by the Public Safety Initiative is the addition of personnel and equipment assigned to investigate crime. These improvements will allow the Police Department

both to investigate more crimes in a more efficient manner and to improve communications with the community regarding the status of investigations.

- *Institute accountability at all levels within the department.* By assigning officers to work with citizens throughout the community, officers will be more accountable to citizens through the development of partnerships. In addition, increased numbers of front line supervisors are provided to ensure that beat teams remain focused on community goals. The personnel evaluation process will be revised to ensure that personnel are responsible for accomplishing performance goals tailored to unique needs of the community they serve.

The Public Safety Initiative required the hiring of 24 police officers and the purchase of new portable radios, marked police vehicles, fax machines, computers, and uniforms. The initiative was to cost $3.3 million in 1996 and $5.7 million in 1997.

The Public Safety Initiative required that the field services division be reorganized. Ninety-one positions were added to this field division: one lieutenant, 19 sergeants, 69 police officers, and two clerks. The structure of the division was revamped: four patrol bureaus (precinct stations) were established, each commanded by a captain. Lieutenants supervised each shift, and sergeants supervised patrol officers in groups of beats, called sectors. The number of officers assigned to each bureau varied based on call load. However, each patrol bureau had nine community policing officers to coordinate police services.

The investigative division was not overlooked by the Public Safety Initiative. Twenty-seven positions were added to it: one lieutenant, 18 detectives, one officer, three chemists, one photo technician, and three clerks. The division received additional personnel to allow them to investigate more crimes.

The third major unit of the Wichita Police Department, the support function, also received additional personnel: three lieutenants, two administrative assistants, one planning analyst, one dispatcher, and one management intern.

The Public Safety Initiative focused on crime prevention, crime analysis, and training. One lieutenant was placed in charge of crime prevention, and one lieutenant was assigned to training to improve the department's ability to hire and train officers. An administrative assistant was added to records, and an analyst was added to crime analysis.

The Public Safety Initiative that laid the foundation for the Wichita community policing strategy is still in operation and seems to be successful. Morale of the department seems high, and most police officers seem to have accepted the concept.

SUMMARY

The training of police officers is going through a transition. The traditional approach, or pedagogical technique, emphasizes information-based learning, primarily through the lecture-driven class. In the new approach, known as the andragogical technique, adult learners are held responsible for their own learning, and the instructor functions as a facilitator who guides and assists them but does not necessarily make learning decisions for them.

In 1990, the Royal Canadian Mounted Police introduced a new approach to training that incorporated many aspects of the andragogical technique because they recognized that their introduction of community policing would require police to be innovative problem solvers with excellent people skills. All police are trained in a problem-solving model, known as CAPRA, that emphasizes understanding the client, analyzing information, working in partnerships, choosing appropriate responses, and assessing for continuous improvement.

The Savannah, Georgia, Police Department recognized that all departmental members had to be trained in the community policing philosophy before it could be implemented. Their seven training modules cover community policing, problem-oriented policing, crime in Savannah, service agencies in Savannah, city codes, working with citizens, developing trust, and conducting crime analysis.

The Los Angeles Sheriff's Department developed a six-hour training course on community policing specifically for supervisors and managers. It covers such topics as the broken windows theory, problem-oriented policing, history of policing, combining investigative and patrol functions, allowing lower-level flexibility in decision making, and permanently assigning patrol officers to neighborhoods.

Other training approaches include regional community policing training centers for both law enforcement officers and civilians, which offer courses at no cost, and citizen police academies where citizens learn about police procedures.

This chapter offered an array of examples of approaches to implementing community policing in police departments, some of which involved extensive strategic planning and others which did not. Astoria's implementation was small scale and slow and did not involve much planning because the department had few resources to spare. The Bellingham Police Department, in contrast, developed a very extensive strategic plan that included principles of community policing, a mission statement, a values statement, and numerous goals and objectives and a time frame for achieving them.

The Madison Police Department adopted total quality management as its management philosophy and incorporated it into the Madison community policing program under the name of quality policing. This philosophy emphasized a focus on the customer (citizen), a problem-solving orientation, a drive toward continuous quality improvement, an openness to employee input, and a commitment to teamwork.

The Portland Police Department initiated community policing in 1988. The chapter offered an outline of major accomplishments and developments in the program over the ensuing decade.

The Savannah, Georgia, Police Department launched community policing in 1991 after an election-time furor over high crime rates. Its plan decentralized patrol service and held captains responsible for crime in their assigned geographical area and for the patrol officers under their command. The Wichita Police Department developed a similar plan, with the exception that they specifically assigned community policing officers to work with patrol officers who were responsible for answering service calls. Wichita's community policing officers were to be a resource to the patrol beat officers and to work in conjunction with beat officers to solve problems on their beats.

KEY TERMS

andragogical approach	Comprehensive	Malcolm Knowles
CAPRA	Community Crime	pedagogical approach
CIP Action Plan	Control Strategy	Public Safety Initiative
citizen police academies	GEOCOM	quality policing
COPS		

REVIEW QUESTIONS

1. Compare andragogical and pedagogical approaches to learning.
2. Discuss the Royal Canadian Mounted Police approach to police training.
3. Discuss at least one innovative community police training program.
4. Discuss the Bellingham Police Department's community policing strategy.
5. Discuss the eight-point action plan of the Eugene, Oregon, Police Department.
6. Discuss quality policing.
7. Discuss the Public Safety Initiative of the Wichita Police Department.

REFERENCES

Couper, D.C., and S.H. Lobitz. 1991. *Quality policing: The Madison experience.* Washington, DC: Police Executive Research Forum.

Cross, P.F. 1981. *Adults as learners.* San Francisco: Jossey-Bass.

Eugene Police Department. 1997. *Community Involved Policing (CIP) Action Plan.* Eugene, OR: Eugene Police Department.

Eugene Police Department. 1998. *Discussion paper on the formation of police/citizen/council committees.* Eugene, OR: Eugene Police Department.

Himelfarb, F. 1997. RCMP learning and renewal: Building on strengths. In *Community policing in a rural setting,* ed. Q. Thurman and E. McGarrel. Cincinnati, OH: Anderson.

Illinois State Police. n.d. Illinois State Police: Community-oriented policing. Unpublished document, Springfield, IL.

Knowles, M. 1990a. *The adult learner: A neglected species.* Houston, TX: Gulf.

Knowles, M. 1990b. The field of human resource development. In *Handbook of human resource development,* 2nd ed., ed. L. Nadler and Z. Nadler. New York: John Wiley.

Meese, E. III. 1991. *Community policing and the police officer.* Perspectives on Policing. Washington, DC: National Institute of Justice.

Palmiotto, M.J., et al. 1998. A model curriculum for police recruit training in community policing. Paper presented at the Academy of Criminal Justice Sciences, March 10–14, Albuquerque, NM.

Portland Police Bureau. 1990. *Community Policing Transition Plan.* Portland, OR: Portland Police Bureau.

Royal Canadian Mounted Police. n.d. *CAPRA problem solving model.*

Wichita Police Department. 1996. *Public Safety Initiative.* Wichita, KS: Wichita Police Department.

Chapter 12

Distinctive Community Policing Programs

CHAPTER OBJECTIVES

1. Be familiar with community policing programs that deal with drugs and drug-related crime.
2. Be familiar with distinctive programs for juveniles.
3. Be familiar with distinctive programs that deal with domestic violence.
4. Be familiar with distinctive programs dealing with the business community.
5. Be familiar with distinctive programs that deal with homelessness.
6. Be familiar with distinctive programs that deal with hate crime.

INTRODUCTION

Although community policing has been explained as a philosophy that should permeate all aspects of policing, this chapter deals with distinctive programs for specific social problems that take a community policing approach. One of the most influential programs, Operation Weed and Seed, was initiated by the U.S. Department of Justice. However, most community policing programs are initiated by local police agencies and sheriff's departments. Many local police agencies have shown creativity and initiative in developing programs to solve their communities' problems.

OPERATION WEED AND SEED

In 1992, the U.S. Department of Justice introduced Operation Weed and Seed to illustrate how comprehensive resources could be coordinated to control drugs and crime, while at the same time improving the quality of life in high-crime neighborhoods. Residents of the neighborhoods targeted were mostly at or below poverty line. The program had a two-pronged strategy: "weed out" violent offenders, and "seed" the neighborhood with treatment, prevention, intervention, and revitalization (Roehl et al. 1996, 2, 9). The first strategy of prevention, intervention, and treatment dealt with substance abuse programs and activities such as health and nutrition, victim assistance, and community crime prevention. Also, Safe Havens Centers were established in most Weed and Seed cities. These were multiservice centers that provided a safe neighborhood facility for locating various youth and adult services.

Many of the seeding activities were present when Operation Weed and Seed was implemented, but some were federally funded. These included Safe Havens, Boys and Girls Clubs, the Race Against Drugs, and Wings of Hope. Also, state and federal funding continued to support existing programs such as HUD renovation projects, D.A.R.E., the Police Athletic League, and Aid to Dependent Children.

The role of community policing in Operation Weed and Seed was that officers were assigned to targeted areas to cultivate relationships with the neighborhood residents and promote efforts to deal with residents' concerns. Community policing officers were not assigned radio service calls so that they could concentrate on problem solving, community contacts, and youth activities, but they still performed law enforcement activities and made arrests where appropriate. Often, they had partners or worked in teams. The program placed community policing officers in contact with prosecutors, probation and parole officers, federal law enforcement agents, neighborhood residents, and governmental and social service personnel and enabled these diverse groups to work together. Through steering committees, various agencies coordinated programs, shared resources, and solved problems.

The extent to which Operation Weed and Seed played a major part in reducing violent crime and drug-related crime has been difficult to evaluate, but Operation Weed and Seed assisted in spreading the community policing philosophy, and the community policing strategy had a positive impact on both the communities involved and the police departments. This program demonstrated that law enforcement can be enhanced by interacting with the community and that law enforcement and community services are not incompatible police functions.

BUSINESS ALLIANCE PROGRAM

The Business Alliance Program grew out of Operation Weed and Seed. Its purpose was to rebuild the economic structure of neighborhoods through community efforts by improving living and working conditions—specifically, to help businesses grow, attract new employment opportunities to neighborhoods, and increase training opportunities for neighborhood residents and the private sector through partnerships between the business community and neighborhood residents. The Business Alliance, when it is successful, can be beneficial to the police, because crime rates are down when the unemployment rate is down. Removing drugs and crime from a neighborhood will not last if the community lacks employment opportunities. Further, community policing has a role to play in Business Alliance programs, because communities and neighborhoods with high crime rates and disorder will not attract or maintain businesses. According to a U.S. Department of Justice publication on Business Alliance (Bureau of Justice Assistance 1994), "Community policing is essential for attracting economic enterprises through which the community, law enforcement, and businesses can support revitalization efforts" (p. 2).

CITIZENS ON PATROL

In 1991, 105 persons from 11 Ft. Worth neighborhoods attended 12 hours of training sessions at the Ft. Worth Police Academy to become members of a group called Citizens on Patrol. Classes covered legal liabilities, patrol procedures, communications, and the criminal code. Upon completing the training, each citizen graduate was given a diploma, cap, T-shirt, and windbreaker bearing the insignia *Citizens on Patrol.* After completing the training, each citizen was required to ride with a patrol officer and to meet with the officer in charge of the Citizens on Patrol program in his or her neighborhood to review regulations. All Citizens on Patrol groups selected a leader who worked the neighborhood patrolling schedule. The group was furnished a portable radio to report suspicious persons or activities to their base station. The base station could be housed either at the local field operations division or at an individual's home. Base units notified the police dispatcher.

Citizens on Patrol interfaced with neighborhood patrol officers (NPOs). Ten NPOs were assigned to each of the four field operations divisions. Each NPO was assigned to a specific geographical neighborhood within the division to address root causes of crime. A variety of nontraditional methods could be used, each with the purpose of including and empowering the community. For example, some NPOs were in charge of sports teams that participated in the City's Youth League. Others were involved with the Explorer Scout groups. NPOs worked closely with Citizens on Patrol groups. They provided crime data and advised Citizens on Patrol groups on when and where to patrol their neighborhood. Also, NPOs worked with Citizens on Patrol to develop or expand problem-solving concepts that fit their specific neighborhood. In one neighborhood, residents were listed in a central data log, and neighbors each had a bumper sticker to identify the neighborhood to which they belonged, allowing for rapid identification of suspicious vehicles in the area. Citizens on Patrol also advised residents when they left themselves open to victimization, such as when they left their garage doors open or left expensive items in an unlocked vehicle parked on the street (Ft. Worth Police Department n.d.).

DOMESTIC VIOLENCE PROGRAMS

Traditional policing considers domestic violence primarily a problem for social services agencies, but community policing agencies recognize that police have an important role to play in reducing it. Domestic violence is often unreported, and victims often lack knowledge about available services. Policing can provide an important link between victims in need of services and domestic violence service agencies in the community. Police officer resistance to domestic violence initiatives is often symptomatic of attitudinal problems.

In 1996, the Cheektowaga Police Department, located in western New York, began participating in the Community Policing to Combat Domestic Violence Program, funded by the U.S. Justice Department.* The Erie County Coalition against Family

---

Source: The next seven paragraphs are adapted from C. Rucinski, et al., *Transitions,* © 1997, Cheektowaga Police Department.

Violence had found that victims of domestic violence were not receiving readily available community services, so the police department sought to change the behavior of police officers who were ineffective in delivery of services to victims of domestic violence and to develop an organizational culture that would be responsive to the domestic violence–related needs of the community.

To combat domestic violence, the Cheektowaga Police Department entered into a partnership with the Haven House, the local battered women's shelter, and the National Conference (formerly the National Conference of Christian and Jews), a human relations organization committed to diversity and to combating prejudice. The domestic violence project had the following objectives:

1. To dispel invalid mythology regarding victims of domestic violence our officers may harbor
2. To increase the interpersonal and communication skills of the responding officers
3. To reduce bias and discrimination by the officer towards victims
4. To increase the skill level of officers in the recognition of violent human dynamics in domestic scenarios
5. To develop data identifying areas for growth and continuous improvement (Rucinski et al. 1997, 1)

Based on the objectives, a two-day domestic violence training program was developed that focused on combating myths about domestic violence, developing interpersonal skills, and overcoming bias and stereotypes that hamper problem solving. Training was open to all police officers, public safety dispatchers, and civilian personnel, including the community partners. Both the community partners and the police had input in selecting training topics and developing vignettes.

In the training, police officers were assisted in identifying risk factors for several types of assaultive behavior, predicting dangerousness of individuals, intervening with effective methods, thereby increasing both officer and victim safety, and working with victims in trauma and crisis situations. They were given practical tips on how to handle domestic violence situations, including defusing the anger of the victim and the aggressor, letting the victim know her options and alternatives, and letting the victim know that shelter was available. The Theater for Change, professional actors, performed three vignettes: one concerning a victim reluctant to press charges, one presenting parties with conflicting stories, and one involving repeated reconciliations between the parties. The National Conference presented a workshop on recognizing diversity, overcoming myths and stereotypes (especially those associated with victims and offenders of domestic violence) that hinder creative problem solving, understanding prejudice, and strengthening communications and interpersonal skills. Other issues addressed by the training were effects of violence on children, signs of domestic violence, resources to help victims, why women stay, and updates on law changes dealing with domestic violence and restraining order protection.

The Cheektowaga Police Department's domestic violence training program is a practical program that police officers can use when they respond to domestic violence calls. Over 300 police officers in Western New York had taken the domestic violence training course as of 1997, and, in 1997, the National League of Cities gave the Cheektowaga Police Department their project award for the program.

Other police agencies throughout America are also taking constructive action to end violence against women. The *Community Police Exchange* ("A Law Enforcement Checklist" 1997) delineated a number of strategies that can be used by police agencies to reduce incidents of violence against women:

- *Create a community roundtable.* Convene a community roundtable bringing together police, prosecutors, judges, child protection agencies, survivors, religious leaders, health professionals, business leaders, defense attorneys, and victim advocate groups; and meet regularly. Create specific plans for needed change, and develop policies among law enforcement, prosecutors and others that will result in a coordinated, consistent response to domestic violence.
- *Record domestic violence.* To help understand and respond to the dimensions of violence against women, develop and require the use of a uniform domestic violence reporting form. It should include an investigative checklist for use in all domestic violence incidents or responses.
- *Continue to educate.* Create informational brochures on domestic violence and sexual assault, which include safety plans and a list of referral services, for distribution in all courthouses, police stations, and prosecutors' offices and in nonlegal settings such as grocery stores, libraries, laundromats, schools, and health centers.
- *Provide clear guidance on responding to domestic violence.* Write new or adapt existing protocol policies for police, courts, and prosecutors regarding domestic violence and sexual assault incidents, and train all employees to follow them. Policies should specify that domestic violence and sexual assault cases must be treated with the highest priority, regardless of the severity of the offense charged or injuries inflicted.
- *Ensure that law enforcement is well informed.* Designate at least one staff member to serve as your agency's domestic violence and sexual assault contact, with responsibility for keeping current on legal development, training resources, availability of services, and grant funds. . . .
- *Reach out to the front lines.* Identify and meet with staff and residents from local battered women's shelters and rape crisis centers to discuss their perceptions of current needs from the law enforcement community. Solicit suggestions for improving the law enforcement response to these crimes.
- *Improve enforcement by implementing a registry of restraining orders and a uniform order of protection.* Implement a statewide registry of restraining orders designed to provide accurate, up-to-date, and easily accessible information on current and prior restraining orders for use by law enforcement and judicial personnel. Develop a uniform statewide protection order for more effective and efficient enforcement.
- *Support and pursue legislative initiatives.* Develop and support legislative initiatives to address domestic violence and sexual assault, including (a) stalking, (b)

death review teams, (c) sentencing guidelines, (d) indefinite restraining orders, and (e) batterer's intervention programs.

- *Conduct training.* Conduct ongoing multidisciplinary domestic violence and sexual assault training for police, prosecutors, judges, advocates, defenders, service providers, child protection workers, educators, and others. . . .
- *Structure courts to respond to domestic violence/create specialized domestic violence courts.* Develop specialized courts that deal exclusively with domestic violence cases in a coordinated, comprehensive manner, where community and court resources can be used together to address domestic violence effectively. (8)

Community police officers have a role to play in many of these recommendations because they involve collaboration with various community groups and agencies to prevent crime and improve the quality of life for community residents. Domestic violence is an area in which community policing can be effective.

DRUG AWARENESS AND RESISTANCE EDUCATION (D.A.R.E.)

The Drug Awareness and Resistance Education (D.A.R.E.) program was initiated by the Los Angeles Police Department in 1983 as an educational program to prevent drug use and violence among young people. In a short time, it made its way to police departments throughout the country. Its emphasis has always been to help students recognize and resist the pressures and influence of their peers so that they refrain from experimenting with alcohol, marijuana, tobacco, and other drugs. Strategies focus on building self-esteem, assertiveness, interpersonal and communication skills, decision-making skills, and awareness of positive alternatives to drug use and gang involvement.

The D.A.R.E. program usually has 17 sessions, with either police officers or deputy sheriffs as classroom instructors. Classes are generally taught from the fourth or fifth grade to the eighth grade. The police officers remain on campus all day to interact with students during lunch and recess.

The combining of D.A.R.E. with community policing can be advantageous. It performs a service to the community, develops a potential partnership with the community's young people, and helps to prevent drug-related and other crime among young people.

GANG RESISTANCE EDUCATION AND TRAINING PROGRAM (G.R.E.A.T.)

The Gang Resistance Education and Training (G.R.E.A.T.) program was designed to

- reduce the incidence of violent youth crime
- reduce gang activity and violence
- provide youth with life skills and strategies to resist gang involvement pressure
- familiarize youth with the means by which to resolve conflicts in a non-violent manner

- provide alternative activities for G.R.E.A.T. graduates during the summer months (Bureau of Alcohol, Tobacco, and Firearms [BATF] 1998b, 2)

It teaches young people to establish goals for themselves, resist pressures, and resolve conflicts without violence and allows them to discover for themselves how gang and youth violence can affect the quality of their lives through structured exercises and interactive approaches to learning.

G.R.E.A.T. began in 1992 through a partnership with the Treasury Department's Bureau of Alcohol, Tobacco, and Firearms and the Phoenix Police Department. The two agencies developed a nine-lesson middle-school curriculum that has become a national model, with more than 800 police agencies involved in it. BATF continues to fund police agencies and manages cooperative agreements and communities' interest in G.R.E.A.T. (BATF 1998a, 1).

Like D.A.R.E., G.R.E.A.T. exposes young people to specific programs that will help them stay out of trouble with legal authorities. It draws on the community policing philosophy in that it involves police partnership with the schools, a federal agency, and the young people.

HATE CRIME PROGRAMS

The federal Hate Crimes Sentencing Enhancement Act of 1994 defines hate crimes as those in which "the defendant intentionally selects a victim, or in the case of a property crime, the property that is the object of the crime, because of the actual or perceived race, religion, national origin, ethnicity, gender, disability, or sexual orientation of any person" (Leadership Conference 1998, 1). The crimes included in hate crimes are murder, forcible rape, robbery, aggravated assault, burglary, larceny, motor vehicle theft, arson, intimidation, and vandalism.

Usually, hate crimes are more violent than other crimes. Generally, their intention is to murder, maim, or injure. Hate crimes are five times more likely than other crimes to involve assaults, and hate crime assaults are twice as likely as other assaults to cause injuries that will result in hospitalization. Offenders of hate crimes are likely to be marauding groups of predators searching for targets at which to direct their hatred.

The attackers or offenders of hate crimes are usually young people, mostly young men in their teens or 20s, who are primarily looking for thrills and respect from friends. Sometimes people carry out hate crimes because they feel insulted: by interracial dating, integration of their neighborhood, their battered wife's decision to leave, their competition with women for jobs, or just the existence of homosexuality, whether male or female. Fanatic ideological racist, religious, or ethnic bigots are the least common offenders (Leadership Conference 1998, 3–4).

Since data have been collected for hate crimes, victims have included whites, blacks, Hispanics, American Indians, Asian Americans, and multiracial people; Jews, Catholics, Muslims, and atheists; and gay men, lesbians, bisexuals, and heterosexuals. As the preceding list indicates, no one is left out: regardless of a person's race, ethnicity, religion, gender, or sexual orientation, there is always someone or some group who might attack him or her simply because of one of those characteristics. Anyone in America could be a victim of a hate crime, and this type of crime should be considered offensive to all Americans with a sense of fair play and justice. Groups and individuals who commit hate crimes pose a threat to our communities.

Both the federal and most state governments have taken actions to stem the tide of hate crimes. In 1990, Congress passed the Hate Crimes Statistical Act to get a handle on the extent of hate crimes occurring in our nation. This act mandates that the Attorney General of the United States obtain data on crimes that manifest evidence of prejudice based on "race, religion, ethnicity, and sexual orientation." In 1992, Congress passed the Juvenile and Delinquency Prevention Act, which requires that each state's juvenile delinquency plan include a component designed to combat hate crime. In 1994, Congress passed the Violence Against Women Act, a comprehensive federal law to address the national problem of violence against women. In 1996, Congress passed the Church Arson Prevention Act to enhance federal jurisdiction over and increase federal penalties for the destruction of houses of worship. This law was passed in response to the numerous burnings of African American churches during the 1990s. In addition, the U.S. Commission on Civil Rights has undertaken to produce radio public service announcements on discrimination and denial of equal protection of the law (Leadership Conference 1998, 11).

Most states have some form of hate crime laws. These state statutes take a variety of forms, which include requiring states to compile statistics on hate crime, holding parents liable for the actions of their children, allowing civil actions against perpetrators of hate crimes, outlawing intimidation of individuals, and outlawing vandalism against houses of worship. Also, several states have passed legislation imposing heavier criminal penalties for hate-motivated crimes. On the local level, a number of police departments have established bias units. The New York City Police Department's special bias unit has been trained to be sensitive to victims, because victims who believe the police are sympathetic to their plight are usually more encouraged to report hate crimes and to be cooperative with investigations and prosecutions. A successful police bias unit develops a working relationship with minority communities, with prosecutors, and with police officers from different law enforcement agencies (Leadership Conference 1998, 12).

Not all police agencies have reported hate crimes. Areas whose police departments report hate crimes comprise 58 percent of the United States population (Leadership Conference 1998, 1–3).

The Chicago Police Department is one department committed to community policing that has addressed hate crimes. It has established policies and procedures for handling criminal and noncriminal incidents motivated by hatred. These policies define hate crimes, outline reporting and notification procedures when an incident occurs, and provide an annual statistical report. They also describe the department's strategies for assessing community tension and responding to potential problems before they escalate. The department's response to hate crimes involves many divisions: the patrol division, the detective division, the youth division, and prevention programs (Chicago Police Department 1998, 1).

Police departments committed to the community policing philosophy must be committed to eliminating hate crimes. A police-community partnership cannot be developed when hate crimes are occurring in a community and the police take no action to prevent them or to arrest the attackers. Police departments need to report hate crimes, encourage reporting, and develop effective intervention strategies. Community policing officers should anticipate potential hate crime problems and engage in problem solving, with the cooperation of residents, the police, and other social and governmental agencies, before threats are put into action.

PROGRAMS FOR THE HOMELESS

In the last 15 to 20 years, the homeless population has grown substantially. According to the National Coalition for the Homeless (NCH), two trends are responsible for homelessness: poverty and a shortage of affordable rental housing. Poverty and homelessness are intimately connected. Poor people frequently cannot afford housing, food, child and health care, and education, and, because housing consumes most of a person's income, housing is often not economically possible. Many people are only one paycheck from poverty and can easily be made homeless by an illness, an accident, or a lost job. By 1996, 14 million people in the United States were living below poverty level. Children made up 40 percent of those living in poverty (NCH 1998, 1).

The eroding labor market, low wages, and decrease in public benefits have all been important factors in increasing poverty. Rates of both unemployment and underemployment have increased since the 1970s. Many people cannot get enough work to live on or have given up looking for work. Low-skilled and poorly educated workers are especially vulnerable as blue-collar jobs continue to disappear. The transition to a global economy has meant that many American corporations are moving their operations overseas, often to countries where workers are paid more cheaply, and are laying off their American workers.

Many poor people, including homeless people, are in the workforce but have very low wages. Many are employed as temporary workers or independent contractors. Generally, these workers receive no benefits and lack job security. Even those working full time may not make enough to get above poverty level. Wages have not kept pace with inflation, and unions have lost much of their power to fight for higher wages (NCH 1998, 1–2).

The availability of public assistance has also decreased in the last decade. Most people who receive public assistance do not receive enough funds to afford rental property. Although the welfare caseloads have decreased since the passage of welfare reform legislation during the 1990s, declining welfare rolls only mean that fewer people are receiving public assistance; they do not mean that people are better off financially. Those individuals terminated from welfare rolls may not be able to find employment or may be able to find only low-paying jobs without benefits. If this is the case, they may not be able to afford housing (NCH 1998, 2–3).

The increasing scarcity of affordable housing must also be considered a contributing factor to homelessness. There are not enough affordable housing units to meet the demands and, in many urban areas, gentrification is resulting in the eviction of low-income tenants and the demolition of low-income housing. Other factors as well have contributed to the homeless problem:

- *Lack of Affordable Health Care:* For families and individuals struggling to pay the rent, a serious illness or disability can start a downward spiral into homelessness, beginning with a lost job, depletion of savings to pay for care, and eventual eviction. . . .
- *Domestic Violence:* Battered women who live in poverty are often forced to choose between abusive relationships and homelessness. . . .
- *Mental Illness:* Approximately 20%-25% of the single adult homeless population suffer from some form of severe and persistent mental illness. . . . [However], many mentally ill homeless people are unable to obtain access to mental health support services such as case management, treatment, and provision of housing.
- *Chemical Dependency:* In recent years, the relationship between substance abuse and homelessness has stirred much debate. While rates of alcohol and drug abuse are disproportionately high among the homeless population, the increase in homelessness in the 1980s cannot be explained by substance abuse. During the 1980s, competition for increasingly scarce low-income housing grew so intense that those with disabilities such as chemical addiction and mental illness were more likely to lose out and find themselves on the streets. (NCH 1998, 4–6)

The homeless problem in America is complex, and the homeless should not be blamed for their situation. Obviously, there are many circumstances that allow people to fall into poverty. In many communities, citizens have chosen to consider this social problem a police problem. But the police cannot handle the homeless problem by using the traditional approach to policing. The community policing approach is far better suited to dealing with homelessness.

An agreement between the City of Miami and the American Civil Liberties Union (ACLU) in December 1997 may be used as a model for other cities that have a homeless population. The agreement, the result of a federal lawsuit, *Pottinger v. Miami,* initiated in 1988, challenged as unconstitutional the city's policy of arresting homeless people for sleeping, eating, and congregating in public and the city's policy of having police confiscate and destroy the possessions of homeless people. Under the ACLU-Miami settlement, the city was required to develop a training program to sensitize Miami police officers to the predicament of homeless people and to ensure that the police do not violate the legal rights of homeless people. The city had to develop guidelines on how the police were to deal with encounters with the homeless. If a police officer observed a homeless person engaging in a violation of a city ordinance, such as sleeping on a park bench, the homeless person would have to be informed of homeless shelters available and given transportation before an arrest could occur. The homeless person could only be arrested if he or she refused available shelter. In addition, the police would not be allowed to destroy the property of homeless people, except under very limited circumstances. Also, an advisory committee composed of homeless people and city representatives would monitor police contacts with the homeless and issue a report every six months (National Law Center 1998, 2).

In 1991, the Santa Monica, California, Police Department set up a Homeless Liaison Program (HLP) that created a four-person police unit to deal directly with the community's concerns about homeless issues. The program goals were to end the cycle of homelessness and improve the quality of life for residents and business owners in the community. The HLP unit works with social service agencies to place homeless people in shelters and long-term housing and to provide basic needs for homeless people (Santa Monica Police Department 1998, 1).

The new approaches to homeless problems in Miami and Santa Monica exemplify the community policing philosophy. Both involve partnership with social service agencies, citizen input, and the use of a broad problem-solving approach.

SCHOOL RESOURCE OFFICERS

The Arvada, Colorado, Police Department started a service resource officer (SRO) program in 1995 to establish high visibility of police officers in and around schools in order to prevent disorder problems. At first, the department assigned one police officer to each of the city's three high schools. In 1996, an SRO was assigned to one middle school as well, and then, as a result of the positive feedback from the middle school, SROs were assigned to all the city's middle schools. SROs function as liaisons between the police department, the school, and the surrounding community and do classroom instruction, student mentoring, crime prevention, problem identification, problem solving, and law enforcement. They deal with issues directly related to the school and its student population and associated problems within the surrounding neighborhood. They are not expected to respond to radio service calls. They are available to students during school hours as well as for many extracurricular activities. The Arvada Police Department's SRO program does not follow rigid guidelines, because schools, students, and police officers all have different personalities. SROs are given the latitude to adapt their distinctive styles to the requirements of each specific issue (Hoffman n.d.).

SENIOR LIAISON OFFICERS

In 1995, the Arvada, Colorado, Police Department initiated a senior liaison officer (SLO) program customized to provide services to citizens 60 years of age and older. SLOs work proactively with seniors to identify concerns and service needs, resolve problems, and obtain appropriate services. The long-range goal of the program is to reduce the fear of crime among seniors and to provide quality police service to senior citizens. The SLOs maintain a high visibility in retired senior facilities and make weekly visits to senior centers. Some of the activities of SLOs are

1. Regular visits to senior housing facilities
2. Security at selected senior activities, such as dances and cultural events
3. Identification of seniors who are abused or neglected, or who can no longer adequately care for themselves, and facilitation in the resolution of these situations so that the best care possible can be provided for them

4. Crime prevention and personal safety classes for seniors
5. Referral to outside resources which assist seniors with concerns and problems, dispute resolution, and other personal services for senior citizens as they are identified
6. Personal contact with seniors who are victims of crime (Hoffman n.d., n.p.)

SUBSTANCE ABUSE NARCOTICS EDUCATION (SANE) PROGRAM

The Substance Abuse Narcotics Education (SANE) program has been in existence in the Los Angeles County Sheriff's Department since the mid-1980s. SANE provides a unique approach in drug prevention that involves the Los Angeles County Sheriff's Department, school districts, schools, and municipalities throughout Los Angeles County. Currently, deputies are working with third- through eighth-grade students in all 54 public school districts of Los Angeles County, juvenile court and community schools, and many private schools.

SANE uses a variety of professionally developed drug prevention education curricula to meet the diverse needs of the various schools and communities. Each curriculum has the components of building self-esteem, teaching coping and decision-making skills and skills for dealing with peer pressure, and basic substance education. Deputies and classroom teachers co-teach the program so that the drug prevention message will come from two role models. All deputies assigned to SANE receive intensive training both in general teaching methodologies and in their special curriculum. Classroom teachers are trained in drug use prevention techniques and recognition, thereby becoming valuable resources for their schools.

The SANE program introduced Curriculum Integrated Drug Abuse Prevention (CIDAP) into the schools. CIDAP is a 10-lesson program whose objectives are to develop an awareness of the effects of drugs on the individual's physical and mental health, increase knowledge of drugs, improve decision-making skills, improve self-esteem, explain peer pressure and offer easy-to-use responses to it, review the social problem of drug abuse, affect students' attitudes toward personal use of drugs, give students a better understanding of themselves, and recommend alternatives to drug abuse. The CIDAP program is very flexible and can be used to supplement an existing curriculum or can function as an independent component of the SANE program.

Because of the increase in gang violence and the connection between gangs and drugs, SANE developed a gang prevention curriculum. This curriculum includes lessons on self-esteem, media influence, alternatives, power, violence, and social values.*

VOLUNTEER PROGRAMS

The Delray Beach Police Department in southern Florida has several volunteer programs that have been incorporated into their community policing strategy. The volunteer programs allow citizens of the community to be involved in policing activities

---

*This information was taken from *Drug Abuse—SANE Program,* a Los Angeles County Sheriff's Department brochure; 1998.

and provide citizens with a better understanding of police work. Volunteers are screened and trained by the police department.

Citizen Observer Patrol

This program consists of approximately 1,000 volunteers from 20 neighborhoods who patrol their own neighborhoods in two-hour shifts over both daytime and nighttime hours. The volunteers are equipped with cellular telephones to call in suspicious activity. In some patrols, volunteers drive their own personal vehicles; in other areas, patrol members use and maintain retired police vehicles that were given to the neighborhood by the police department.

Roving Patrol

The Roving Patrol consists of 15 volunteers. It is used by the Delray Beach Police Department in areas that are experiencing property crime problems and areas that require a high-profile type of patrolling. Members use retired police department vehicles equipped with yellow light bars and spotlights. They are trained in the use of a police radio as well as a cell phone. They do three-hour shifts between 6:00 p.m. and 11:00 p.m. throughout the week. They are also used in conjunction with detectives in auto theft ring operations.

Haitian Roving Patrol

The Haitian Roving Patrol has 18 members and patrols with retired police vehicles in the Haitian community as well as on other assignments throughout the city. They assist patrol officers in translations on radio calls for service. Haitian volunteers also walk the downtown business section during the evening hours, providing extra security to merchants and shoppers alike. The Haitian volunteers are dedicated to the Roving Patrol program and the Delray Beach Police Department. Three of them hold the rank of volunteer sergeants. They are trained in the use of the cell phone and binoculars.

Downtown Walking Patrol

The Downtown Walking Patrol has eight volunteer members. They patrol in pairs and receive training in the use of the police radio and cell telephone. They walk a beat in the downtown Delray Beach area and provide a sense of security to both merchants and shoppers. They give directions to tourists and answer questions for the business owners. They also watch out for suspicious activity and call the police on the radio when they are needed.

Parking Enforcement Specialist

There are 51 volunteer Parking Enforcement Specialists. These volunteers provide parking enforcement for the downtown, beach, business district, and shopping center areas throughout the city. They use retired city vehicles with a yellow flashing light and golf carts that allow them to get around easily in traffic. They receive 40 hours of instruction on traffic laws and ticket writing.

Traffic Monitoring Patrol

The Traffic Monitoring Patrol has 14 volunteers. Traffic monitors are trained in the operation and setup of the mobile radar trailer, which is used in problem areas of the city where there are speed problems. The trailer has a self-contained radar unit and a large display board that displays the speed of approaching vehicles. If these vehicles exceed the posted speed limits, the members write down the tag numbers of the suspected vehicles, and the city sends the owners of the vehicles a warning letter. The traffic monitoring program provides training and safety for citizens in the City of Delray Beach and is a good public relations tool for the police department. The volunteers also assist in the collection of data for the department's traffic unit for developing plans to resolve traffic-related problems, such as traffic facilities, accidents, and speed enforcement.

Volunteer Front Lobby

The Volunteer Front Lobby team has 16 volunteers. Volunteers are responsible for greeting all visitors who enter the police department and assisting them in contacting the appropriate police personnel. Also, the Front Lobby team assists in administrative duties.

Volunteer Administrative Support Team

The Administrative Support Team has 25 members. These volunteers help sworn and support personnel in a variety of administrative duties, such as victim services, tracking auto thefts, tracking pawn shop tickets, and filing paperwork (R.G. Overman, Chief, Delray Beach Police Department, personal communication, 1998).

YOUTH PROGRAMS

A joint study by the National Institute of Justice and the Carnegie Corporation of New York (Chaiken 1998) found that many popular approaches to dealing with juvenile violence have not worked. The main approach used has been that of locking up juveniles in state and local facilities. The juveniles are supervised when in custody, but once they are released nothing seems to change. Even boot camps provide only temporary supervision. Many teenage programs available after school are ineffective. Many gang prevention approaches that provide information about the wickedness of delinquency backfire, with the juveniles becoming more delinquent. The approaches that appear to be the most promising for preventing delinquency and violence are long term, continuous, and comprehensive. They involve adult tutors and mentors who instruct young people in social skills and give them opportunities to practice these skills. Many national youth organizations follow this approach. They provide a variety of activities appropriate for a young person's age and period of development.

- An *environment* in which kids are valued and are considered resources rather than problems for their community
- *Activities* that present teens with real challenges and experiences in planning, preparing for, and publicly presenting projects they and their communities truly value
- Ongoing *outreach* to teens and adults in the community, with messages that are understandable (Chaiken 1998, xiv)

The National Institute of Justice study found that partnerships between the police and youth service agencies take many forms, from police officer involvement in youth advisory boards to police coaching of sports teams to youth involvement in crime prevention programs. Several youth service organizations have requested police to provide programs for young people to make them aware of dangerous situations, such as threats and pressures from gang members, violence among family members, and rape attempts. The police have devoted hours as volunteers to Boy Scouts and Girl Scouts in high-crime areas. They have also been involved in juvenile diversion programs (i.e., programs that provide sentencing alternatives to locking up offenders) (Chaiken 1998, xv). All of these are aspects of community policing.

Bristol Family Center

The Bristol Family Center in Bristol, Connecticut, is a comprehensive facility for boys and girls. The center opens at 6:30 a.m. and closes at 9:00 p.m. Its facilities have child care, swimming, gymnastics, dance classes, and other activities. The center offers special events for adolescents in specific grades: for example, there are sixth-grade Friday night socials.

The Bristol Police Department and the Youth Service Bureau are leaders in the coalition that implements the juvenile offender diversion program in which the Bristol Family Center collaborates. Also involved in the coalition are the Bristol public schools, the department of probation, and agencies that provide clinical services for children.

A juvenile review board composed of representatives from the above-named agencies meets on a regular basis to review juvenile cases. The discussion at the board meeting is confidential. Initially, the meeting opens with a juvenile officer giving information concerning the incident leading to the arrest of a juvenile. The board members may provide information about the family and children involved in the incident. Because the board is composed not only of police officers and probation officers but also of school counselors, after-school activities directors, and psychological counselors, a multifaceted view of the juvenile offender can be obtained. The board explores alternative options to locking up juvenile offenders, such as having them write an essay about the consequences of their actions or perform community service. Juveniles who perform community service in the Family Center sign a contract as to the number of hours they will be working. If they fail to live up to the contract, their case is referred to the juvenile review board again for action. Family Center staff members are not made aware of why a youngster is performing community service.

In addition to serving on the juvenile review board, the Bristol Police often work directly with children on projects sponsored by youth organizations. These projects can include "antidrug poster coloring contests," at which officers have the opportunity to talk with children, and training for youth organization participants on topics such as gang awareness and dealing with babysitting emergencies. The community policing "Walk-and-Talk" approach encourages police officers to stop at youth centers as part of community outreach (Chaiken 1998, 22–24).

Spokane Police Department

Spokane's community policing approach has encouraged police officers to address problems involving the children and teens of Spokane and to develop a wide variety of programs to create an environment of safety for young people. One officer developed "Every 15 Minutes," a program to prevent alcohol-related deaths of teenagers on prom nights. Also, a two-day, one-night multimedia program was developed for junior and senior high school students to alert them to the dangers of drinking and driving.

The COPS-N-Kids annual truck and car event held every August is an attempt to deal with teenagers' "cruising." The Spokane officers support cruising by teenagers if it is properly monitored and directed. Police officers have convinced automobile business owners and adults with automobile hobbies to come to the cruising area and to assist the teenagers in maintaining their vehicles. The August truck and car event, which includes free food and drinks, is promised to the teenagers if they remain trouble free.

A COPS (community-oriented policing) West Ministation, staffed by community volunteers and a neighborhood resource officer, has several programs directed at or including young people, such as McGruff safe houses, responses to calls involving conflicts between community children and teenagers, and adult patrolling of the streets before and after school to ensure children's safety and to keep older children from harassing younger children (Chaiken 1998, 35–37).

ADDITIONAL COMMUNITY POLICING PROGRAMS

To cover all community policing programs in one chapter would be an impossible task. Below are listed a few more community policing programs. Several of these have been around for several decades and were initiated for improving police-community relations or for crime prevention:

- *Basic Car Plan:* A city is subdivided by computer-generated crime occurrence data, and teams of nine officers are given 24-hour responsibility for a specific subdivision on permanent assignment to develop proprietary interest through formal and informal meetings with the public.

- *Basic Commander System:* A police sergeant is given command of 20 officers, including detectives, who only investigate crimes in the beat command area on permanent assignment and who are closely monitored for job satisfaction and efficiency at improving citizen satisfaction.
- *Bilingual Programs:* These include such efforts as bilingual public announcements, bilingual education programs, and incentive pay for dual-language officers.
- *Compstat:* Weekly community policing meetings are held. Precinct commanders and other police executives are rotated on a regular basis, and immediate accountability is required every five weeks from the executives in front of the others at staff meetings to see if crime goes down under their watch.
- *Neighborhood Advisory Councils:* Community leaders are invited or elected to explore community needs. They meet with assigned police officers to consider specific patrol tactics and manpower scheduling and to voice complaints and commend personnel.
- *Split Force Policing:* One portion of the police force is used for responding to calls regularly and another portion is assigned to directed patrol. (North Carolina Wesleyan College 1998, 1–10)

SUMMARY

Many police departments and federal and state law enforcement agencies have established distinctive programs that deal with specific problems and take a community policing approach. They address a variety of issues, such as drugs, gangs, juvenile delinquency, domestic violence, hate crimes, and homelessness.

In 1992, the U.S. Department of Justice introduced Operation Weed and Seed to illustrate how comprehensive resources could be coordinated to control drugs and crime, while at the same time improving the quality of life in high-crime neighborhoods. The Weed and Seed program had a two-pronged strategy: "weed out" violent and drug offenders, and "seed" the neighborhood with treatment, prevention, intervention, and revitalization. Community policing officers were assigned to neighborhoods to participate in both weeding and seeding strategies.

Police have an important role to play in addressing domestic violence effectively, and several police departments are initiating proactive approaches in this area. For example, the Cheektowaga, New York, Police Department entered a partnership with a local battered women's shelter and a human relations organization committed to diversity and to combating prejudice to design a comprehensive domestic violence training program that police officers could use when they responded to domestic violence calls. The program dispelled myths about domestic violence, taught effective intervention methods, helped officers to work with victims in trauma and crisis situations, and presented resources to help victims.

Drug Awareness and Resistance Education (D.A.R.E.) and Gang Resistance Education and Training (G.R.E.A.T.) are programs to help elementary- and middle-school children stay away from drugs and gangs. These programs have police officers functioning as instructors in the classroom. This allows the police officer to be a role model for children and also gives children a chance to see the police officer as a person.

Hate crimes are crimes in which the defendant intentionally selects a victim or his or her property because of the victim's actual or perceived race, religion, national origin, ethnicity, gender, disability, or sexual orientation. Hate crimes are usually more violent than other crimes and disrupt communities by creating fear and encouraging further hostilities. Community policing programs can assess potential hate crime problems and address them before they escalate through community cooperation.

Homelessness is a social problem created by poverty and a lack of affordable rental housing. It is frequently referred to the police to deal with by enforcing city ordinances. But tactics that force homeless people to leave a public area or that confiscate or destroy their possessions often violate homeless people's legal rights, as well as being inhumane. Community policing programs have found more constructive approaches to dealing with homelessness that involve cooperation with social service agencies.

Volunteer programs have played a big role in community policing. These programs can involve citizens in neighborhood patrols, traffic monitoring, parking enforcement, or receptionist or administrative support services to the police department. Because of tight budgets, police departments cannot provide all the services that the citizens of the community may want. The use of citizen volunteers can assist the police in providing resources that they normally could not provide. Community policing departments have found a valuable community resource.

KEY TERMS

Business Alliance	hate crimes	SANE
Citizens on Patrol	juvenile diversion	school resource officers
D.A.R.E.	programs	senior liaison officers
G.R.E.A.T.	Operation Weed and Seed	

REVIEW QUESTIONS

1. What role should the police play in the homeless problem?
2. Why should local police agencies be involved in efforts to reduce hate crimes?
3. How can community policing be involved in the Business Alliance?
4. What are the major advantages of Operation Weed and Seed to community policing?
5. Why should police officers be involved in D.A.R.E. and G.R.E.A.T.?

REFERENCES

Bureau of Alcohol, Tobacco, and Firearms. 1998a. G.R.E.A.T. National Policy Board. http://www.atf.treas.gov/great/phisosy.htm. Accessed July 27.

Bureau of Alcohol, Tobacco, and Firearms. 1998b. G.R.E.A.T. (Gang Resistance Education and Training). http://www.pbso.org/html/gang_resistance_education_and_htm. Accessed July 27.

Bureau of Justice Assistance. 1994. *Business Alliance: Planning for business and community partnerships.* Washington, DC: U.S. Department of Justice.

Chaiken, M.R. 1998. *Kids, cops, and communities.* Washington, DC: National Institute of Justice.

Chicago Police Department. 1998. How does the Chicago Police Department respond to hate crimes? http://www.ci.chi.il.us/communitypolicing/hatecrimes1994/response.html. Accessed July 30.

Ft. Worth Police Department. n.d. Citizens on Patrol. *Blue Code.* Ft. Worth, TX: Ft. Worth Police Department.

Hoffman, D. n.d. *Community Service Unit information.* Arvada, CO: Arvada Police Department.

A law enforcement checklist: Important steps to end violence against women. 1997. *Community Policing Exchange.* November/December, p. 8.

Leadership Conference Education Fund Online. 1998. Cause for Concern: Hate crime in America. http:/civilrights.org/lcef/hate. Accessed July 29.

National Coalition for the Homeless. 1998. Why are people homeless? NCH Fact Sheet no. 1. http://nch.ari.net/causes.html. Accessed May.

National Law Center on Homelessness and Poverty. 1998. Recent Legal Development. http://www.nichp.org/legal.htm. Accessed March.

North Carolina Wesleyan College. 1998. List of community policing programs. http://www.ncwc.edu/~toconnor/comlist.htm.

Roehl, J.A., et al. 1996. *National Process of Operation Weed and Seed.* Washington, DC: National Institute of Justice.

Rucinski, C., et al. 1997. *Transitions.* Cheektowaga, NY: Cheektowaga Police Department.

Santa Monica Police Department. 1998. Homeless Liaison Program. www.santamonicapd.org/units/hip_team/hipteam.htm. Accessed July 26.

Chapter 13

The Future of Community Policing

CHAPTER OBJECTIVES

CHAPTER OBJECTIVES

1. Have an understanding of the challenges to community policing that Taylor et al. (1998) describe.
2. Be familiar with Taylor et al.'s (1998) list of five questions that indicate whether a department is doing community policing.
3. Be familiar with the Community Policing Strategies Conferences' recommendations for community policing.
4. Be familiar with possible future trends in community policing.

INTRODUCTION

In the latter part of the twentieth century, police practitioners realized that the professional or traditional policing approach was not working to solve and prevent crimes and that a new strategy would be necessary. The new approach became known as "community policing." It was based on the recognition that police could be successful in solving crimes only when the public fully cooperated with them by providing information that could lead to the arrest of law violators.

Community-oriented policing grew out of earlier experiments, such as team policing, ministations, programs to improve police-community relations, directed patrol, and foot patrols, that had been conducted in the 1960s and 1970s. One especially important concept to the development of community policing was problem-oriented policing, first described by Herman Goldstein, a noted police scholar. It involved encouraging police officers to take a proactive approach to solving recurrent problems on their beat rather than merely having them respond to 911 calls and do random patrol. Community policing builds on problem-oriented policing by adding the crucial ingredient of community participation in both problem identification and problem solving.

One driving force behind "community policing" has been citizens' desire to improve and maintain the quality of life in their neighborhoods. Average citizens have indicated their concern about disorderly behavior that, if unchecked, can lead to neighborhood deterioration. Because quality of life can only be improved or maintained when the neighborhood residents play a major role, the role of police has been to assist the community in this endeavor. As policing has evolved more and more toward a service orientation, police departments have focused more on addressing crime problems and enforcement priorities as the community defines them.

CHALLENGES TO COMMUNITY POLICING

In 1998, Robert Taylor and colleagues authored an article that assesses where community policing is today. Taylor et al. stated that problems with community policing lie not so much with the concept itself as with the definition and implementation of the concept. It would be foolish to debate an approach that emphasizes development of a working relationship between the police and the community. But given how much federal money is available to police departments that wish to do community policing, everyone is claiming to be doing it—even when the changes made are superficial. Taylor et al. list five questions, all of which police departments should be able to say yes to if they are truly doing community policing:

1. Do we really see implemented most of the changes that community policing advocates have recommended?
2. Have the entrance requirements for new officers been changed to reflect changes in the police role?
3. Has the department changed recruit training from a military oriented academy to curriculum more in tune with the new role demanded by community policing?
4. Has the department flattened its organizational pyramid and placed more decision making in the hands of the officer? Has it implemented these structural changes on a citywide basis?
5. Has the chief turned the organization "upside down" and become committed to participatory dialogue with officers as a major part of his management style? (3)

According to Taylor et al., community policing faces the following challenges:

1. Community policing has been insufficiently evaluated. Research has mostly focused on specific programs and not on the holistic implementation of the community policing concept.
2. Community policing requires system change in all of city government.
3. Community policing is often not fully implemented and exists more in plans on paper than in actual policy.
4. Community policing has become too "politicized." It has been shielded from criticism in the departments that adopt it and the agencies that promote it.
5. Nationally, the rates of crime in general and violent crime in particular are declining. But the effect, if any, of community policing on this trend is difficult to disentangle from the effects of other factors, such as an increasingly aging population. Community policing may be claiming successes for which it is not responsible. (3–5)

RECOMMENDATIONS

The following are recommendations for improving community policing. They came out of the 1997 Community Policing Strategies Conference sponsored by the Southwestern Law Enforcement Institute.

1. Police departments should sustain community involvement in the community policing effort.
2. Organizational structure must change if community policing is to be sustained. It must be decentralized, flattened, and streamlined.
3. Police departments, in their effort to sustain community policing, need to learn how to harness the power of new technology.
4. Management styles should emphasize motivation and coaching and deemphasize control over employees.
5. Policing strategies should change in response to the changing dynamics of the community.
6. Police agencies should see a community as a group of neighborhoods that are different.
7. Supervisors should be trained on how to evaluate community policing officers.
8. Training should focus on diversity issues.
9. Separate community policing units must be abolished. Every police officer is a community policing officer.
10. All the criminal justice components—courts, probations, prosecution, corrections—should become community oriented.
11. Community-oriented policing should evolve into community-oriented government. Other governmental agencies should implement the tenets of the community policing philosophy and help take control of the neighborhoods.
12. Police agencies should develop strategies to sustain community partnerships.

PROSPECTS OF COMMUNITY POLICING

The outlook for community policing is positive. Insightful police chiefs are questioning and searching for ways to improve community policing. The 12 recommendations will eventually become a reality. Already, many police departments have made great strides in implementing them. Further, the intellectual quality of police executives and line personnel is much superior collectively today than at any time in the past, and we can expect this trend to continue. Police recruits will be better educated, with most having a college degree, and their training will be expanded to include problem solving as well as law enforcement.

We can expect that in the future the community will play a larger role in the community policing effort. The community will decide which problems and crimes have priority and will be involved in the decisions on who gets hired as a police officer to work in their community. They will take on more responsibilities in crime prevention and take control not only of crime and disorder problems but also of other problems affecting the community, such as those involving streets and zoning.

Community advisory councils will have a larger role in community policing. They will meet on a regular basis and work with operational commanders and police department policy makers to decrease the crime rate, disorder, and the fear of crime. They will identify and prioritize problems that a majority of residents want solved, will devise strategies to solve them, and will evaluate the success of the solutions devised. They will print a newsletter to inform residents of recent crime trends and provide crime prevention techniques to help citizens avoid becoming victims. They will also organize citizen patrols with the supervision of the police department. Working with the police, they will have graffiti and abandoned cars removed and will strive to clear the neighborhood of any appearance of disorganization.

Eventually, community-oriented policing will evolve into community-oriented government. Realistically, the police are not equipped to handle all of a community's problems, and all too often they have been saddled with problems caused by the incompetence or inefficiency of other government agencies. In the future, all the components of the criminal justice system—courts, probation, parole, prosecution, and corrections—will be community oriented.

By now, community policing is so incorporated into police departments' operations that it is unlikely to disappear. Further, once communities have bought into the concept, they are unlikely to endorse a return to the traditional model of policing. When residents and business owners feel empowered to safeguard their communities, it will be difficult to turn the clock back.

Community policing will survive if it changes as society changes. Its concepts and strategies must continue to evolve.

When community policing becomes fully implemented in all the ways that Taylor et al. described, the term *community policing* will no longer be used because all policing will be based on the assumption that the community has a crucial role in crime prevention and crime solving.

SUMMARY

In the latter part of the twentieth century, police practitioners realized that the professional policing approach was not working and that the public needed to play a larger role in solving and preventing crime. They began to experiment with a new approach that emphasized community participation in identifying, preventing, and solving crime and disorder problems. The new strategy became known as "community policing." The approach addressed not only law enforcement issues but also broader issues of improving and maintaining neighborhoods' quality of life.

The outlook for community policing is positive. We can expect that police training will be expanded to include community problem solving and that educational requirements for recruits will rise. Also, the community will play a larger role in making decisions about the policing they want for their neighborhood and the problems that have priority. They will do more in crime prevention and will take control of not only crime and disorder problems but also other problems affecting the community.

REVIEW QUESTIONS

1. Describe the changes that Taylor et al. consider necessary for the full implementation of community policing.
2. Describe the five "core challenges" that community policing faces, according to Taylor et al.
3. What does the future hold for community policing?

REFERENCE

Taylor, R.W., et al. 1998. Core challenges facing community policing: The emperor still has no clothes. *Academy of Criminal Justice Services Today* 17(1):1–5.

Index